Evolutionary Developmental Biology of Invertebrates 1

Andreas Wanninger

Editor

Evolutionary Developmental Biology of Invertebrates 1

Introduction, Non-Bilateria, Acoelomorpha, Xenoturbellida, Chaetognatha

Springer

Editor
Andreas Wanninger
Department of Integrative Zoology
University of Vienna
Faculty of Life Sciences
Wien
Austria

ISBN 978-3-7091-1985-3   ISBN 978-3-7091-1862-7   (eBook)
DOI 10.1007/978-3-7091-1862-7

Springer Wien Heidelberg New York Dordrecht London
© Springer-Verlag Wien 2015
Softcover reprint of the hardcover 1st edition 2015

Cover illustration: Scanning electron micrograph of a larva of the demosponge, *Amphimedon queenslandica*. Image courtesy of Sally Leys, Alberta. See Chapter 4 for details.

Printed on acid-free paper

Springer-Verlag GmbH Wien is part of Springer Science+Business Media (www.springer.com)

Preface

The evolution of life on Earth has fascinated mankind for many centuries. Accordingly, research into reconstructing the mechanisms that have led to the vast morphological diversity of extant and fossil organisms and their evolution from a common ancestor has a long and vivid history. Thereby, the era spanning the nineteenth and early twentieth century marked a particularly groundbreaking period for evolutionary biology, when leading naturalists and embryologists of the time such as Karl Ernst von Baer (1792–1876), Charles Darwin (1809–1882), Ernst Haeckel (1834–1919), and Berthold Hatschek (1854–1941) realized that comparing ontogenetic processes between species offers a unique window into their evolutionary history. This revelation lay the foundation for a research field today commonly known as Evolutionary Developmental Biology, or, briefly, EvoDevo.

While for many of today's EvoDevo scientists the principle motivation for studying animal development is still in reconstructing evolutionary scenarios, the analytical means of data generation have radically changed over the centuries. The past two decades in particular have seen dramatic innovations with the routine establishment of powerful research techniques using micro-morphological and molecular tools, thus enabling investigation of animal development on a broad, comparative level. At the same time, methods were developed to specifically assess gene function using reverse genetics, and at least some of these techniques are likely to be established for a growing number of so-called emerging model systems in the not too distant future. With this pool of diverse methods at hand, the amount of comparative data on invertebrate development has skyrocketed in the past years, making it increasingly difficult for the individual scientist to keep track of what is known and what remains unknown for the various animal groups, thereby also impeding teaching of state-of-the-art Evolutionary Developmental Biology. Thus, it appears that the time is right to summarize our knowledge on invertebrate development, both from the classical literature and from ongoing scientific work, in a treatise devoted to EvoDevo.

Evolutionary Developmental Biology of Invertebrates aims at providing an overview as broad as possible. The authors, all renowned experts in the field, have put particular effort into presenting the current state of knowledge as comprehensively as possible, carefully weighing conciseness against level of detail. For issues not covered in depth here, the reader may consult additional textbooks, review articles, or web-based resources, particularly on well-established model systems such as *Caenorhabditis elegans* (www.wormbase.org) or *Drosophila melanogaster* (www.flybase.org).

Evolutionary Developmental Biology of Invertebrates is designed such that each chapter can stand alone, and most chapters are dedicated to one phylum or phylum-like taxonomic unit. The main exceptions are the hexapods and the crustaceans. Due to the vast amount of data available, these groups are treated in their own volume each (Volume 4 and Volume 5, respectively), which differ in their conceptual setups from the other four volumes. In addition to the taxon-based parts, chapters on embryos in the fossil record, homology in the age of genomics, and the relevance of EvoDevo for reconstructing evolutionary and phylogenetic scenarios are included in Volume 1 in order to provide the reader with broader perspectives of modern-day EvoDevo. A chapter showcasing developmental mechanisms during regeneration is part of Volume 2.

Evolutionary Developmental Biology of Invertebrates aims at scientists that are interested in a broad comparative view of what is known in the field but is also directed toward the advanced student with a particular interest in EvoDevo research. While it may not come in classical textbook style, it is my hope that this work, or parts of it, finds its way into the classrooms where Evolutionary Developmental Biology is taught today. Bullet points at the end of each chapter highlight open scientific questions and may help to inspire future research into various areas of Comparative Evolutionary Developmental Biology.

I am deeply grateful to all the contributing authors that made *Evolutionary Developmental Biology of Invertebrates* possible by sharing their knowledge on animal ontogeny and its underlying mechanisms. I warmly thank Marion Hüffel for invaluable editorial assistance from the earliest stages of this project until its publication and Brigitte Baldrian for the chapter vignette artwork. The publisher, Springer, is thanked for allowing a maximum of freedom during planning and implementation of this project and the University of Vienna for providing me with a scientific home to pursue my work on small, little-known creatures.

This volume starts off with three chapters that set the stage for the entire treatise by covering general aspects of EvoDevo research, including its relevance for animal phylogeny, homology issues in the age of developmental genomics, and embryological data in the fossil record. These are followed by taxon-based chapters on the animals that are commonly considered to have branched off the animal tree of life before the evolution of the Bilateria: the Porifera, Placozoa, Cnidaria (with the Myxozoa being treated separately), and Ctenophora. In addition, the Acoelomorpha, Xenoturbellida, and Chaetognatha are included, all with currently hotly debated phylogenetic affinities.

Tulbingerkogel, Austria   Andreas Wanninger
January 2015

Contents

EvoDevo and Its Significance for Animal Evolution and Phylogeny

Alessandro Minelli

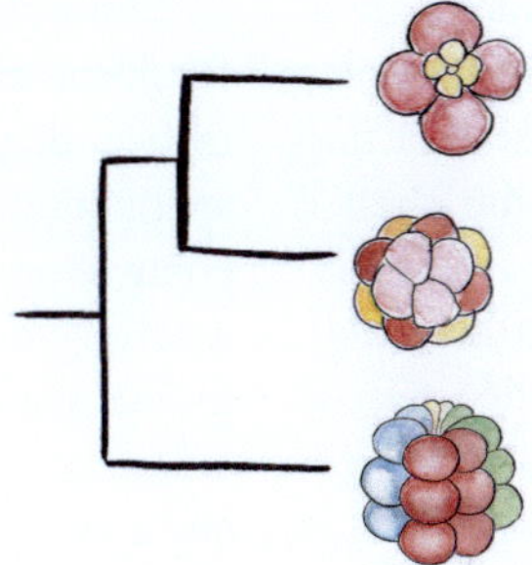

Chapter vignette artwork by Brigitte Baldrian.
© Brigitte Baldrian and Andreas Wanninger.

A. Minelli
Department of Biology, University of Padova,
Via Ugo Bassi 58 B, I 35131 Padova, Italy
e-mail: alessandro.minelli@unipd.it

A. Wanninger (ed.), *Evolutionary Developmental Biology of Invertebrates 1:
Introduction, Non-Bilateria, Acoelomorpha, Xenoturbellida, Chaetognatha*
DOI 10.1007/978-3-7091-1862-7_1, © Springer-Verlag Wien 2015

ONTOGENY VS PHYLOGENY

Despite Steinböck's (1963, p. 49) dismissive statement that "ontogeny has only a very limited value for phylogenetic questions," successful attempts to infer phylogenetic relationships from comparative information about the developmental schedules of animal species are numerous, beginning with two well-known, eighteenth-century examples. One is Thompson's (1830) discovery of the crustacean nature of barnacles, based on his observation of nauplius larvae metamorphosing into sessile adults (see Vol. 4, Chapter 5) whose morphology deviates so strongly from the arthropod ground plan that Linné (1758) placed *Lepas* (inclusive of barnacles) in his Vermes Testacea (i.e., the shelled mollusks) rather than in his Insecta (a "class" broadly equivalent to present-day Arthropoda). The other example is Kowalewski's (1866) discovery of the affinities between vertebrates and ascidians, revealed by the presence of the notochord in the larva of the latter (Vol. 6, Chapter 4). This does not imply, however, that the relationships between ontogeny and phylogeny are always easy to discover or that these follow simple and perhaps universal principles such as Haeckel's (1866) "biogenetic law." Haeckel's recapitulationist views, indeed, have never been again much in favor since Garstang (1922) demonstrated that many larval adaptations are recent and independent; and a further strong blow to the theory was de Beer's (1930, 1940) demonstration of the pervasiveness of heterochrony. However, new opportunities to extract phylogenetic information from ontogenetic data have been emerging since the advent of evolutionary developmental biology (Telford and Budd 2003; Cracraft 2005; Minelli 2007, 2009; Minelli et al. 2007).

EVOLUTIONARY DEVELOPMENTAL BIOLOGY

Evolutionary developmental biology, or EvoDevo, is one of the most active frontiers of the life sciences, despite the fuzzy definition of its scope and its sometimes problematic boundaries in respect to the parent disciplines – evolutionary biology and developmental biology. Comprehensive overviews of origins, aims, and methods of evolutionary developmental biology can be found in Hall (1998) and Hall and Olson (2003); other useful book-size accounts, although more selective in their approach, are Wilkins (2001), Minelli (2003a), Carroll et al. (2005), and Minelli and Fusco (2008).

As one should expect for a newly established, or reestablished, field of study, EvoDevo is still struggling to define its own identity; short introductions to the internal debate have been provided by Arthur (2002) and Müller (2008).

Many researchers (e.g., Carroll et al. 2005) view EvoDevo essentially as comparative developmental genetics, that is, as the comparative study of the spatial and temporal expression patterns of genes controlling the establishment of body architecture: anterior-posterior and dorsoventral polarity; longitudinal patterning of the main body axis; segmentation, production, and patterning of appendages; and so on, down to details such as the differentiation of eyespots on butterfly wings or the rows of specialized bristles forming the sex combs on the forelegs of *Drosophila* males. This comparative approach to the study of gene expression pattern has produced the positive effect of rapidly increasing the number of organisms used in the lab as model species. In turn, the expanding taxonomic coverage of these studies has helped generating results of potentially high relevance for phylogenetic research. At the level of the genetic mechanisms controlling development, it has become meaningful, and operationally feasible, to address questions of homology between features of vastly divergent taxa.

Arguably, simply broadening the scope of comparison beyond the traditional bunch of model species, such as *Caenorhabditis elegans*, *Drosophila melanogaster*, and *Mus musculus*, would hardly justify the recognition of a new, distinct discipline. EvoDevo, however, is characterized by a problem agenda that could not be satisfactorily fulfilled within the premises of either evolutionary or developmental biology in isolation. This is true, for example, for the origin of evolutionary novelties (Müller 1990; Müller and Wagner 1991, 2003; Wagner 2000, 2011; Galis 2001; Müller and Newman 2003, 2005; Minelli and Fusco 2005;

Love 2008; Moczek 2008; Pigliucci 2008; Shubin et al. 2009; Brigandt and Love 2010, 2012; Hall and Kerney 2012; Peterson and Müller 2013). The most discipline-specific problem addressed by EvoDevo is the nature and the properties of *evolvability*, defined by Hendrikse et al. (2007) as "the capacity of a developmental system to evolve." This means that EvoDevo characteristically focuses on the *arrival* of the fittest rather than on the *survival* of the fittest. How far this shift of focus should be considered an extension, either marginal or substantial, of the evolutionary synthesis paradigm, or a radical alternative to the same, is still a matter of dispute (e.g., Laubichler 2010; Minelli 2010; Pigliucci and Müller 2010), but this is not relevant to our subject.

To introduce, instead, an overview of the possible significance of EvoDevo in the context of phylogenetic analysis, it is fair to repeat, at the outset, the comment made 10 years ago by Wiens et al. (2005) that up to now the overall contribution of EvoDevo to phylogenetics has been quite small. But this is arguably due to the limited awareness of EvoDevo by a large majority of phylogeneticists, and vice versa, rather than to the exiguity of the potential intersection between the two disciplines. Eventually discovering mutual foreignness between EvoDevo and phylogenetics would be ironic, indeed: let's recall that Gould's magisterial introduction to one of the roots of EvoDevo, namely, the study of heterochrony, was published in 1977 under the title *Ontogeny and Phylogeny*. There are instead several important areas to which EvoDevo can contribute to progress in phylogenetics. I will articulate these areas in the following sections, mainly taking examples from invertebrates.

RECAPITULATION VS CLADISTIC ASSESSMENTS OF CHARACTER POLARITY

In its earliest steps, long before getting its current name, EvoDevo contributed substantially to a critical revisitation of Haeckel's recapitulationism, the principle according to which ontogeny recapitulates phylogeny. As mentioned before, de Beer's books (1930, 1940) dissected the possible relationship between ontogeny and phylogeny in such a way that these eventually revealed the wealth of alternative patterns, recapitulation being only one among several possible scenarios and, arguably, not necessarily the most common among them. De Beer's analysis eventually resulted in the birth of the modern studies on heterochrony, especially after this area was popularized by Gould's (1977) book.

In the meantime, debates about the phylogenetic signal contained in ontogenetic sequences developed in cladistic circles. Some cladists, like Rieppel (1979), were critical of the independence of ontogenetic information from the morphological data used in outgroup comparisons. Others, however, thought otherwise.

Among the criteria to be used for polarizing characters, i.e., to distinguish the plesiomorphic from the apomorphic state of a character, Hennig (1966) had suggested ontogenetic character precedence. Somehow echoing Haeckel's biogenetic principle, this criterion postulated that the derived character states are to be found in late developmental stages, whereas similarities shared at earlier stages are generally symplesiomorphies (shared primitive character states) that cannot be used to infer phylogenetic relationships. The ontogenetic character precedence was regarded by some authors (e.g., Fink 1982) as reliable as the outgroup comparison, whereas others (e.g., Kluge 1985) pointed to its lack of general applicability and still others (e.g., Nelson 1978; de Queiroz 1985) suggested different reformulations of the principle, effectively taking distance from the original recapitulationist flavor of Hennig's principle. For example, Nelson (1978, p. 327) reformulated the "biogenetic law" in the following terms: "given an ontogenetic character transformation, from a character observed to be more general to a character observed to be less general, the more general character is primitive and the less general advanced." A different formulation was recently suggested by Martynov (2012, p. 833) as the main principle of his *ontogenetic systematics*, which should be based on "progressive (addition of stages and characters) or regressive (reduction of already existing stages and structures) modification of ancestral taxon, the diagnosis of which corresponds to the model of its ontogenetic cycle."

An unusual extension of the recapitulationist paradigm into the area of animal behavior has been recently proposed by Barrantes and Eberhard (2010) with a comparative study of the web-spinning behavior in spiders. These authors found that the design of the web spun by adults of three *Latrodectus* species is more divergent than the design of those spun by juveniles of the same species and more similar to those of young spiders of the genus *Steatoda* than to those of the adult of the latter genus.

But let's move to more explicit suggestions and recent examples of their application.

DEVELOPMENTAL GENES AND PHYLOGENETIC INFERENCE

Gene-Based Homology

The first step toward a phylogenetic analysis is getting informative data. How can EvoDevo contribute to filling a matrix?

The contribution of EvoDevo to the assessment of homology is controversial (see Chapter 2). On the one hand, it is right from early works in what was still to be named EvoDevo that biologists realized that traits firmly regarded as homologous by comparative morphologists can have quite a different developmental origin. On the other hand, one of the most visible successes of EvoDevo has been the discovery that homologous genes are often involved in building equivalent structures in the most disparate animals, although this equivalence has been generally regarded as nonhomologous by comparative morphologists. On homology, see Minelli and Fusco (2013), Wagner (2014), and Chapter 2 herein.

The Genotype→Phenotype Map

One of the most far-reaching results of EvoDevo studies is the growing awareness of the complexity (and, to a very large extent, unpredictability) of the genotype→phenotype map, that is, of the cascade of processes through which a given phenotypic trait is controlled by the expression of a given gene (e.g., Alberch 1991; Altenberg 1995; Mezey et al. 2000; Kell 2002; West-Eberhard 2003; Pigliucci 2010; Wagner and Zhang 2011). To put it in simple terms, this mapping is rarely, if ever, a one-to-one function (one gene→one phenotypic trait), but it is generally one-to-many (pleiotropy; e.g., Wagner and Zhang 2011, 2013; Paaby and Rockman 2013) or many-to-one (convergence and or redundancy) and eventually many-to-many.

In *Drosophila*, some 50 genes are the direct targets of transcription factors encoded by Hox genes (Pearson et al. 2005): some of these genes are involved in apoptosis and others in the control of cell cycle, cell motility, intercellular signaling, or cell adhesion (Davidson 2006), and there are hundreds of genes whose expression is downstream of the expression of one or more of the Hox genes (Mastick et al. 1995; Botas and Auwers 1996).

Convergence

A nice example of the intricacies of the genotype→phenotype map is the fact that the same genes can regulate the development of homologous structures through significantly different cellular processes. A recently studied example is offered by the sex combs of male *Drosophila* species. The key regulatory genes involved in the production of these rows of specialized bristles are the same in the different species that have been investigated, but the cellular mechanisms through which they operate are different, not only between members of different subgenera (*Sophophora* vs *Lordiphosa*) (Atallah et al. 2012) but also between quite closely related species that are classified in the same subgenus (*Sophophora*: species of the *obscura* and *melanogaster* species groups; Barmina and Kopp 2007; Tanaka et al. 2009, 2011).

Developmental genes, Hox genes included, are as prone to convergence as morphological characters are. An example is offered by the mechanisms controlling leg repression in one of the body regions (abdomen or opisthosoma) of some arthropod clades. In insects, the Hox genes *Ultrabithorax* (*Ubx*) and *abdominal-A* mediate leg repression, thus providing the most obvious difference between a leg-bearing thorax and a legless abdomen. Things are different in spiders, so far as the recent findings of Khadjeh et al. (2012) on *Achaearanea tepidariorum* will hold for the whole clade. Here, the gene *Antennapedia* (*Antp*) represses leg formation in the first segment of the opisthosoma, whereas both *Antp* and *Ubx* show their (redundant) effect in repressing leg formation in the following segment.

From Gene Phylogeny to a Comparison of Gene Expression Patterns

Most of the total output of EvoDevo research has been a growing knowledge of the identity, sequence, patterns of expression, and relative position in developmental control cascades of "developmental genes," i.e., of genes demonstrably involved in the control of specific ontogenetic events or in the deployment of specific traits of body architecture. Among these genes are those involved in body segmentation and those (the Hox genes) that specify positions along the anterior-posterior body axis. Indeed, the discovery of the high degree of conservation of these genes across the animal kingdom was one of the main successes that contributed to establishment of EvoDevo as a promising new biological discipline.

The potential phylogenetic signal contained in these genes can be studied at different levels, as shown here briefly on the example of the Hox genes.

A first level of analysis is the reconstruction of gene phylogeny, a necessary step, not only required to reveal duplications and thus to disentangle orthologous from paralogous sequences but also to establish relationships between gene families that may evolve either in concert or in divergent manner and eventually to polarize gene changes. The literature on the phylogenetic relationships of Hox genes is extensive (e.g., Finnerty and Martindale 1998; Kourakis and Martindale 2000; Ferrier and Holland 2001; Ferrier and Minguillon 2003; Garcia-Fernàndez 2005a, b; Duboule 2007; Ferrier 2007, 2010; Butts et al. 2008). A recent review by Holland (2012) highlights the phylogenetic relationships between the Hox family and other gene families also involved in development (*ParaHox, Evx, Dlx, En, NK4, NK3, Msx*, and *Nanog*), all together forming the ANTP class.

The second step is to use gene sequences to reconstruct the phylogeny of the organisms from which the sequences have been obtained. Will EvoDevo suggest to give preferences to selected gene families? In the past, several biologists looked at chromosome structure as at privileged morphological traits, insofar as chromosomes contain genes and genes are involved in determining the phenotype. In the same vein, some authors have looked at Hox genes – genes controlling aspects of the overall body architecture – as to privileged genes, possibly carrying important phylogenetic signal. Of course, the phylogenetic information potentially carried by the highly conserved homeobox sequence (the "morphological signature" of this gene class) will be very different from the phylogenetic information potentially carried by the remaining of the molecule, especially by regions distant from the homeobox. Eventually, Hox gene sequences have been used in reconstructing the mutual relationships of bilaterian phyla (de Rosa et al. 1999; Balavoine et al. 2002; Hueber et al. 2013) or to investigate phylogeny within large phyla such as Arthropoda (Cook et al. 2001). Other studies have contributed to fix affinities, e.g., of bryozoans (ectoprocts) as lophotrochozoans (Passamaneck and Halanych 2004). Apart from the Hox genes, specific signatures have been found in many other developmental genes. An example is provided by the *bone morphogenetic protein* genes, which are represented in all insects by *decapentaplegic* and *glass bottom boat* (*gbb*); a third gene, *screw* (*scw*), is found in *Drosophila melanogaster* and other flies, and recent comparative studies have placed the *gbb/scw* duplication in the interval between the origin of the Brachycera and the origin of the Cyclorrhapha, that is, between 200 and 150 Ma ago (Wotton et al. 2013).

The last step is to compare gene expression patterns in order to trace homologies, especially in cases where morphological evidence does not seem to allow a definitive assessment. This approach has been followed, for example, by Hughes and Kaufman (2002), Copf et al. (2003), and Angelini and Kaufman (2005) in comparing body regions of different arthropod groups and by Lichtneckert and Reichert (2005) in delineating homologies between vertebrate and arthropod brains. Jager et al. (2006) (see also Manuel et al. 2006) used the expression patterns of Hox genes to align the anterior appendages of sea spiders with those of other arthropods, thus yielding results that are in contrast with the morphological (neuroanatomical) evidence obtained by Maxmen et al. (2005). There are, however, examples of developmental genes whose expression patterns confirm the homologies suggested by morphology. One of these is *Brachyury* (*bra*): its expression in the notochord of chordates has been fittingly chosen by Ferrier (2011) as a good example of a homologous gene with a homologous function in a homologous morphological character, a far from marginal example, the presence of a notochord being an apomorphy of a phylum.

miRNA and Phylogeny

According to Wheeler et al. (2009), a substantial increase in morphological complexity along the evolutionary history of metazoans is linked to a corresponding increase in the number and specificity of action of miRNAs. The same authors stress the high phylogenetic value of these molecules, confirm the previously established (Hertel et al. 2006; Sempere et al. 2006; Prochnik et al. 2007) major expansion of the miRNA family at the base of the nephrozoan clade, and identify the presence of 34 miRNA families in the last common ancestor of protostomes and deuterostomes, to the exclusion of acoels. A few miRNAs have been discovered in sponges (Robinson et al. 2013) but none of these is shared with eumetazoans.

Conservation of Gene Function and Developmental System Drift

The value of detailed patterns of Hox gene expression as a base to establish homology of segments or positions along the main body axis has probably been overestimated (Abzhanov et al. 1999; Brenneis et al. 2008), because of the observable evolutionary shifts of the anterior boundary of expression of many Hox genes within arthropods, especially *Antennapedia*, *Ultrabithorax*, *abdominal-A*, and *abdominal-B* (Hughes and Kaufman 2002).

Shifts corresponding to a positional inversion along the main body axis are very unlikely. Morphologically, the *fore*wings of male strepsipterans (the females are wingless and mostly vermiform) are quite similar to the halteres of dipterans – their characteristically modified *hind*wings. However, whatever the mechanisms specifying the peculiar structure of the strepsipteran forewings, it is quite unlikely that these evolved from the dipteran condition, through a "macromutation" switching the haltere specification from the meta- to the mesothorax, in turn restoring the metathoracic wings to a more conventional morphology. This unconventional hypothesis was suggested by Whiting and Wheeler (1994) as a tentative EvoDevo counterpart of their phylogeny of holometabolous insects, in which the Strepsiptera turned out to be the sister group of the Diptera (for the putative Diptera+Strepsiptera monophylum, the name Halteria was also proposed) (Whiting et al. 1997). The need to demonstrate the actual occurrence of such a macromutation eventually vanished, as soon as subsequent phylogenetic analyses (e.g., Rokas et al. 1999; Wiegmann et al. 2009) refuted the monophyly of the Halteria, thus showing the independent evolution, in the two clades, of morphologically similar but positionally nonequivalent "halteres."

Evolving Gene Functions

Comparative developmental genetics has revealed many examples of evolutionary changes in gene function. For example, *fushi tarazu* and *oskar* may have initially functioned in the central nervous system but later became involved in the patterning of the early embryo, as seen today in *Drosophila* (Ewen-Campen et al. 2012; Heffer et al. 2013). The evolution of new functional roles has been documented to occur even if the phenotypic traits previously controlled by a gene are subjected to strong stabilizing selection; this is why the evolution of new functions in a lineage of orthologous genes (i.e., independent from gene duplications) has been called *developmental system drift* (DSD) (True and Haag 2001; Haag 2014). Even organs that are identical at the cellular level, because they are produced through an identical cell lineage, can experience rapid DSD. This has been shown by Verster et al. (2014) by comparing over 20 species of *Caenorhabditis* where functional divergence has been found in orthologous genes regulating sex determination, early embryonic patterning, vulva development, and excretory physiology.

The phylogenetically widespread involvement of *Pax6/ey* homologs in eye morphogenesis (e.g., Halder et al. 1995; Tomarev et al. 1997; Glardon et al. 1998; Kmita-Cunisse et al. 1998; Chow et al. 1999; Pineda et al. 2000) has led to the hypothesis of a monophyletic origin of bilaterian eyes (e.g., Gehring and Ikeo 1999; Gehring 2000), contrary to a well-entrenched opinion, based on gross morphological differences between ciliary- and rhabdomeric-type eyes, suggesting an at least diphyletic origin of eyes. More cautiously, Wagner (2001) suggested that ancestrally *Pax-6* homologs may have been involved in initiating the development of light-sensitive epithelia, eventually a key component of subsequently evolved eye types such as the compound eye of arthropods and the camera eye of squids, but the hypothesis of a monophyletic origin of the eye has been strongly rejected by others, among which Harris (1997) and Meyer-Rochow (2000). In addition to the arguments provided by comparative morphology, Harris (1997) remarked that the expression of *Pax-6* is not restricted to the eyes. For example, in vertebrates this gene is also expressed in the nasal placodes, the

diencephalon, the latero-ventral hindbrain, and the spinal cord (Li et al. 1994; Amirthalingam et al. 1995). In *Drosophila*, its homolog *ey* is also expressed in the brain and the ventral nerve cord, and in the squid, *Pax6* expression extends to the brain and the arms (Tomarev et al. 1997). Even more intriguing is the fact that *Pax6* homologs are also present in eyeless animals. In the nematodes, for example, *vab-3* is involved in the differentiation of the cephalic body end and *mab-18* is expressed in the precursors of peripheral sense organs (Chisholm and Horvitz 1995; Harris 1997). In the sea urchins, a *Pax6* homolog is expressed in the tube feet (Czerny and Busslinger 1995). Summing up, *Pax6* is likely a patterning gene, expressed in the head, which has been repeatedly involved (or, better, co-opted; see below) in the regulation of eye development.

Two arthropod genes of the Hox family have undergone dramatic functional changes. In selected branches of the arthropod tree, both of them have lost their original function as specifiers of position along the main body axis. One of these genes is *fushi tarazu*, which is involved in segmentation and, in insects only, in neurogenesis. The other gene is *zerknuellt* (*zen*), which is involved in dorsoventral patterning. In the Diptera, a duplication of *zen* has given rise to *bicoid*, whose functional role has continued to evolve rapidly: in *Drosophila*, it is required for the normal development of the head and thorax, and in the phorid *Megaselia abdita*, it is additionally required for the development of four abdominal segments (Stauber et al. 2000).

Gene Regulatory Networks and Their Evolution

Eventually, following the rapidly increasing knowledge on gene control cascades, research focus has shifted from the evolution of individual genes, and of their expression, to the evolution of whole gene regulatory networks (Davidson 2006; see also Davidson et al. 2002, 2003; Davidson and Erwin 2006; see also Chapter 2). From the perspective of phylogenetic reconstruction, this means moving from the limited evidence of homology provided by single genes, whose involvement in a given developmental process is prone to convergent evolution (multiple independent co-option events), to the more robust evidence provided by whole sets of functionally integrated genes (Ferrier 2011).

Comparative developmental genetics is able to reveal the intricate nature of gene networks such as those underlying the architectural design of the nervous system of bilaterians (Denes et al. 2007), the segmented body of arthropods (Dray et al. 2010), and the notochord of chordates (Kugler et al. 2011), a kind of synapomorphy packages for the corresponding clades.

Davidson (2006) described the developmental regulatory genome as something like a computer, with four classes of subcircuits: (i) batteries of genes involved in cell differentiation, (ii) little invariant subcircuits repeatedly involved in less specific functions, (iii) switches, and (iv) "kernels," complex and highly conserved networks responsible for specifying morphogenetic fields from which particular body parts arise. One of those kernels, for example, would be responsible for the specification of the endoderm. Kernels would be most robust to change and are thus likely to be shared by distantly related clades. Davidson envisaged a phylogenetic hierarchy of regulatory networks, e.g., bilaterian kernels, protostome kernels, and ecdysozoan kernels.

However, selected parts of a gene regulatory network may show unequal rate of evolution. For example, within the gene regulatory network (GRN) controlling the specification of endomesoderm in nematodes, a preliminary analysis of genome sequences of *Haemonchus contortus* and *Brugia malayi* suggests that evolution is most rapid for some zygotic genes involved in the specification of blastomere identity (Maduro 2006).

If we accept that development is controlled by GRNs, it follows that the evolution of development and form is due to changes within GRNs (Carroll 2008), but this is arguably an excessive generalization.

An exceptional example of the evolvability of developmental gene networks has been revealed by Kugler et al. (2011) with a comparison of notochord development between the pelagic urochordate *Oikopleura* and the ascidian *Ciona intestinalis* (Vol. 6, Chapter 4). In the latter, some 50 genes are known to be activated downstream of *bra*, but 24 of them do not have a homolog in the small, very compact genome of *Oikopleura*. Some of the latter have undergone a lineage-specific duplication, but less than a half of them are apparently expressed in the context of notochord formation. For an extensive discussion on gene regulatory networks and their bearings on character identity and evolution, see Chapter 2.

Gene Loss and Character Loss

From the perspective of phylogenetic reconstruction, character loss is a frequent cause of problems.

In an important study of salamander phylogeny, Wiens et al. (2005) have shown the misleading effects of paedomorphosis on phylogenetic analysis, because of which a previous analysis by Gao and Shubin (2001), based on morphological data, had placed most paedomorphic families in a single clade. As demonstrated by the new analysis, problems are not solved by simply excluding from the data matrix the characters suspected to be paedomorphic and by taking into account the parallel evolution of adaptive changes associated with the aquatic habitat typical of salamander larvae generally and definitely retained in the paedomorphic lineages. A possibly more disturbing problem is the absence, in the paedomorphic lineages, of those synapomorphies that in non-paedomorphic taxa develop at metamorphosis.

In respect to regressive changes, EvoDevo has much to offer beyond a conceptual framework, especially in those cases in which a regressive change is apparently due to gene loss. This has been tentatively suggested (Aboobaker and Blaxter 2003a; Minelli 2009) as a possible explanation for the relatively simple organization of the nematodes, compared to most ecdysozoans, which possibly correlates with a reduction in the number of Hox genes (which is coupled, however, with a very high rate of evolution of the surviving members of this gene family; Aboobaker and Blaxter 2003b). The most intriguing example of a likely correlation between the loss of a gene and the loss of a body part is, however, the extreme reduction of the abdomen in the parasitic crustacean *Sacculina carcini*, matched by the loss of the Hox gene *abdominal-A* (Blin et al. 2003; cf. Vol. 4, Chapter 5).

Gene and Gene Network Co-option vs Paramorphism

Since the last years of the past century, it has become fashionable to interpret major events in the evolution of the genetic control of development in terms of *co-option* of individual genes or even of whole gene regulatory networks. Gene co-option would be usually dependent on previous gene duplication. Following the latter event, neofunctionalization of a duplicate gene would add a new trait to the phenotypic features under its control. Co-option, for example, would explain the evolution of arthropod and vertebrate appendages (Tabin et al. 1999). According to Pires-daSilva and Sommer (2003), all developmental processes involved in the generation of new structures would necessarily depend on co-option.

However, we should probably advocate gene co-option only when an existing gene gets a new role in a developmental process in which it was not previously involved or in a body part where it was previously not expressed, only when the developmental process or the body part with which it now becomes involved was already in existence (Minelli 2009). This is the case of the wing eye spots of many butterflies, which are centered on a group of *Distal-less*-expressing cells (Carroll et al. 1994). *Distal-less* has a much older and phylogenetically much more general role in animal development, as an early marker of the sites where appendages will form, including insect legs, "polychaete" parapodia, vertebrate limbs, and sea urchin podia (Panganiban et al. 1997). In butterflies, *Distal-less* has been co-opted to mark the position of new "virtual axes," but the presence of wings does not depend on this novel expression of the gene.

The concept of co-option does not apply, however, when a novel pattern of expression of a gene, or of a whole gene regulative network, coincides with the origin of a new body part. It is possible, indeed, that the evolving phenotypic outcome of that gene's expression is a story of exaptation rather than one of co-option. This is arguably the case of *nanos*, originally a determinant of the posterior end of the trunk (cf. Rabinowitz et al. 2008), subsequently turned into a specifier of germ cell identity, and also of *Pax6*, perhaps exapted from pigment specifier to specifier of the eye (Kozmik 2005).

Genes involved in patterning the main body axis may have also a role in the proximo-distal patterning of appendages. This secondary expression is unlikely the result of co-option of these genes' function in patterning a new body feature (the appendage) that supposedly evolved prior to, and independent of, these genes' expression. If, on the contrary, this new gene expression evolved together with the origin of the appendage, this would be a case of paramorphism (Minelli 2000). With time, the patterning role of these genes in the appendage will likely diverge from the corresponding role in the trunk; nevertheless the appendage is likely to behave like a duplicate of the main body axis and thus to retain some characteristic traits of the latter. This may explain why the appendages of segmented animals are frequently segmented, while those of unsegmented animals never are. If we accept the hypothesis of axis paramorphism, we shall perhaps revise some popular interpretation of character polarity.

For example, is the arthropod (first) antenna a specialized leg, or vice versa? Dong et al. (2001) favored the antenna-first hypothesis, whereas Casares and Mann (1998) initially supported the "leg-first" hypothesis, but in a later paper (Casares and Mann 2001) they accepted that the appendages may have been already different (and segmented) since their very first expression. However, if the relationship between the (segmented) appendages and the (also segmented) main body axis of arthropods is one of paramorphism, the whole question of the primacy of the leg versus the antenna would become meaningless (Minelli 2003b; Minelli and Fusco 2005), and no scheme of character transition from one form to the other would be applicable (Minelli et al. 2007).

SEGMENTATION: GENES AND BILATERIAN PHYLOGENY

The first suggestion that arthropod and annelid segmentation may have evolved independently, thus shaking the solidity of one of the oldest "supraphyletic" assemblages – the one closely corresponding to Cuvier's (1812) old *embranchement* of the Articulata – was based on rudimentary EvoDevo arguments (Minelli and Bortoletto 1988). Shortly thereafter, the Articulata hypothesis was rejected by a phylogenetic analysis based on a for the time extensive matrix of morphological data (Eernisse et al. 1992). Eventually, a molecular analysis (Aguinaldo et al. 1997) confirmed the lack of close affinities between arthropods and annelids and revealed the existence of a clade of molting animals, segmented and unsegmented, which received the now popular name Ecdysozoa.

In the following years, the Articulata vs Ecdysozoa debate (e.g., Schmidt-Rhaesa et al. 1998; Wägele et al. 1999; Zrzavý 2001; Scholtz 2002, 2003; Giribet 2003; Nielsen 2003a, b; Schmidt-Rhaesa 2004, 2006; Pilato et al. 2005; Ivanova-Kazas 2013) was mostly centered on steadily revised interpretations of morphological evidence (including descriptive embryology), in the light of a growing set of phylogenetic analyses. The need of a contribution from EvoDevo, however, became increasingly important, insofar as a growing detail of segmentation processes was understood at the level of gene expression, in a few model organisms at least. It became thus critically important to determine what comparative developmental genetics could say about the single or multiple origin of segmentation. Eventually, the newly emerging phylogeny (e.g., Adoutte et al. 2000; Halanych 2004; Bourlat et al. 2008; Dunn et al. 2008; Telford and Littlewood 2009; Edgecombe et al. 2011; Mallatt et al. 2012), strongly based on molecular evidence, provided a background against which the problem of the evolution of segmentation could be framed in the following alternative terms: (i) segmentation evolved before the split between Ecdysozoa and Lophotrochozoa and perhaps even before the split between Protostomia and Deuterostomia, i.e., essentially, at the base of the Bilateria – if so, segmentation would have been secondarily lost several times – or (ii) segmentation evolved independently in the arthropod, annelid, and vertebrate lineages, from unsegmented ancestors, which were also the last common ancestor of all Bilateria and the last common ancestor of Ecdysozoa and Lophotrochozoa.

Discussions about the mono- vs polyphyletic origin of segmentation are far from settled. Comparative studies of the genetic control of segmentation have played an increasing role in the dispute. In the 1980s, the presence of regularly spaced stripes of *engrailed* (*en*) expression along the elongating main axis of the embryo emerged as a potentially reliable proof in favor of a segmentation mechanism shared by all segmented metazoans. In arthropods, indeed, *en* is expressed in transversal rows of cells immediately anterior to the future segmental margin. It is also expressed in a series of transversal stripes in the embryos of leeches, polyplacophoran mollusks, onychophorans, as well as in amphioxus and in the vertebrates (Jacobs et al. 2000). This does not mean, however, that in all these metazoans *en* is actually involved in segmentation. In *Drosophila*, *en* expression is limited to the ectoderm, where it marks compartment boundaries, besides being involved in the patterning of the nervous system. In the leech, its expression extends to the mesoderm but in the ectodermal derivatives it is not involved in patterning the nervous system into segmental units (Shankland 2003). Besides these spatial (germ layer or tissue level) differences, *en* expression is also diverse temporally. In vertebrates, *en* homologs are expressed in the segmental mesodermal units (somites), but only after these are formed (Holland and Holland 1998). Moreover, homologs of *en* are present and expressed during the embryonic development, also in non-segmented animals such as mollusks (*Patella*: Nederbragt et al. 2002; Vol. 2, Chapter 7). In a variety of segmented and unsegmented animals including arthropods, annelids, mollusks, and echinoderms, the ectodermal expression of *en* is associated with skeletal development (Jacobs et al. 2000). In polychaetes, *en* is regularly expressed in the chaetal sacs (Seaver et al. 2001). In mollusks, *en-expressing* cells surround the ectodermal cells producing shell material (Moshel et al. 1998; Wanninger and Haszprunar 2001). In ophiuroid echinoderms, *en*-expressing ectodermal cells delimit the areas where the ossicles are produced (Lowe and Wray 1997). It is thus quite possible that the association of segmentation with *en* expression is only an indirect one rather than evidence of a common origin of segmentation.

More recently, the idea of a single origin of segmentation in bilaterians has been floated anew, based on the common involvement, shared between arthropods and vertebrates, of a periodic, oscillatory behavior in the expression of genes involved in the Notch/Delta signaling pathway (Stollewerk et al. 2003). To be more precise, this oscillating behavior is now firmly established as central to the segmentation process in vertebrates (e.g., Jiang et al. 2000; Holley et al. 2002; Mara et al. 2007; Özbudak and Lewis 2008; Lewis et al. 2009; Oates et al. 2012). In annelids, there is some positive evidence for the involvement of Notch signaling in segmentation in the leech *Helobdella robusta* (Rivera and Weisblat 2009), but not in the polychaete *Capitella* sp. 1 (Thamm and Seaver 2008). In arthropods, where it has been detected in several lineages (e.g., in the spider *Cupiennius salei*: Stollewerk et al. 2003; the cockroach *Periplaneta americana*: Pueyo et al. 2008; the flour beetle *Tribolium*: Sarrazin et al. 2012), this mechanism does not seem to be universally present or, at least, universally required for segmentation (Kainz et al. 2011: *Gryllus*), but this condition might well be secondary. However, the recent discovery of oscillatory transcription in *Arabidopsis*, with patterning effect on the positioning of the lateral root primordia (Moreno-Risueno et al. 2010), suggests that a "segmentation clock" is a general principle governing patterning in growing tissues, but this also suggests its multiple evolution in multicellulars (Richmond and Oates 2012); even among the metazoans, it has possibly evolved multiple times through the parallel co-option of ancestral gene regulatory networks (Chipman 2010).

RETHINKING EMBRYOLOGICAL EVIDENCE OF PHYLOGENETIC RELATIONSHIPS

The Phylogenetic Signal of Cleavage Patterns

Acoels (Chapter 9) are characterized by duet spiral cleavage; hydrozoans and other cnidarians (Chapter 6) have variable (Beklemishev 1963), unstable cleavage patterns, but this character is not easily coded in a matrix.

Synapomorphies of annelids, mollusks, entoprocts, nemerteans, and rhabditophorans are quite likely their quartet spiral cleavage, with the typical orientation of the mitotic spindles during the earliest mitoses and their characteristic cell lineage (reviewed in Nielsen 2008; cf. Vol. 2, Chapters 3, 6, 7, 8, and 9). The phylogenetic value of sharing spiral cleavage is likely strengthened by the low probability of multiple independent transitions to such an idiosyncratic cleavage pattern. The opposite transition (spiral to radial cleavage) is possibly quite easier, as shown by the coexistence of both patterns in a member of an otherwise typical spiralian group, the Rhabditophora. At the eight-cell stage, some embryos of the lecithoepitheliate *Prorhynchus stagnalis* have eight blastomeres of equal size, but others have four macromeres and four micromeres, as in radial and spiral cleavage, respectively (Steinböck and Ausserhofer 1950).

Germ Layer Homology

Rather than on objective morphological or molecular evidence, germ layers have being often identified in terms of their prospective fate. This theory-laden approach (Hall 1998) has invited comparisons even between embryos with clearly distinguishable germ layers as individualized cell sheets and embryos where germ layers are not distinguishable as morphological units. As a consequence, what had been called germ layers became the initial pools of cells eventually fated to produce specific tissues or organ systems rather than objectively recognizable morphological units in the embryo before organogenesis. Eventually, however, comparative developmental genetics has led to the identification of genes selectively expressed in one or the other of the germ layers, thus suggesting a more objective criterion upon which to compare features of embryos with morphologically identifiable germ layers with those without. For example, in their effort to homologize endomesoderm across eumetazoans – diplo- as well as triploblastic ones – Technau and Scholz (2003) have focused on *GATA 4-6*, *twist*, *snail*, and *brachyury*.

Interestingly, endoderm-specific genes have been found in *Caenorhabditis elegans*, where the distinct germ "layers" are not discernible, due to the very small total number of cells in the embryo (Maduro and Rothman 2002). Other genes, such as *snail* and *twist*, are characteristically expressed in the mesoderm. Eventually, a *snail* homolog has been found in the coral *Acropora millepora* (Hayward et al. 2004) and in the sea anemone *Nematostella vectensis* (Martindale et al. 2004), where it arguably contributes to the specification of the endoderm in respect to the ectoderm (Ball et al. 2004; Martindale et al. 2004; Chapter 6). A *twist* homolog has been found in the hydrozoan *Podocoryne carnea* (Spring et al. 2000). This is potentially of interest in respect to the repeatedly floated question of the possible presence of mesoderm in the Cnidaria, which are traditionally described as diploblastic (but see Boero et al. 1998; Seipel and Schmid 2005, 2006; Burton 2008; Chapter 6).

Persisting difficulties in finding reliable homologies between cnidarian germ layers and those of bilaterians are deepened by the diverse behavior of hydrozoans, whose germ cells generally differentiate from ectodermal interstitial cells, but in *Protohydra* and *Boreohydra*, germ cells originate instead from the endoderm (Van de Vyver 1993). Moreover, nervous cells originate from the endoderm in the hydrozoan *Phialidium gregarium* (Thomas et al. 1987) but from the ectoderm in scyphozoans (Nielsen 2001). Problems, however, are not restricted to Cnidaria. Malpighian tubules are ectodermal in insects but endodermal in chelicerates, and in tardigrades the midgut is of mesodermal origin (Kristensen 2003) rather than endodermal, as it would be expected to be.

PRIMARY VS SECONDARY LARVAE

Among the synapomorphies of clades such as the Holometabola among the Insecta and the Epimorpha among the Chilopoda are characters of their postembryonic development, holometaboly ("complete metamorphosis"), and epimorphosis (postembryonic development without addition of segments or appendages). Other "higher" taxa have been tentatively characterized by the presence of specific larval types, e.g., the trochophore or the tornaria. Larval morphology is however liable to profound and even rapid change, up to complete disappearance. EvoDevo can thus offer a valuable contribution to phylogenetics, insofar as it can provide reliable scenarios of the evolvability of larvae and determine the degree to which larval and adult traits can actually evolve independently – a property likely to be different in different major clades of metazoans.

Quite long ago, Steinböck (1963) argued that the phylogenetic significance of the larvae has been considerably overestimated. Today, in the context of cladistic methods and language, we can say that even coding larval characters in matrices intended for the reconstruction of "higher" group relationships is fraught with problems. First, we have not even a satisfactory definition of larva (for a discussion, see Minelli 2009). Second, across the metazoans, larvae certainly evolved several times. Third, the widely accepted distinction between primary and secondary larvae is far from obvious and perhaps unwarranted. This is briefly discussed here.

When proposing a distinction between primary and secondary larvae, it is necessary to specify the node(s) of the phylogenetic tree corresponding to ground plans we credit with possessing either larval type. In the literature it seems often to be implicitly accepted that the last common ancestor of all recent metazoans, the Urbilateria, was an indirect developer. This does not rule out, however, the possibility that some clades re-evolved a secondary larva after having lost the primary one.

According to the phylogenetic scenario proposed by (Davidson 1991; see also Peterson et al.

1997; Cameron et al. 1998; Peterson and Davidson 2000), ancestral bilaterians would have lacked the later evolved genetic circuitry responsible for the complex body structure of their modern descendants. Their simpler genetic networks were only capable to produce little animals with a bodily organization directly comparable to that of the larva of many living invertebrates. In this scenario, the modern bilaterian adult is interpreted as an evolutionary novelty, a terminal addition grafted onto the original body plan, which is eventually conserved in the larva. As a consequence, larvae such as the trochophore and the tornaria would be primary because they would be older and recapitulative in respect to the corresponding adults. An often implied corollary is their supposed monophyletic origin.

However, there are problems with phylogeny (Valentine et al. 1999; Jenner 2000; Sly et al. 2003). Mollusk veligers are probably homoplastic (Ponder and Lindberg 1997; Waller 1998; Lindberg et al. 2004). Transitions from one larval type to another are frequent and often reversible. Planktotrophic larvae corresponding to the "primary" larva of Davidson and others are often lost and acquired again (Haszprunar et al. 1995; McEdward and Janies 1997; McHugh and Rouse 1998). Independent transition from planktonic to non-planktonic larvae occurred many times even within one genus, as in the case of *Conus* (Duda and Palumbi 1999).

The opposite idea that all larvae are secondary has been championed by many authors (e.g., Garstang 1922; de Beer 1954; Hadži 1955; Steinböck 1963; Conway Morris 1998; Valentine and Collins 2000; Collins and Valentine 2001; Hadfield et al. 2001), although often without a precise reference to a specific node in the metazoan tree.

The most serious difficulty with Davidson's scenario is the implied polyphyletic origin of the "zootype," that is, of the anterior-posterior patterning of the main axis of the bilaterians controlled by the Hox genes (Slack et al. 1993). Nothing like a zootype organization is found in any of the putative "primary" larvae. For example, in the pluteus of the sea urchin *Strongylocentrotus purpuratus*, Hox gene expression is limited to the adult rudiment (Arenas-

Mena et al. 2000). Similarly, in the trochophore of the polychaete *Chaetopterus*, Hox gene expression is limited to future adult tissues, while it does not show up in any of the larval structures that are fated to disappear at metamorphosis (Peterson et al. 2000). Things are broadly similar in other polychaetes, although in the late trochophore of *Platynereis dumerilii*, *Hox1* is expressed in the apical tuft cells (Kulakova et al. 2007).

Nielsen (2003a, b) regarded the lack of Hox gene expression in these larvae as an argument in favor of their primary nature. However, it is also possible (Minelli 2009) that the anterior-posterior patterning of the main body axis is a very old feature. If so, the lack of Hox gene expression in the larval tissues may indicate that the larva has been secondarily intercalated in the developmental schedule, in correspondence to an early developmental phase where Hox genes were still silent. We should not rule out, however, that other larvae may correspond to a later, Hox-expressing developmental phase. Let's remark in this context that trochophore-like larvae may have evolved repeatedly (Haszprunar et al. 1995).

TEMPO AND MODE IN EVOLUTION

Heterochrony in Phylogenetics: Noise or Data?

From the perspective of Haeckelian recapitulation, heterochrony is exception to the rule; in inferring phylogeny from ontogeny, it turns straight into noise. Indeed, it was right by showing the pervasiveness of heterochrony throughout the animal kingdom that de Beer, as mentioned before, was able to refute the "biogenetic law." However, de Beer was also able to provide a first classification of the possible kinds of change in ontogenetic sequences, thus remotely introducing two ideas that could be subsequently exploited in phylogenetics.

On the one side, de Beer's analysis suggested at least some degree of modularity of ontogenetic sequences. Anticipation, postponement, and changes in relative speed can only be predicated of "units," be these individual developmental processes or individual developmental stages.

This could suggest that homologs of which we can trace the evolution are not necessarily the organs – more generally, the structural features – of adult animals but perhaps also those of earlier stages or, better, (i) ontogenetic stages as such (e.g., the gastrula or the germband stage in arthropod embryonic development) and (ii) developmental processes as such (e.g., gastrulation, or a particular sequence of cell lineage).

On the other hand, the very possibility to classify heterochronies could invite a search for the phylogenetic signal possibly present in heterochronies as such. Patterns of heterochrony may contain useful phylogenetic signal, as demonstrated by, e.g., Guralnick and Lindberg (2001), who produced a phylogenetic tree of several lophotrochozoan taxa based on the timing of cell lineage events and found that the phylogenetic hypothesis thus obtained replicated patterns found in more traditional analyses. In another study, patterns of heterochrony in the developmental sequences of Branchiopoda were used to identify the origin of Cladocera (Fritsch et al. 2013).

Growth Heterochrony vs Sequence Heterochrony

At the beginning of this century, a decisive enhancement of the use of heterochrony as a source of data for phylogenetic reconstruction was obtained following a shift of focus from growth heterochrony to sequence heterochrony, to use a terminology introduced by Smith (2001). Virtually all of the traditional literature on heterochrony (e.g., Gould 1977; Alberch et al. 1979; McNamara 1986, 1995; McKinney 1988; McKinney and McNamara 1991) refers to *growth heterochrony*, i.e., to developmental changes in size and shape relationships.

However, many interesting evolutionary changes in developmental schedules are not changes in either size or shape. This is why Smith (1996, 2001, 2002, 2003) and Velhagen (1997) have suggested a different approach, termed *sequence heterochrony*, in which heterochrony is identified in the changes in the position of a developmental event relative to other events in

the same ontogenetic sequence. Several techniques have been proposed to analyze sequence heterochronies. Any two events A and B in a developmental sequence occur in one of the following orders: (i) A occurs before B, (ii) A and B are simultaneous, or (iii) A occurs after B. These timing relationships, or event pairs, are given a numerical score. Data are thus assembled in a matrix that can be analyzed under maximum parsimony. In these efforts, the major problem to be addressed is how to dissect ontogeny into reasonably independent units, as required by a cladistic analysis. This difficulty was acknowledged since the earliest studies in this area (e.g., Velhagen 1997; Bininda-Emonds et al. 2002). Schulmeister and Wheeler (2004) remarked that the optimization of developmental event sequences on a given cladogram based on event pairing may lead to unacceptable results because event pairing treats interdependent features as if they were independent. To overcome this problem, they suggested a method of character optimization treating the entire developmental sequence as a single character and aiming to determine the transformation cost between pairs of character states. Parsimov, another method for examining heterochronies in a phylogenetic framework, was introduced by Jeffery et al. (2005). In this parsimony-based method, the least number of event displacements (heterochronies) that explains all the observed event-pair changes is identified for each branch of the tree, thus eventually obtaining all alternative, equally parsimonious explanations, out of which a consensus is derived that contains the developmental changes that form part of every equally most parsimonious explanation.

Hot Points of Change Along the Developmental Schedule

One of the reasons to abandon von Baer's (1828) scenario of morphological divergence regularly increasing with the embryos progressing along their developmental trajectory and Haeckel's recapitulationist view according to which the evolutionary novelties are essentially terminal additions to the largely invariant earlier developmental stages is the fact that some developmental stages are more conservative (or more variable) than others, although not in a monotonic relationship with developmental age.

It is now fashionable to describe embryonic development in terms of the so-called hourglass model, to signify that the earliest stages (especially, but not exclusively, those under exclusive or prevailing control of maternal genes) are more extensively and easily divergent than later embryonic stages (Duboule 1994; Raff 1996; Hall 1997; Galis and Metz 2001). From initially different starting points (first discussed for insects by Sander 1976), developmental trajectories converge toward a much more conserved stage, often recognizable as characteristic for an individual phylum, which is called the phylotypic stage (Sander 1983) or at least a largely conserved segment of the developmental trajectory that has been termed the phylotypic period (Richardson et al. 1997). As expected, gene expression is maximally conserved around the phylotypic period (*Drosophila*: Kalinka et al. 2010).

Early-stage divergence, especially between closely related species, is often a direct consequence of the different amount of yolk stored in the female gamete during oogenesis; for example, thus is the case of two sea urchin species, the lecithotrophic *Heliocidaris erythrogramma* and the planktotrophic *H. tuberculata* (e.g., Parks et al. 1988; Wray and Raff 1991; Henry et al. 1992). More interesting, however, are other examples of early-stage divergence that cannot be explained in such a simple "mechanistic" way. The most dramatic case is the nematodes, among which the pattern of cleavage, the spatial arrangement, and the differentiation of cells have diverged dramatically during the history of the phylum, without producing corresponding changes in the adult phenotype (Schierenberg and Schulze 2008; Schulze and Schierenberg 2011).

Early divergence is sometimes noticeable even at intraspecific level, as shown by Tills et al. (2011) for the pond snail *Radix balthica*.

Heterochrony is not limited to the embryonic segment of the developmental schedule, but its occurrence along the postembryonic development is not frequently studied and is still less used to infer phylogenetic relationship. A promising example is the crustacean genus *Niphargus*: a preliminary study by Fišer et al. (2008) has revealed extensive sequence heterochrony along the postembryonic development, independence between events being more pronounced in mid-aged instars.

Saltational Evolution and Discontinuous Variation

Continuous variation is notoriously difficult to handle when we are confronted with the problem of partitioning it into bins to be differently coded in a data matrix used in a phylogenetic analysis. However, from the perspective of evolutionary change, continuous variation fits well within a gradualistic neo-Darwinian paradigm. The opposite is true when the observed character states are widely separated. In this case, there is no problem in partitioning our set into unambiguously distinct classes (unless the differences are so big that we may have problems recognizing two states as homologous). However, from an evolutionary point of view, we would not expect closely related taxa to be separated by an apparently unbridgeable gap. In other terms, we do not expect evolution to be saltational. However, this expectation is due for revision, in the light of facts that are possibly intractable in a traditional evolutionary scenario, but may become reasonable in the light of EvoDevo.

Major phenotypic differences may not necessarily depend on major changes or even rearrangements, at the genetic or genomic level. As mentioned above, the genotype→phenotype map is not necessarily simple or obvious, and a single instance of saltational evolution may require a reassessment of the phylogenetic signal carried by a given character. For example, the presence of 21 or 23 pairs of legs in the adult was long regarded as a reliable synapomorphy of the Scolopendromorpha, the other "higher" clades among the Chilopoda having instead either 15 (Scutigeromorpha, Craterostigmomorpha, Lithobiomorpha) or at least 27 (Geophilomorpha) pairs of legs. Recently, a scolopendromorph species with either 39 or 43 pairs of legs has been described (Chagas et al. 2008). What most matters (Minelli et al. 2009) besides the obvious need to reformulate the diagnosis of the clade Scolopendromorpha is that the newly discovered species (*Scolopendropsis duplicata*) is not the sister group to all remaining scolopendromorphs, or at least to a substantial subclade within them, but a very close relative of a "normal" species (*Scolopendropsis bahiensis*), to the same genus of which it has been thus assigned. The nature of the change in developmental mechanisms that in this case has broken a long entrenched phenotypic stability (the Carboniferous *Mazoscolopendra* had 21 pairs of legs; Mundel 1979) is not known, but it is not difficult to hypothesize a point mutation potentially responsible for this one-shot duplication of segment number.

Patterns of "saltational" variation are perhaps less rare than our gradualistic tradition has thus far invited to expect. EvoDevo is the obvious tool for accommodating them within our growing hypotheses of phylogenetic relationships.

References

Aboobaker A, Blaxter M (2003a) *Hox* gene evolution in nematodes: novelty conserved. Curr Opin Genet Dev 13:593–598

Aboobaker AA, Blaxter ML (2003b) *Hox* gene loss during dynamic evolution of the nematode cluster. Curr Biol 13:37–40

Abzhanov A, Popadić A, Kaufman TC (1999) Chelicerate *Hox* genes and the homology of arthropod segments. Evol Dev 1:77–89

Adoutte A, Balavoine G, Lartillot N, Lespinet O, Prud'homme B, de Rosa R (2000) The new animal phylogeny: reliability and implications. Proc Natl Acad Sci U S A 97:4453–4456

Aguinaldo AMA, Turbeville JM, Lindford LS, Rivera MC, Garey JR, Raff RA, Lake JA (1997) Evidence for a clade of nematodes, arthropods and other moulting animals. Nature 387:489–493

Alberch P (1991) From genes to phenotype: dynamical systems and evolvability. Genetica 84:5–11

Alberch P, Gould SJ, Oster G, Wake D (1979) Size and shape in ontogeny and phylogeny. Paleobiology 5:296–317

Altenberg L (1995) Genome growth and the evolution of the genotype-phenotype map. In: Banzhaf W, Eeckman FH (eds) Evolution and biocomputation. Computational models of evolution. Springer, Berlin, pp 205–259

Amirthalingam K, Lorens JB, Saetre BO, Salaneck E, Fjose A (1995) Embryonic expression and DNA-binding properties of zebrafish Pax-6. Biochem Biophys Res Commun 215:122–128

Angelini DR, Kaufman TC (2005) Comparative developmental genetics and the evolution of arthropod body plans. Annu Rev Genet 39:95–119

Arenas-Mena C, Cameron AR, Davidson EH (2000) Spatial expression of *Hox* cluster genes in the ontogeny of a sea urchin. Development 127:4631–4643

Arthur W (2002) The emerging conceptual framework of evolutionary developmental biology. Nature 415:757–764

Atallah J, Watabe H, Kopp A (2012) Many ways to make a novel structure: a new mode of sex comb development in *Drosophila*. Evol Dev 14:476–483

Balavoine G, de Rosa R, Adoutte A (2002) *Hox* clusters and bilaterian phylogeny. Mol Phylogenet Evol 24:366–373

Ball EE, Hayward DC, Saint R, Miller DJ (2004) A simple plan – cnidarians and the origins of developmental mechanisms. Nat Rev Genet 5:567–577

Barmina O, Kopp A (2007) Sex-specific expression of a *Hox* gene associated with rapid morphological evolution. Dev Biol 311:277–286

Barrantes G, Eberhard WG (2010) Ontogeny repeats phylogeny in *Steatoda* and *Latrodectus* spiders. J Arachnol 38:485–494

Beklemishev VN (1963) On the relationships of the Turbellaria to other groups of the animal kingdom. In: Dougherty EC (ed) The lower metazoa. Comparative biology and phylogeny. University of California Press, Berkeley, pp 234–244

Bininda-Emonds ORP, Jeffery JE, Coates MI, Richardson MK (2002) From Haeckel to event-pairing: the evolution of developmental sequences. Theory Biosci 121:297–320

Blin M, Rabet N, Deutsch JS, Mouchel-Vielh E (2003) Possible implication of *Hox* genes *Abdominal-B* and *abdominal-A* in the specification of genital and abdominal segments in cirripedes. Dev Genes Evol 213:90–96

Boero F, Gravili C, Pagliara P, Piraino S, Bouillon J, Schmid V (1998) The Cnidarian premises of metazoan evolution: from triploblasty, to coelom formation, to metamery. Ital J Zool 65:5–9

Botas J, Auwers L (1996) Chromosomal binding sites of ultrabithorax homeotic proteins. Mech Dev 56:129–138

Bourlat SJ, Nielsen C, Economou AD, Telford MJ (2008) Testing the new animal phylogeny: a phylum level molecular analysis of the animal kingdom. Mol Phyl Evol 49:23–31

Brenneis G, Ungerer P, Scholtz G (2008) The chelifores of sea spiders (Arthropoda, Pycnogonida) are the appendages of the deutocerebral segment. Evol Dev 10:717–724

Brigandt I, Love AC (2010) Evolutionary novelty and the evo-devo synthesis: field notes. Evol Biol 37:93–99

Brigandt I, Love AC (2012) Conceptualizing evolutionary novelty: moving beyond definitional debates. J Exp Zool (Mol Dev Evol) 318B:417–427

Burton PM (2008) Insights from diploblasts; the evolution of mesoderm and muscle. J Exp Zool (Mol Dev Evol) 310B:5–14

Butts T, Holland PWH, Ferrier DE (2008) The urbilaterian Super-Hox cluster. Trends Genet 24:259–262

Cameron RA, Peterson KJ, Davidson EH (1998) Developmental gene regulation and the evolution of large animal body plans. Am Zool 38:609–620

Carroll SB (2008) Evo-devo and an expanding evolutionary synthesis: a genetic theory of morphological evolution. Cell 134:25–36

Carroll SB, Gates J, Keys D, Paddock SW, Panganiban GE, Selegue JE, Williams JA (1994) Pattern formation and eyespot determination in butterfly wings. Science 265:109–114

Carroll SB, Grenier JK, Weatherbee SD (2005) From DNA to diversity: molecular genetics and the evolution of animal design, 2nd edn. Blackwell, Malden

Casares F, Mann RS (1998) Control of antennal versus leg development in *Drosophila*. Nature 392:723–726

Casares F, Mann RS (2001) The ground state of the ventral appendage in *Drosophila*. Science 293:1477–1480

Chagas A Jr, Edgecombe GD, Minelli A (2008) Variability in trunk segmentation in the centipede order Scolopendromorpha: a remarkable new species of *Scolopendropsis* Brandt (Chilopoda: Scolopendridae) from Brazil. Zootaxa 1888:36–46

Chipman AD (2010) Parallel evolution of segmentation by co-option of ancestral gene regulatory networks. Bioessays 32:60–70

Chisholm AD, Horvitz HR (1995) Patterning of the *Caenorhabditis elegans* head region by the Pax-6 family member *vab-3*. Nature 377:52–55

Chow RL, Altmann CR, Lang RA, Hemmati-Brivanlou A (1999) *Pax6* induces ectopic eyes in a vertebrate. Development 126:4213–4222

Collins AG, Valentine JW (2001) Defining phyla: evolutionary pathways to metazoan body plans. Evol Dev 3:432–442

Conway Morris S (1998) Metazoan phylogenies: falling into place or falling to pieces? A paleontological perspective. Curr Opin Genet Dev 8:662–666

Cook CE, Smith ML, Telford MJ, Bastianello A, Akam M (2001) *Hox* genes and the phylogeny of the arthropods. Curr Biol 11:759–763

Copf T, Rabet N, Celniker SE, Averof M (2003) Posterior patterning genes and the identification of a unique body region in the brine shrimp *Artemia franciscana*. Development 130:5915–5927

Cracraft J (2005) Phylogeny and evo-devo: characters, phylogeny and historical analysis of the evolution of development. Zoology 108:345–356

Cuvier G (1812) Sur un nouveau rapprochement à établir entre les classes qui composent le règne animal. Ann Mus Nat Hist Nat Paris 19:73–84

Czerny T, Busslinger M (1995) DNA-binding and transactivation properties of Pax-6: three amino acids in the paired domain are responsible for the different sequence recognition of Pax-6 and BSAP (Pax-5). Mol Cell Biol 15:2858–2871

Davidson EH (1991) Spatial mechanisms of gene regulation in metazoan embryos. Development 113:1–26

Davidson EH (2006) The regulatory genome: gene regulatory networks in development and evolution. Academic Press, Oxford

Davidson EH, Erwin DH (2006) Gene regulatory networks and the evolution of animal body plans. Science 311:796–800

Davidson EH, Rast JP, Oliveri P, Ransick A, Calestani C, Yuh C-H, Minokawa T, Amore G, Hinman V, Arenas-Mena C, Otim O, Brown CT, Livi CB, Lee PY, Revilla R, Rust AG, Zj P, Schilstra MJ, Clarke PJC, Arnone MI, Rowen L, Cameron RA, McClay DR, Hood L, Bolouri H (2002) A genomic regulatory network for development. Science 295:1669–1678

Davidson EH, McClay DR, Hood L (2003) Regulatory gene networks and the properties of the developmental process. Proc Natl Acad Sci U S A 100:1475–1480

de Beer GR (1930) Embryology and evolution. Clarendon Press, Oxford

de Beer GR (1940) Embryos and ancestors. Clarendon Press, Oxford

de Beer GR (1954) The evolution of the metazoa. In: Huxley J, Hardy AC, Ford EB (eds) Evolution as a process. Allen and Unwin, London, pp 24–33

de Queiroz K (1985) The ontogenetic method for determining character polarity and its relevance to phylogenetic systematics. Syst Zool 34:280–299

de Rosa R, Grenier JK, Andreeva T, Cook CE, Adoutte A, Akam M, Carroll SB, Balavoine G (1999) Hox genes in brachiopods and priapulids and protostome evolution. Nature 399:772–776

Denes AS, Jekely G, Steinmetz PR, Raible F, Snyman H, Prud'homme B, Ferrier DEK, Balavoine G, Arendt D (2007) Molecular architecture of annelid nerve cord supports common origin of nervous system centralization in Bilateria. Cell 129:277–288

Dong PDS, Chu J, Panganiban G (2001) Proximodistal domain specification and interactions in *Drosophila* developing appendages. Development 128:2365–2372

Dray N, Tessmar-Raible K, Le Gouar M, Vibert L, Christodoulou F, Schipany K, Guillou A, Zantke J, Snyman H, Béhague J, Vervoort M, Arendt D, Balavoine G (2010) Hedgehog signaling regulates segment formation in the annelid *Platynereis*. Science 329:339–342

Duboule D (1994) Temporal colinearity and the phylotypic progression: a basis for the stability of a vertebrate Bauplan and the evolution of morphologies through heterochrony. Dev 1994 Suppl 135–142

Duboule D (2007) The rise and fall of *Hox* gene clusters. Development 134:2549–2560

Duda TF, Palumbi SR (1999) Developmental shifts and species selection in gastropods. Proc Natl Acad Sci U S A 96:10272–10277

Dunn CW, Hejnol A, Matus DQ, Pang K, Browne WE, Smith SA, Seaver E, Rouse GW, Obst M, Edgecombe GD, Sørensen MV, Haddock SHD, Schmidt-Rhaesa A, Okusu A, Kristensen RM, Wheeler WC, Martindale MQ, Giribet G (2008) Broad phylogenomic sampling improves resolution of the animal tree of life. Nature 452:745–749

Edgecombe GD, Giribet G, Dunn CW, Hejnol A, Kristensen RM, Neves RC, Rouse GW, Worsaae K, Sørensen MV (2011) Higher-level metazoan relationships: recent progress and remaining questions. Org Divers Evol 11:151–172

Eernisse DJ, Albert JS, Anderson FE (1992) Annelida and Arthropoda are not sister taxa: a phylogenetic analysis of spiralian metazoan morphology. Syst Biol 41:305–330

Ewen-Campen B, Srouji JR, Schwager EE, Extavour CG (2012) Oskar predates the evolution of germ plasm in insects. Curr Biol 22:2278–2283

Ferrier DEK (2007) Evolution of *Hox* gene clusters. In: Papageorgiou S (ed) *Hox* gene expression. Springer, New York, pp 53–67

Ferrier DEK (2010) Evolution of *Hox* complexes. Adv Exp Med Biol 689:91–100

Ferrier DEK (2011) Tunicates push the limits of animal evo-devo. BMC Biol 9:3

Ferrier DE, Holland PW (2001) Ancient origin of the *Hox* gene cluster. Nat Rev Genet 2:33–38

Ferrier DE, Minguillon C (2003) Evolution of the *Hox/ParaHox* gene clusters. Int J Dev Biol 47:605–611

Fink WL (1982) The conceptual relationship between ontogeny and phylogeny. Paleobiology 8:254–264

Finnerty JR, Martindale MQ (1998) The evolution of the *Hox* cluster: insights from outgroups. Curr Opin Genet Dev 8:681–687

Fišer C, Bininda-Emonds ORP, Blejec A, Sket B (2008) Can heterochrony help explain the high morphological diversity within the genus *Niphargus* (Crustacea: Amphipoda)? Org Divers Evol 8:146–162

Fritsch M, Bininda-Emonds ORP, Richter S (2013) Unraveling the origin of Cladocera by identifying heterochrony in the developmental sequences of Branchiopoda. Front Zool 10:35

Galis F (2001) Key innovations and radiations. In: Wagner GP (ed) The character concept in evolutionary biology. Academic Press, San Diego, pp 58–605

Galis F, Metz JA (2001) Testing the vulnerability of the phylotypic stage: on modularity and evolutionary conservation. J Exp Zool (Mol Dev Evol) 291:195–204

Gao K-Q, Shubin NH (2001) Late Jurassic salamanders from northern China. Nature 410:574–577

Garcia-Fernàndez J (2005a) *Hox*, *ParaHox*, Proto*Hox*: facts and guesses. Heredity 94:145–152

Garcia-Fernàndez J (2005b) The genesis and evolution of homeobox gene clusters. Nat Rev Genet 6:881–892

Garstang W (1922) The theory of recapitulation: a critical restatement of the biogenetic law. J Linnean Soc Zool 35:81–101

Gehring WJ (2000) Reply to Meyer-Rochow. Trends Genet 16:245

Gehring WJ, Ikeo K (1999) *Pax 6*: mastering eye morphogenesis and eye evolution. Trends Genet 15:371–377

Giribet G (2003) Molecules, development and fossils in the study of metazoan evolution: articulata versus Ecdysozoa revisited. Zoology 106:303–326

Glardon S, Holland LZ, Gehring WJ, Holland ND (1998) Isolation and developmental expression of the amphioxus *Pax-6* gene (*AmphiPax-6*): insights into eye and photoreceptor evolution. Development 125:2701–2710

Gould SJ (1977) Ontogeny and phylogeny. The Belknap Press of Harvard University Press, Cambridge, MA

Guralnick RP, Lindberg DR (2001) Reconnecting cell and animal lineages: what do cell lineages tell us about the evolution and development of Spiralia? Evolution 55:1501–1519

Haag ES (2014) The same but different: worms reveal the pervasiveness of developmental system drift. PLoS Genet 10:e1004150

Hadfield MG, Carpizo-Ituarte EJ, Del Carmen K, Nedved BT (2001) Metamorphic competence, a major adaptive convergence in marine invertebrate larvae. Am Zool 41:1123–1131

Hadži J (1955) K diskusiji o novi sistematiki živalstva (Zur Diskussion über das neue zoologische System.) Razprave, Slovenska Akademija Znanosti in Umetnosti, Razrad za Prirodoslovne in Medicinske Vede, Ljubljana 3:175–207 [In Slovenian with German summary]

Haeckel E (1866) Generelle Morphologie der Organismen. Allgemeine Grundzüge der organischen Formen-Wissenschaft, mechanisch begründet durch die von Charles Darwin reformirte Descendenz-Theorie, vol 1, Allgemeine Anatomie der Organismen. Reimer, Berlin

Halanych KM (2004) The new view of animal phylogeny. Annu Rev Ecol Evol Syst 35:229–256

Halder G, Callaerts P, Gehring WJ (1995) Induction of ectopic eyes by targeted expression of the *eyeless* gene in *Drosophila*. Science 267:1788–1792

Hall BK (1997) Phylotypic stage or phantom, is there a highly conserved embryonic stage in vertebrates? Trends Ecol Evol 12:461–463

Hall BK (1998) Evolutionary developmental biology, 2nd edn. Chapman & Hall, London

Hall BK, Kerney R (2012) Levels of biological organization and the origin of novelty. J Exp Zool (Mol Dev Evol) 318B:428–437

Hall BK, Olson WM (2003) Keywords and concepts in evolutionary developmental biology. Harvard University Press, Cambridge, MA

Harris WA (1997) *Pax-6*: where to be conserved is not conservative. Proc Natl Acad Sci U S A 94:2098–2100

Haszprunar G, von Salvini-Plawen L, Rieger RM (1995) Larval planktotrophy. Acta Zool 76:141–154

Hayward DC, Miller DJ, Ball EE (2004) *snail* expression during embryonic development of the coral *Acropora*: blurring the diploblast/triploblast divide? Dev Genes Evol 214:257–260

Heffer A, Xiang J, Pick L (2013) Variation and constraint in *Hox* gene evolution. Proc Natl Acad Sci U S A 110:2211–2216

Hendrikse JL, Parsons TE, Hallgrímsson B (2007) Evolvability as the proper focus of evolutionary developmental biology. Evol Dev 9:93–401

Hennig W (1966) Phylogenetic systematics. University of Illinois Press, Urbana

Henry JJ, Klueg KM, Raff RA (1992) Evolutionary dissociation between cleavage, cell lineage and embryonic axes in sea urchin embryos. Development 114:931–938

Hertel J, Lindemeyer M, Missal K, Fried C, Tanzer A, Flamm C, Hofacker IL, Stadler PF, The Students of Bioinformatics Computer Labs 2004 and 2005 (2006) The expansion of the metazoan microRNA repertoire. BMC Genomics 7:25

Holland LZ, Holland ND (1998) Developmental gene expression in amphioxus: new insights into the evolutionary origin of vertebrate brain regions, neural crest, and rostrocaudal segmentation. Am Zool 38:647–658

Holland PWH (2012) Evolution of homeobox genes. WIREs Dev Biol. doi:10.1002/wdev.78

Holley SA, Jülich D, Rauch GJ, Geisler R, Nüsslein-Volhard C (2002) *her1* and the notch pathway function within the oscillator mechanism that regulates zebrafish somitogenesis. Development 129:1175–1183

Hueber SD, Rauch J, Djordjevic MA, Gunter H, Weiller GF, Frickey T (2013) Analysis of central Hox protein types across bilaterian clades: on the diversification of central Hox proteins from an Antennapedia/Hox7-like protein. Dev Biol 383:175–185

Hughes CL, Kaufman TC (2002) *Hox* genes and the evolution of the arthropod body plan. Evol Dev 4:459–499

Ivanova Kazas OM (2013) Origin of arthropods and of the clades of Ecdysozoa. Russ J Dev Biol 44:221–231

Jacobs DK, Wray CG, Wedeen CJ, Kostriken R (2000) Molluscan *engrailed* expression, serial organization, and shell evolution. Evol Dev 2:340–347

Jager M, Murienne J, Clabaut C, Deutsch J, Le Guyader H, Manuel M (2006) Homology of arthropod anterior appendages revealed by *Hox* gene expression in a sea spider. Nature 441:506–508

Jeffery JE, Bininda-Emonds ORP, Coates MI, Richardson MK (2005) A new technique for identifying sequence heterochrony. Syst Biol 54:230–240

Jenner RA (2000) Evolution of animal body plans: the role of metazoan phylogeny at the interface between pattern and process. Evol Dev 2:208–221

Jiang Y-J, Aerne BL, Smithers L, Haddon C, Ish-Horowicz D, Lewis J (2000) Notch signalling and the synchronization of the somite segmentation clock. Nature 408:475–479

Kainz F, Ewen-Campen B, Akam M, Extavour CG (2011) Notch/Delta signalling is not required for segment generation in the basally branching insect *Gryllus bimaculatus*. Development 138:5015–5026

Kalinka AT, Varga KM, Gerrard DT, Preibisch S, Corcoran DL, Jarrells J, Ohler U, Bergman CM, Tomancak P (2010) Gene expression divergence recapitulates the developmental hourglass model. Nature 468:811–814

Kell DB (2002) Genotype–phenotype mapping: genes as computer programs. Trends Genet 18:555–559

Khadjeh S, Turetzek N, Pechmann M, Schwager EE, Wimmer EA, Damen WGM, Prpic N-M (2012) Divergent role of the *Hox* gene *Antennapedia* in spiders is responsible for the convergent evolution of abdominal limb repression. Proc Natl Acad Sci U S A 109:4921–4926

Kluge AG (1985) Ontogeny and phylogenetic systematics. Cladistics 1:13–27

Kmita-Cunisse M, Loosli F, Bierne J, Gehring WJ (1998) Homeobox genes in the ribbonworm *Lineus sanguineus*: evolutionary implications. Proc Natl Acad Sci U S A 95:3030–3035

Kourakis MJ, Martindale MQ (2000) Combined-method phylogenetic analysis of *Hox* and *ParaHox* genes of the metazoa. J Exp Zool 288:175–191

Kowalewski A (1866) Entwicklungsgeschichte der einfachen Ascidien. Mém Acad Imp Sci St Petersburg (7)10(15):1–19

Kozmik Z (2005) *Pax* genes in eye development and evolution. Curr Opin Genet Dev 15:430–438

Kristensen RM (2003) Comparative morphology: do the ultrastructural investigations of Loricifera and Tardigrada support the clade Ecdysozoa? In: Legakis A, Sfenthourakis S, Polymeni R, Thessalou-Legaki M (eds) The new panorama of animal evolution. Proceedings of the 18th international congress of Zoology. Pensoft, Sofia-Moscow, pp 467–477

Kugler JE, Kerner P, Bouquet J-M, Jiang D, Di Gregorio A (2011) Evolutionary changes in the notochord genetic toolkit: a comparative analysis of notochord genes in the ascidian *Ciona* and the larvacean *Oikopleura*. BMC Evol Biol 11:21

Kulakova M, Bakalenko N, Nivikova E, Cook CE, Eliseeva E, Steinmetz PRH, Kostyuchenko RP, Dondua A, Arendt D, Akam M, Andreeva T (2007) *Hox* gene expression in larval development of the polychaetes *Nereis virens* and *Platynereis dumerilii* (Annelida, Lophotrochozoa). Dev Genes Evol 217:39–54

Laubichler MD (2010) Evolutionary developmental biology offers a significant challenge to the Neo-Darwinian paradigm. In: Ayala FJ, Arp R (eds) Contemporary debates in philosophy of biology. Wiley-Blackwell, New York, pp 199–212

Lewis J, Hanisch A, Holder M (2009) Notch signaling, the segmentation clock, and the patterning of vertebrate somites. J Biol 8:44

Li HS, Yang JM, Jacobson RD, Pasko D, Sundin O (1994) *Pax-6* is first expressed in a region of ectoderm anterior to the early neural plate: implications for stepwise determination of the lens. Dev Biol 162:181–194

Lichtneckert R, Reichert H (2005) Insights into the urbilaterian brain: conserved genetic patterning mechanisms in insect and vertebrate brain development. Heredity 94:465–477

Lindberg DR, Ponder WF, Haszprunar G (2004) The Mollusca: relationships and patterns from their first half-billion years. In: Cracraft J, Donoghue MJ (eds) Assembling the tree of life. Oxford University Press, Oxford, pp 252–278

Linné C (1758) Systema Naturae per regna tria Naturae secundum classes, ordines, genera, species, cum charateribus, differentiis, synonymis, locis. Editio decima. Tomus I. Apud Laurentium Salvium, Holmiae

Love AC (2008) Explaining evolutionary innovations and novelties: criteria of explanatory adequacy and epistemological prerequisites. Philos Sci 75:874–886

Lowe CJ, Wray GA (1997) Radical alterations in the roles of homeobox genes during echinoderm evolution. Nature 389:718–721

Maduro MF (2006) Endomesoderm specification in *Caenorhabditis elegans* and other nematodes. Bioessays 28:1010–1022

Maduro MF, Rothman JH (2002) Making worm guts: the gene regulatory network of the *Caenorhabditis elegans* endoderm. Dev Biol 246:68–85

Mallatt J, Craig CW, Yoder MJ (2012) Nearly complete rRNA genes from 371 animalia: updated structure-based alignment and phylogenetic analysis. Mol Phylogenet Evol 63:604–617

Manuel M, Jager M, Murienne J, Clabaut C, Le Guyader H (2006) *Hox* genes in sea spiders (Pycnogonida) and the homology of arthropod head segments. Dev Genes Evol 216:481–491

Mara A, Schroeder J, Chalouni C, Holley SA (2007) Priming, initiation and synchronization of the segmentation clock by *delta D* and *delta C*. Nat Cell Biol 9:523–530

Martindale MQ, Pang K, Finnerty JR (2004) Investigating the origins of triploblasty: 'mesodermal' gene expression in a diploblastic animal, the sea anemone *Nematostella vectensis* (Phylum, Cnidaria; Class, Anthozoa). Development 131:2463–2474

Martynov AV (2012) Ontogenetic systematics: the synthesis of taxonomy, phylogenetics, and evolutionary developmental biology. Paleontol J 46:833–864

Mastick GS, McKay R, Oligino T, Donovan K, Lopez AJ (1995) Identification of target genes regulated by homeotic proteins in *Drosophila melanogaster* through genetic selection of Ultrabithorax protein-binding sites in yeast. Genetics 139:349–363

Maxmen A, Browne WE, Martindale MQ, Giribet G (2005) Neuroanatomy of sea spiders implies an appendicular origin of the protocerebral segment. Nature 437:1144–1148

McEdward LR, Janies DA (1997) Relationships among development, ecology and morphology in the evolution of echinoderm larvae and life cycles. Biol J Linn Soc 60:381–400

McHugh D, Rouse GW (1998) Life history evolution of marine invertebrates: new views from phylogenetic systematics. Trends Ecol Evol 13:182–186

McKinney ML (ed) (1988) Heterochrony in evolution: a multidisciplinary approach. Plenum, New York

McKinney ML, McNamara KJ (1991) Heterochrony. The evolution of ontogeny. Plenum, New York

McNamara KJ (1986) A guide to the nomenclature of heterochrony. J Paleontol 60:4–13

McNamara KJ (ed) (1995) Evolutionary change and heterochrony. Wiley, Chichester

Meyer-Rochow VB (2000) The eye: monophyletic, polyphyletic or perhaps biphyletic? Trends Genet 16:244–245

Mezey JG, Cheverud JM, Wagner GP (2000) Is the genotype-phenotype map modular?: a statistical approach using mouse quantitative trait loci data. Genetics 156:305–311

Minelli A (2000) Limbs and tail as evolutionarily diverging duplicates of the main body axis. Evol Dev 2:157–165

Minelli A (2003a) The development of animal form. Cambridge University Press, Cambridge

Minelli A (2003b) The origin and evolution of appendages. Int J Dev Biol 47:573–581

Minelli A (2007) Invertebrate taxonomy and evolutionary developmental biology. Zootaxa 1668:55–60

Minelli A (2009) Perspectives in animal phylogeny and evolution. Oxford University Press, Oxford

Minelli A (2010) Evolutionary developmental biology does not offer a significant challenge to the Neo-Darwinian paradigm. In: Ayala FJ, Arp R (eds) Contemporary debates in philosophy of biology. Wiley-Blackwell, New York, pp 213–226

Minelli A, Bortoletto S (1988) Myriapod metamerism and arthropod segmentation. Biol J Linn Soc 33:323–343

Minelli A, Fusco G (2005) Conserved vs. innovative features in animal body organization. J Exp Zool (Mol Dev Evol) 304B:520–525

Minelli A, Fusco G (eds) (2008) Evolving pathways. Key themes in evolutionary developmental biology. Cambridge University Press, Cambridge

Minelli A, Fusco G (2013) Homology. In: Kampourakis K (ed) The philosophy of biology: a companion for educators. Springer, Dordrecht, pp 289–322

Minelli A, Negrisolo E, Fusco G (2007) Reconstructing animal phylogeny in the light of evolutionary developmental biology. In: Hodkinson TR, Parnell JAN (eds) Reconstructing the tree of life: taxonomy and systematics of species rich taxa. Taylor and Francis – CRC Press, Boca Raton, pp 177–190

Minelli A, Chagas-Júnior A, Edgecombe GD (2009) Saltational evolution of trunk segment number in centipedes. Evol Dev 11:318–322

Moczek AP (2008) On the origins of novelty in development and evolution. Bioessays 30:432–447

Moreno-Risueno MA, Van Norman JM, Moreno A, Zhang J, Ahnert SE, Benfey PN (2010) Oscillating gene expression determines competence for periodic *Arabidopsis* root branching. Science 329:1306–1311

Moshel SM, Levine M, Collier JR (1998) Shell differentiation and *engrailed* expression in the *Ilyanassa* embryo. Dev Genes Evol 208:135–141

Müller GB (1990) Developmental mechanisms at the origin of morphological novelty: a side-effect hypothesis. In: Nitecki MH (ed) Evolutionary innovations. University of Chicago Press, Chicago, pp 99–130

Müller GB (2008) Evo-devo as a discipline. In: Minelli A, Fusco G (eds) Evolving pathways. Key themes in evolutionary developmental biology. Cambridge University Press, Cambridge, pp 5–30

Müller GB, Newman SA (eds) (2003) Origination of organismal form: beyond the gene in developmental and evolutionary biology. MIT Press, Cambridge, MA

Müller GB, Newman SA (2005) The innovation triad: an EvoDevo agenda. J Exp Zool (Mol Dev Evol) 304B:487–503

Müller GB, Wagner GP (1991) Novelty in evolution: restructuring the concept. Annu Rev Ecol Syst 22:229–256

Müller GB, Wagner GP (2003) Innovation. In: Hall BK, Olson WM (eds) Keywords and concepts in evolutionary developmental biology. Harvard University Press, Cambridge, MA, pp 218–227

Mundel P (1979) The centipedes (chilopoda) of the Mazon Creek. In: Nitecki MH (ed) Mazon Creek fossils. Academic, New York, pp 361–378

Nederbragt AJ, van Loon AE, Dictus WJAG (2002) Expression of *Patella vulgata* orthologs of *engrailed* and *dpp-BMP2/4* in adjacent domains during molluscan shell development suggests a conserved compartment boundary mechanism. Dev Biol 246:341–355

Nelson GJ (1978) Ontogeny, phylogeny, paleontology and the biogenetic law. Syst Zool 27:324–345

Nielsen C (2001) Animal evolution: interrelationships of the living phyla, 2nd edn. Oxford University Press, Oxford

Nielsen C (2003a) Defining phyla: morphological and molecular clues to metazoan evolution. Evol Dev 5:386–393

Nielsen C (2003b) Proposing a solution to the Articulata-Ecdysozoa controversy. Zool Scr 32:475–482

Nielsen C (2008) Ontogeny of the spiralian brain. In: Minelli A, Fusco G (eds) Evolving pathways: key themes in evolutionary developmental biology. Cambridge University Press, Cambridge, pp 399–416

Oates AC, Morelli LG, Ares S (2012) Patterning embryos with oscillations: structure, function and dynamics of the vertebrate segmentation clock. Development 139:625–639

Özbudak EM, Lewis J (2008) Notch signalling synchronizes the zebrafish segmentation clock but is not needed to create somite boundaries. PLoS Genet 4:e15

Paaby AB, Rockman MV (2013) The many faces of pleiotropy. Trends Genet 29:66–73

Panganiban G, Irvine SM, Lowe C, Roehl H, Corley LS, Sherbon B, Grenier JK, Fallon JF, Kimble J, Walker M, Wray GA, Swalla BJ, Martindale MQ, Carroll SB (1997) The origin and evolution of animal appendages. Proc Natl Acad Sci U S A 94:5162–5166

Parks AL, Parr BA, Chin J-E, Leaf DS, Raff RA (1988) Molecular analysis of heterochronic changes in the evolution of direct developing sea urchins. J Evol Biol 1:27–44

Passamaneck YJ, Halanych KM (2004) Evidence from *Hox* genes that bryozoans are lophotrochozoans. Evol Dev 6:275–281

Pearson JC, Lemons D, McGinnis W (2005) Modulating *Hox* gene functions during animal body patterning. Nat Rev Genet 6:893–904

Peterson KJ, Cameron RA, Davidson EH (1997) Set-aside cells in maximal indirect development: evolutionary and developmental significance. BioEssays 19:623–631

Peterson KJ, Davidson EH (2000) Regulatory evolution and the origin of the bilaterians. Proc Natl Acad Sci U S A 97:4430–4433

Peterson KJ, Irvine SQ, Cameron RA, Davidson EH (2000) Quantitative assessment of *Hox* complex expression in the indirect development of the polychaete annelid *Chaetopterus* sp. Proc Natl Acad Sci U S A 97:4487–4492

Peterson T, Müller GB (2013) What is evolutionary novelty? Process versus character based definitions. J Exp Zool (Mol Dev Evol) 320B:345–350

Pigliucci M (2008) What, if anything, is an evolutionary novelty? Philos Sci 75:887–898

Pigliucci M (2010) Genotype→phenotype mapping and the end of the 'genes as blueprint' metaphor. Phil Trans R Soc B 365:557–566

Pigliucci M, Müller G (eds) (2010) Evolution: the extended synthesis. MIT Press, Cambridge, MA

Pilato G, Binda MG, Biondi O, D'Urso V, Lisi O, Marletta A, Maugeri S, Nobile V, Rappazzo G, Sabella G, Sammartano F, Turrisi G, Viglianisi F (2005) The clade Ecdysozoa, perplexities and questions. Zool Anz 244:43–50

Pineda D, Gonzalez J, Callaerts P, Ikeo K, Gehring WJ, Saló E (2000) Searching for the prototypic eye genetic network: *Sine oculis* is essential for eye regeneration in planarians. Proc Natl Acad Sci U S A 97:4525–4529

Pires-daSilva A, Sommer RJ (2003) The evolution of signalling pathways in animal development. Nat Rev Genet 4:39–49

Ponder WF, Lindberg DR (1997) Towards a phylogeny of gastropod molluscs: an analysis using morphological characters. Zool J Linn Soc 119:83–265

Prochnik SE, Rokhsar DS, Aboobaker AA (2007) Evidence for a microRNA expansion in the bilaterian ancestor. Dev Genes Evol 217:73–77

Pueyo JI, Lanfear R, Couso JP (2008) Ancestral Notch-mediated segmentation revealed in the cockroach *Periplaneta americana*. Proc Natl Acad Sci U S A 105:16614–16619

Rabinowitz JS, Chan XY, Kingsley EP, Duan Y, Lambert JD (2008) *nanos* is required in somatic blast cell lineages in the posterior of a mollusc embryo. Curr Biol 18:331–336

Raff RA (1996) The shape of life: genes, development and the evolution of animal form. University of Chicago Press, Chicago

Richardson MK, Hanken J, Gooneratne ML, Pieau C, Raynaud A, Selwood L, Wright GM (1997) There is no highly conserved embryonic stage in the vertebrates: implications for current theories of evolution and development. Anat Embryol 196:91–106

Richmond DL, Oates AC (2012) The segmentation clock: inherited trait or universal design principle? Curr Opin Genet 22:600–606

Rieppel O (1979) Ontogeny and the recognition of primitive character states. Zschr zool Syst Evol Forsch 17:57–61

Rivera AS, Weisblat DA (2009) And Lophotrochozoa makes three: Notch/Hes signaling in annelid segmentation. Dev Genes Evol 219:37–43

Robinson JM, Sperling EA, Bergum B, Adamski M, Nichols SA, Adamska M, Peterson KJ (2013) The identification of microRNAs in calcisponges: independent evolution of microRNAs in basal metazoans. J Exp Zool (Mol Dev Evol) 320B:84–93

Rokas A, Kathirithamby J, Holland PWH (1999) Intron insertion as a phylogenetic character: the *engrailed* homeobox of Strepsiptera does not indicate affinity with Diptera. Insect Mol Biol 8:527–530

Sander K (1976) Specification of the basic body plan in insect embryogenesis. Adv Insect Physiol 12:125–238

Sander K (1983) The evolution of patterning mechanisms: gleanings from insect embryogenesis and spermatogenesis. In: Goodwin BC, Holder N, Wylie CC (eds) Evolution and development. Cambridge University Press, Cambridge, pp 137–159

Sarrazin AF, Peel AD, Averof M (2012) A segmentation clock with two-segment periodicity in insects. Science 336:338–341

Schierenberg E, Schulze J (2008) Many roads lead to Rome: different ways to construct a nematode. In: Minelli A, Fusco G (eds) Evolving pathways: key themes in evolutionary developmental biology. Cambridge University Press, Cambridge, pp 261–280

Schmidt-Rhaesa A (2004) Ecdysozoa versus Articulata. Sitzungsber Ges Naturforsch Freunde Berl 43:35–49

Schmidt-Rhaesa A (2006) Perplexities concerning the Ecdysozoa: a reply to Pilato et al. Zool Anz 244:205–208

Schmidt-Rhaesa A, Bartolomaeus T, Lemburg C, Ehlers U, Garey JR (1998) The position of the Arthropoda in the phylogenetic system. J Morphol 238:263–285

Scholtz G (2002) The Articulata hypothesis – or what is a segment? Org Divers Evol 2:197–215

Scholtz G (2003) Is the taxon Articulata obsolete? Arguments in favour of a close relationship between annelids and arthropods. In: Legakis A, Sfenthourakis S, Polymeni R, Thessalou-Legaki M (eds) The new panorama of animal evolution. Proceedings of the 18th international congress of zoology, Athens 2000. Pensoft, Sofia, pp 489–501

Schulmeister S, Wheeler WC (2004) Comparative and phylogenetic analysis of developmental sequences. Evol Dev 6:50–57

Schulze J, Schierenberg E (2011) Evolution of embryonic development in nematodes. EvoDevo 2:18

Seaver EC, Paulson D, Irvine SQ, Martindale MQ (2001) The spatial and temporal expression of *Ch-en*, the *engrailed* gene in the polychaete *Chaetopterus* does not support a role in body axis segmentation. Dev Biol 236:195–209

Seipel K, Schmid V (2005) Evolution of striated muscle: jellyfish and the origin of triploblasty. Dev Biol 282:14–26

Seipel K, Schmid V (2006) Mesodermal anatomies in cnidarian polyps and medusa. Int J Dev Biol 50:589–599

Sempere LF, Cole CN, McPeek MA, Peterson KJ (2006) The phylogenetic distribution of metazoan microRNAs: insights into evolutionary complexity and constraint. J Exp Zool (Mol Dev Evol) 306B:575–588

Shankland M (2003) Evolution of body axis segmentation in the bilaterian radiation. In: Legakis A, Sfenthourakis S, Polymeni R, Thessalou-Legaki M (eds) The new panorama of animal evolution. Proceedings of the 18th international congress of zoology. Pensoft, Sofia, pp 187–195

Shubin N, Tabin C, Carroll S (2009) Deep homology and the origins of evolutionary novelty. Nature 457:818–823

Slack JMW, Holland PWH, Graham CF (1993) The zootype and the phylotypic stage. Nature 361:490–492

Sly BJ, Snoke MS, Raff RA (2003) Who came first – larvae or adults? Origins of bilaterian metazoan larvae. Int J Dev Biol 47:623–632

Smith KK (1996) Integration of craniofacial structures during development in mammals. Am Zool 36:70–79

Smith KK (2001) Heterochrony revisited: the evolution of developmental sequences. Biol J Linn Soc 73:169–186

Smith KK (2002) Sequence heterochrony and the evolution of development. J Morphol 252:82–97

Smith KK (2003) Time's arrow: heterochrony and the evolution of development. Int J Dev Biol 47:613–621

Spring J, Yanze N, Middel AM, Stierwald M, Gröger H, Schmid V (2000) The mesoderm specification factor twist in the life cycle of jellyfish. Dev Biol 228:363–375

Stauber M, Taubert H, Schmidt-Ott U (2000) Function of *bicoid* and *hunchback* homologs in the basal cyclorrhaphan fly *Megaselia* (Phoridae). Proc Natl Acad Sci U S A 97:10844–10849

Steinböck O (1963) Origin and affinities of the lower metazoa: the "aceloid" ancestry of the Eumetazoa. In: Dougherty EC (ed) The lower metazoa. Comparative biology and phylogeny. University of California Press, Berkeley, pp 40–54

Steinböck O, Ausserhofer B (1950) Zwei grundverschiedene Entwicklungsabläufe bei einer Art (*Prorhynchus stagnatilis* M. Sch., Turbellaria). Arch Entw Mech 144:155–177

Stollewerk A, Schoppmeier M, Damen WGM (2003) Involvement of *Notch* and *Delta* genes in spider segmentation. Nature 423:863–865

Tabin CJ, Carroll SB, Panganiban G (1999) Out on a limb: parallels in vertebrate and invertebrate limb patterning and the origin of appendages. Am Zool 39:650–663

Tanaka K, Barmina O, Kopp A (2009) Distinct developmental mechanisms underlie the evolutionary diversification of *Drosophila* sex combs. Proc Natl Acad Sci U S A 106:4764–4769

Tanaka K, Barmina O, Sanders LE, Arbeitman MN, Kopp A (2011) Evolution of sex-specific traits through changes in HOX dependent *doublesex* expression. PLoS Biol 9:e1001131

Technau U, Scholz C (2003) Origin and evolution of endoderm and mesoderm. Int J Dev Biol 47:531–539

Telford MJ, Budd GE (2003) The place of phylogeny and cladistics in Evo-Devo research. Int J Dev Biol 47:479–490

Telford ML, Littlewood DTJ (eds) (2009) Animal evolution: genomes, fossils, and trees. Oxford University Press, Oxford

Thamm K, Seaver EC (2008) Notch signaling during larval and juvenile development in the polychaete annelid *Capitella* sp I. Dev Biol 320:304–318

Thomas MB, Freeman G, Martin VJ (1987) The embryonic origin of neurosensory cells and the role of nerve cells in metamorphosis of *Phialidium gregarium* (Cnidaria, Hydrozoa). Int J Invertebr Reprod Dev 11:265–287

Thompson JV (1830) Zoological researches. Memoir IV. On the cirripedes or barnacles; demonstrating their deceptive character; the extraordinary metamorphosis they undergo, and the class of animals to which they undisputably belong. King and Ridings, Cork, pp 69–85

Tills O, Rundle SD, Salinger M, Haun T, Pfenninger M, Spicer JI (2011) A genetic basis for intraspecific differences in developmental timing? Evol Dev 13:542–548

Tomarev SI, Callaerts P, Kos L, Zinovieva R, Halder G, Gehring W, Piatigorsky J (1997) Squid *Pax-6* and eye development. Proc Natl Acad Sci U S A 94:2421–2426

True JR, Haag ES (2001) Developmental system drift and flexibility in evolutionary trajectories. Evol Dev 3:109–119

Valentine JW, Collins AG (2000) The significance of moulting in ecdysozoan evolution. Evol Dev 2:152–156

Valentine JW, Jablonski D, Erwin DH (1999) Fossils, molecules and embryos: new perspectives on the Cambrian explosion. Development 126:851–859

Van de Vyver G (1993) [Classe des Hydrozoaires] Reproduction sexuée – Embryologie. In: Grassé PP (ed) Traité de Zoologie, 3(2). Masson, Paris, pp 417–473

Velhagen WA Jr (1997) Analyzing developmental sequences using sequence units. Syst Biol 46:204–210

Verster A, Ramani A, McKay S, Fraser A (2014) Comparative RNAi screens in *C. elegans* and *C. briggsae* reveal the impact of developmental system drift on gene function. PLoS Genet 10:e1004077

von Baer KE (1828) Über Entwicklungsgeschichte der Thiere: beobachtung und Reflexion, I. Bornträger, Königsberg

Wägele JW, Erikson T, Lockhart P, Misof B (1999) The Ecdysozoa: artifact or monophylum? J Zool Syst Evol Res 37:211–223

Wagner GP (2000) What is the promise of developmental evolution? Part I: why is developmental biology necessary to explain evolutionary innovations? J Exp Zool (Mol Dev Evol) 288:95–98

Wagner GP (ed) (2001) The character concept in evolutionary biology. Academic Press, San Diego

Wagner A (2011) The origins of evolutionary innovations. A theory of transformative change in living systems. Oxford University Press, Oxford

Wagner GP (2014) Homology, genes, and evolutionary innovation. Princeton University Press, Princeton

Wagner GP, Zhang J (2011) The pleiotropic structure of the genotype-phenotype map: the evolvability of complex organisms. Nat Rev Genet 12:204–213

Wagner GP, Zhang J (2013) On the definition and measurement of pleiotropy. Trends Genet 29:383–384

Waller TR (1998) Origin of the molluscan class Bivalvia and a phylogeny of the major groups. In: Johnston PA, Haggard JW (eds) Bivalves: an eon of evolution. University of Calgary Press, Calgary, pp 1–45

Wanninger A, Haszprunar G (2001) The expression of an engrailed protein during embryonic shell formation of the tusk-shell, *Antalis entails* (Mollusca, Scaphopoda). Evol Dev 3:312–321

West-Eberhard MJ (2003) Developmental plasticity and evolution. Oxford University Press, New York

Wheeler BM, Heimberg AM, Moy VN, Sperling EA, Holstein TW, Heber S, Peterson KJ (2009) The deep evolution of metazoan microRNAs. Evol Dev 11:50–68

Whiting MF, Wheeler WC (1994) Insect homeotic transformation. Nature 368:696

Whiting M, Carpenter JC, Wheeler QD, Wheeler WC (1997) The Strepsiptera problem: phylogeny of the holometabolous insect orders inferred from 18S and 28S ribosomal DNA sequences and morphology. Syst Biol 46:1–68

Wiegmann B, Trautwein M, Kim JW, Cassel B, Bertone M, Winterton S, Yeates D (2009) Single-copy nuclear genes resolve the phylogeny of the holometabolous insects. BMC Biol 7:34

Wiens JJ, Bonett RM, Chippindale PT (2005) Ontogeny discombobulates phylogeny: paedomorphosis and higher-level salamander relationships. Syst Biol 54:91–110

Wilkins A (2001) The evolution of developmental pathways. Sinauer Associates, Sunderland

Wotton KR, Alcaine Colet A, Jaeger J, Jimenez-Guri E (2013) Evolution and expression of BMP genes in flies. Dev Genes Evol 223:335–340

Wray GA, Raff RA (1991) The evolution of developmental strategy in marine invertebrates. Trends Ecol Evol 6:45–50

Zrzavý J (2001) Ecdysozoa versus Articulata: clades, artifacts, prejudices. J Zool Syst Evol Res 39:159–163

Homology in the Age of Developmental Genomics

Günter P. Wagner

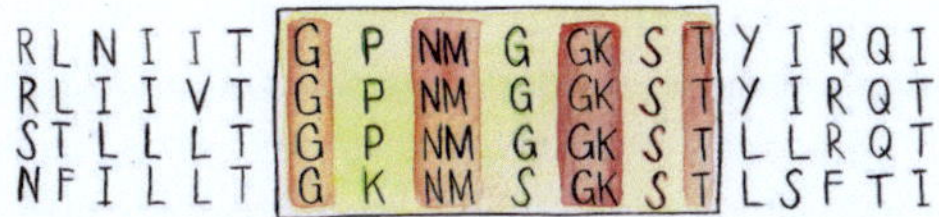

Chapter vignette artwork by Brigitte Baldrian.
© Brigitte Baldrian and Andreas Wanninger.

G.P. Wagner
Department of Ecology and Evolutionary Biology,
Yale Systems Biology Institute,
New Haven, CT, USA

Department of Obstetrics, Gynecology and
Reproductive Sciences, Yale University,
New Haven, CT, USA
e-mail: gunter.wagner@yale.edu

A. Wanninger (ed.), *Evolutionary Developmental Biology of Invertebrates 1:
Introduction, Non-Bilateria, Acoelomorpha, Xenoturbellida, Chaetognatha*
DOI 10.1007/978-3-7091-1862-7_2, © Springer-Verlag Wien 2015

INTRODUCTION

The homology concept was introduced into pre-Darwinian evolutionary biology by Richard Owen as referring to "the same organ in different animals regardless of form and function" (Owen 1848). Since then, it has played not only a fundamental role as an organizing idea in comparative anatomy but also an important role in preparing the way for evolutionary biology (Donoghue 1992; Amundson 2005). Homology is the primary evidence for phylogenetic relationships among organisms, and whenever we project experimental results from a model organism onto humans, we assume homology among the mechanisms in humans and the model organism. Homology was fully integrated into the Darwinian tradition through Lankester's redefinition as an organ in two species that is derived from the same organ in the most recent common ancestor of the two species (Lankester 1870). Nevertheless, the homology concept remains controversial primarily because it seems to escape a simple rigorous definition. Homology shares this attribute with other fundamental concepts like that of a species or a gene. In addition, homology is hard to pin down mechanistically. Apparently, homology is among the concepts biologists have a hard time living with but certainly can't live without. This situation often leads to considerable frustration among biologists, and some have suggested abandoning the concept altogether (Wake 2003), a move that is hardly feasible.

Morphological evidence for phylogenetic relationships among extant organisms is increasingly replaced with molecular data, which seems to make controversies around morphological homology obsolete. I think, however, that abandoning the homology concept would be counterproductive, since it still has an important role to play in both evolutionary and developmental biology and also in other branches of organismal biology (Wagner 2014).

Why, then, is homology still necessary and important in the twenty-first-century biology? There are two broad reasons why homology is still central to evolutionary and developmental biology. First, homology reflects a broad pattern of biological diversity, and, second, a deeper understanding of the nature of homology (i.e., character identity) is essential for research into the origin of evolutionary novelties (Müller and Wagner 1991; Müller and Newman 1999; Wagner and Lynch 2010) and thus is essential for understanding of how complex organisms (and characters) arose in evolution.

The fact that we can identify parts of organisms, e.g., brains, wings, and shells, and find corresponding parts in other, sometimes distantly related, organisms is a fundamental fact about biological diversity. This fact shows that animals and plants consist of quasi-independent building blocks that can have historical continuity over considerable stretches of time. For instance, the insect eye is at least as old as the crown insects, i.e., more than 400 Mio years. Thus, homology reflects the truism that animals can be highly structured and that the organizational features of organisms can be highly conserved even in species that live in radically different environments and are leading radically different lives. An evolutionary biology that does not accommodate these facts into its conceptual outline is missing its goal, namely, to provide a rational explanation of biological diversity.

Homology is about the historical continuity of body parts and about the nature of character identity. As such, homology is about the nature and conservation of organizational building blocks of multicellular organisms. A complementary problem is to explain how novel body parts originate in evolution, i.e., the problem of evolutionary novelties. Novel body parts originate in evolution, and derived lineages can have parts that are not present in any of the ancestral lineages, like the wings of pterygot insects and others. We cannot even begin to investigate how novel body parts originated if we do not understand what body part identity is in the first place.

The claim that a deeper understanding of homology is necessary for a productive research program on the origin of novelties in evolution can be supported by an analogy argument about species and the origin of species. Following the publication of Darwin's *The Origin of Species* in 1859, many researchers attempted to study species origins. Much work failed to reach its goal, because it was based on the mistaken idea that the origin of novel species is synonymous with the origin of a distinguishing, defining characteristic (i.e., an autapomorphy), like the color of petals, or the presence or absence of a wing marking in a butterfly. Understanding character change and adaptation is an important research program in its own right, but in itself does not solve or even address the problem of species origins. It required an understanding of population biology to make clear that the task of investigating speciation is synonymous with the task of investigating how gene pools become separated (Dobzhansky 1937; Mayr 1942), leading to two independent lineages of evolutionary change. Similarly, the origin of novel body parts cannot be addressed if we do not have some degree of clarity what body part identity really is. Hence, homology and novelty are complementary concepts (Müller and Wagner 1991). Homology is about what is the same in different animals, and novelty is what is different, i.e., not homologous, among different animals.

For most of its history, homology and its complement, novelty, have been hard to pin down. However, the situation has changed dramatically in recent decades, in particular due to the maturation of phylogenetic inference methods (Felsenstein 2003) and comparative methods (Maddison et al. 1984; Donoghue 1989), as well as due to the growth of comparative developmental biology (Raff 1996; Wilkins 2002; Carroll 2008), aka developmental evolution. For the first time there is a realistic chance to connect our mechanistic understanding of development with a sophisticated understanding of evolutionary biology and phylogenetics (Wagner 2014). In this chapter I want to outline how the new tools of developmental genomics can be harnessed to elucidate the genetic basis of character identity and how evolutionary understanding can be used to guide the experimental analysis of animal development. Before we can proceed to the technical aspects of a research program on the origin of novelties, we need to address a number of conceptual challenges.

CHALLENGES

There are many issues that can and have to be raised when we talk about homology, but these can be boiled down to two types of questions: the question of *character individuation* and the question of *what the mechanistic cause of character identity is*. When we propose that two characters in two species are homologous, then we have to be prepared to define what is "the thing that is the same" in these two species. For instance, if we say that the pectoral fin of a teleost and the tetrapod forelimb are homologous, it raises the question, what is the unit that is "the same" in this case? Is the shoulder girdle part of the homologous body part or just the elements distal to the shoulder girdle? This is a question of individuality: can the body be divided into clearly demarcated building blocks that objectively demarcate quasi-independent body parts? The other question is, what is the mechanistic basis for the individuation of these body parts? Can we propose an experimentally testable model that explains body part individuation?

To address these questions, I want to develop my argument in four steps: first, I want to assert the fact that there is a lower limit to character individuality, below which comparison and individual identification of physical body parts are meaningless (Riedl 1978). This argument will establish that homology only applies to individualized parts of the body. Second, I will introduce and defend the distinction between character identity and character states (Wagner 2007). The point will be that these reflect different aspects of biological reality and thus need to be distinguished. Third, I will address Patterson's claim that homology is coextensive and therefore identical in meaning to apomorphy, i.e., any shared derived trait (Patterson 1982). Patterson's idea is a reasonable position, given that both character identities (the presence and absence of parts) and character states (how body parts are shaped) can contain phylogenetic information. However, given the importance of character identity, I will argue that apomorphy is a more general term than homology, such that apomorphy can stand for both shared derived character identities and shared derived character states. Finally, I will argue that character identity is caused by a distinct gene regulatory network upstream of the genes that determine character states (Wagner 2007, 2014). This model provides a mechanistic interpretation to the conceptual distinction between character identity and character states. In addition, this proposal gives an explanation how two body parts in two different species can be "the same" under "every variety of form and function," because character identity is mechanistically decoupled from character states. This model implies an agenda for a research program on the origin of novel characters, i.e., to understand that the origin of a novel character is equal to understanding the origin of a novel character identity network.

Individuality

In the case of clear-cut homology assessments, there is often little doubt about the individuality of the compared structures. Say, the cerebellum of a cow is clearly homologous to that of a dog, and in turn there is no question that in each of these animals the cerebellum is a distinct developmental individuality compared, say, to the olfactory bulb or any other brain region or any other body part. The individuality of the cerebellum is reflected in its distinct location with respect to other brain regions, the fact that it can be lost as a discrete unit without major effects of the rest of the brain, its connections to the rest of the brain, as well as its tissue architecture and the nature of its cell types, e.g., the Purkinje cells that are the characteristic cell type of the cerebellum. However, it is also clear that not every physically separated part of the body has developmental individuality. For instance, for the 20–30 trillion erythrocytes that exist in my bloodstream at any moment, there is no one-to-one correspondence to the 20–30 trillion erythrocytes that exist in the body of another person, not to speak of members of another species. These cells are obviously multiple copies of the same thing, human erythrocytes. The same is true for different instances

of the same protein that are produced from the same gene. Even though each molecule is a physical individual, biologically they are instances of the same kind, say, alcohol dehydrogenase (ADH) or aromatase. Hence, it is meaningless to ask which single molecule of ADH in one animal corresponds to which single ADH molecule in another animal. It is, however, meaningful to ask which ADH gene corresponds to which ADH sequence in another animal's genome. The reason is that only one copy of an orthologous gene is passed on from parent to offspring, so that there is a unique lineage of descent over time. Homology is about the existence of a unique lineage of descent of things that are passed on over generations (Ghiselin 2005). This is also true for morphological characters, where homology is clearest where there is only one copy of a body part instantiated in each individual.

In this context it is worthwhile to remember the classical definition of homology by Owen: "Homologue is the same organ in different animals under every variety of form and function" (Owen 1848). What is noteworthy here is that Owen only speaks of organs or body parts rather than resemblances or attributes. Clearly, at its origin, the concept is intended to capture the notion of character identity. It is also important to note that there is an explicit distinction made between the character itself and its various realizations, aka character states: "any variety of form and function." This focus on the identity of body parts is still retained in Lankester's evolutionary reinterpretation of homology (Lankester 1870). The original quote is: "Without doubt the majority of evolutionists would agree that by asserting that an <u>organ A</u> in an animal α to be homologous to and <u>organ B</u> in an animal β, they mean that in some common ancestor κ the <u>organs A and B</u> were represented by an <u>organ C</u>, and that α and β have inherited their organs A and B from κ" (Lankester 1870, p. 36; underlines added by this author). This precision, namely, what homology is actually about, i.e., body part identity, got lost in the twentieth century among the leaders of the new synthesis biology. For instance, G. G. Simpson defines homology as "<u>resemblance</u> due to inheritance from a common ancestry" (Simpson 1961, p. 78). Similarly, Ernst Mayr writes of "attributes" as being homologous to "characteristics" in a common ancestor (Mayr 1982, p. 465). It seems plausible to speculate that this loss of specificity in the understanding of homology is in great part responsible for the confusion surrounding the homology concept in the twentieth century and the frustration it engendered.

Character Identity and Character States

The language we use to describe organismal structure and diversity does not naturally convey whether we speak of the same body part in different shapes or whether we speak of different body parts as such. We use different words to describe paired appendages of fishes and tetrapods (paired fins and limbs) or different forms of wings like the elytra of beetles, even though they actually are the same body parts (fins and limb as well as forewing and elytra). In the sciences, however, we have to be clear what is a body part identity and what refers to different states of the same character.

A clear illustration of the difference between terms that name character identities and those that are names for character states is the comparison of various insect "wings," or better dorsal appendages of the second and third thorax segments. Most clades of pterygot insects have four wings, meaning that there are two pairs of dorsal appendages that are shaped so that they can aid in flying. However, there are some lineages with highly modified dorsal appendages, like Diptera, whose name suggests that they have only two wings or one pair of wings. From a functional point of view, it is true that there is only one pair of dorsal appendages that is dedicated to lift production, but from the developmental evolutionary point of view, even dipterans still have four wings. The point is that the second pair of dorsal appendages, "wings" sensu *lato*, are highly modified and are called halteres (Fig. 2.1). Halteres are not "wings" in the functional sense but gyroscope-like sensory organs that aid in flight, but do not produce lift.

There is broad consensus that the haltere is derived from the hindwing of four-winged ancestors and thus is the "same" body part as the hindwing of a butterfly, even though with different shapes and functions. Clearly, the haltere is a hindwing that has assumed an extreme character state, that of a haltere. Hence, *hindwing* is the term we should use to name the *character identity* of the dorsal appendage on the metathorax of a fly, and *haltere* is a term that describes a *character state* of the hindwing.

Similarly, in beetles, the forewing is modified from a lift-producing blade to a protective cover called elytra (Fig. 2.1). Nevertheless, it is clear that elytra are derived from lift-producing wing blades in ancestral four-winged insects, and thus even the beetles have four wings, even though only the hindwing is still used in flying (in those beetles that actually are able to fly). Hence, the body part identity of the dorsal appendages on the mesothorax of beetles is that of a *forewing*, but their *character state* is called *elytra*.

Homology Is a Narrower Concept Than Apomorphy

Both character identities and character states can convey phylogenetic information, meaning that there is inheritance and historical continuity of character states such that corresponding character states can be used to characterize monophyletic clades. At the same time, the presence or absence of a particular character identity can also characterize monophyletic clades (e.g., mandibles in Mandibulata). Diptera share a haltere-shaped hindwing, and the clade of Coleoptera can be characterized by having "elytracized" forewings. Hence, for the purposes of phylogenetic reconstruction, the distinction between characters and character states is not of fundamental importance; the distinction between shared derived attributes (apomorphies) and shared ancestral attributes (plesiomorphies), however, is of importance (Hennig 1966). This realization led Patterson to equate homology to apomorphy in an influential paper from 1982 (Patterson 1982), adding to the confusion about the nature of homology.

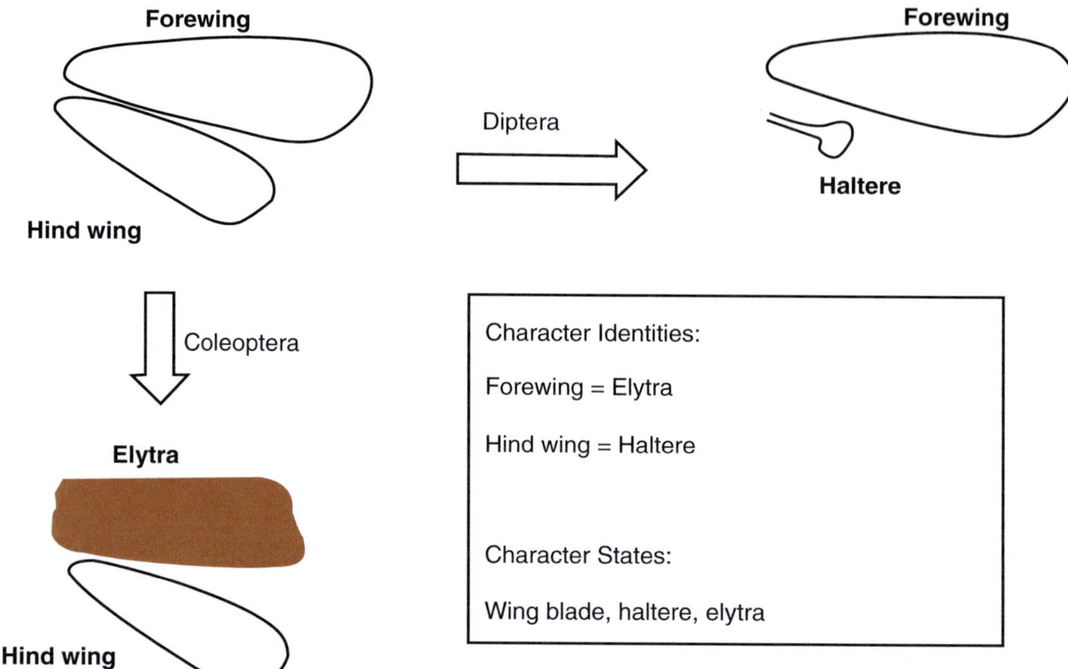

Fig. 2.1 Illustration of the difference between character states and character identities. Ancestrally, pterygot insects have two wings, a forewing and a hindwing. In dipterans the hindwing is transformed into a club-shaped appendage called haltere. The haltere acts as a gyroscopic sense organ in flight. Clearly, there is historical continuity between hindwings and halteres, which means that they are homologous but have different character states, either as a wing blade or as a haltere. In coleopterans, in contrast, the forewing is replaced by a protective cover called elytra. The elytra is clearly homologous to forewings but is a derived character state

Given that the homology concept never was meant to describe corresponding character states (see above) and that body part identity and body part phenotype reflect different dimensions of organismal structure, it seems more productive to recognize that the term "apomorphy" is the broader concept than homology. Apomorphy applies to any inherited biological attribute that has historical continuity. In contrast, homology is better reserved for what it was intended, to describe character identity, i.e., "the same organs under any variety of form and function." In that sense apomorphy is the broader concept since it encompasses both shared derived character identities and shared derived character states. In contrast, homology, as originally conceived by Owen and Lankester, only refers to shared derived character identities.

In formal terms one can say that

$$\text{Apomorphy} = \{\text{shared derived character identities, shared derived character states}\}$$

And thus homology is not equal to apomorphy, as Patterson proposed, but rather part of or a special case of apomorphy:

$$\text{Homology} \subset \text{Apomorphy}$$

Homologies are a subset of apomorphies, or homology implies apomorphy but not the other way around:

$$\text{Homology} \rightarrow \text{Apomorphy}$$

I think this modification of Patterson's proposal leaves intact what he intended, namely, that apomorphy is the broader and in a way more fundamental concept, and it recognizes the need to distinguish between body part identities and body part phenotypes (character states).

The Genetic Basis of Character Identity

All the discussion in this section so far is useful to sort out unnecessary confusion around the homology concept but does not address the most fundamental problem of the homology concept: What do we really mean when we talk of "the same organ under any variety of form and function"? How can anything be "the same" in spite of any degree of morphological dissimilarity and difference in biological role? What do we mean by "the same"? This problem is not even addressed by the evolutionary reinterpretation of homology by Lankester, since it asks us to know what he means when he says that two organs A and B are derived from the *same* organ C in a common ancestor. All the evolutionary notion of homology adds is a historical dimension, namely, that of evolutionary continuity, but does nothing to answer questions of what sameness really means and how to recognize it (Wagner 1994).

The situation is not helped by the well-documented fact that clearly homologous structures can have quite different developmental genetic underpinnings (Roth 1988; Wagner and Misof 1993; Sommer and Sternberg 1994; Wray and Abouheif 1998; Hall 2003; Hallgrimsson et al. 2009). Celebrated examples are vulva development in nematodes (Sommer and Sternberg 1994) and segmentation among insects (Damen 2007). Advances in developmental genetics have clearly documented developmental variation of homologous characters, but this insight is as old as experimental developmental biology itself. For instance, Spemann and Mangold documented differences in the requirement for inductive signals for lens development in different species of anurans. Spemann made an attempt to reconcile his experimental work with his experience as a comparative anatomist, and his conclusion was negative. There was no way for him to relate his anatomical work to his experimental developmental work (Spemann 1915). In 1971 Gavin de Beer published a small monograph with the telling title *Homology: An Unsolved Problem* (de Beer 1971), where he also compiled evidence

that homology cannot be synonymous with identity of genetic information.

An important hint of how to solve this conundrum came from the comparative developmental genetics of the *Ubx* gene in insects (Deutsch 2005). The *Ubx* gene was discovered because of the effect on the haltere of a loss-of-function mutation. Loss of *Ubx* function leads to a transformation of the haltere into a more wing-like appendage. Subsequently, it was shown that *Ubx* suppresses a number of genes that have been shown to be important in the development of a wing blade (Weatherbee et al. 1998). A reasonable interpretation of these results was that *Ubx* is perhaps an "anti-wing blade" gene. This idea was tested by documenting *Ubx* gene expression in a four-winged insect, the butterfly *Junonia*. Thereby, *Ubx* was found to be expressed in the developing hindwings of *Junonia*, clearly showing that *Ubx* is not an "anti-wing blade" gene (Weatherbee et al. 1999). The decisive evidence that *Ubx* is really involved in character identity, rather than wing shape development, came from a study of *Ubx* function in the beetle *Tribolium castaneum* (Tomoyasu et al. 2005). A knockdown of *Ubx* during *Tribolium* development leads to two pairs of elytra. Since elytra are only formed in forewings and two pairs of elytra do not make any functional sense, it is clear that the knockdown of *Ubx* leads to a loss of hindwing identity and the establishment of a default identity, namely, that of a forewing (Deutsch 2005). In other words, the role of the *Ubx* gene is to determine hindwing identity or more broadly metathorax identity. *Ubx* function is not tied to a particular hindwing phenotype, or hindwing character state, but is necessary for hindwing identity, regardless what the shape or the function of the hindwing is in the respective species. Accordingly, *Ubx* in wing development plays the role of a *character identity gene*, which led to the hypothesis that character identity is determined by character identity networks (ChIN, (Wagner 2007)).

The second observation was that putative ChINs are much more conserved than the inductive signals that initiate the development of a body part of a cell type. Examples are the eye gene regulatory network (Friedrich 2006), the segment polarity gene network which could be understood as the ChIN for insect segments, and the gene regulatory network underlying the development of paired vertebrate appendages (fins and limbs). From these three observations derives the hypothesis that character identity is rigidly linked to the activity of ChINs, while character states are determined by the set of realizer genes that are regulated by the genes in the ChIN. Realizer genes are genes that code for proteins that do physiological work, e.g., enzymes, extracellular matrix proteins, and cytoskeletal molecules. A concept that is closely allied to the ChIN model is that of core regulatory networks of cell types (Graf and Enver 2009) or the terminal selector genes of cell types (Hobert 2011). Core regulatory networks and terminal selector gene networks are the ChINs for cell type identity.

If the phenotype of a character is not determined by the ChIN genes, what is? Since the same character can have different phenotypes but still be determined by the same ChIN genes, the role of the ChIN is "abstract," i.e., not bound to and does not determine a certain phenotype. *The role of ChIN is to enable the expression of different sets of realizer genes.* In this model, the separation of gene regulatory networks responsible for character identity and genes responsible for the phenotype of the character explains how a character can be "the same regardless of any variety of form and function."

The role of ChINs can be understood by placing them into the basic hierarchy of development. There are three functional roles to be distinguished in the development of any body part: positional information, character identity determination, and execution of the phenotype (Fig. 2.2). Positional information signals tell the cell where it is in the embryo and what "it is supposed to do." The examples cited above show that this level of the developmental hierarchy can be highly variable even for homologous characters. These signals in turn activate the character identity network which translates the positional information into a distinct gene regulatory network state that determines the developmental and

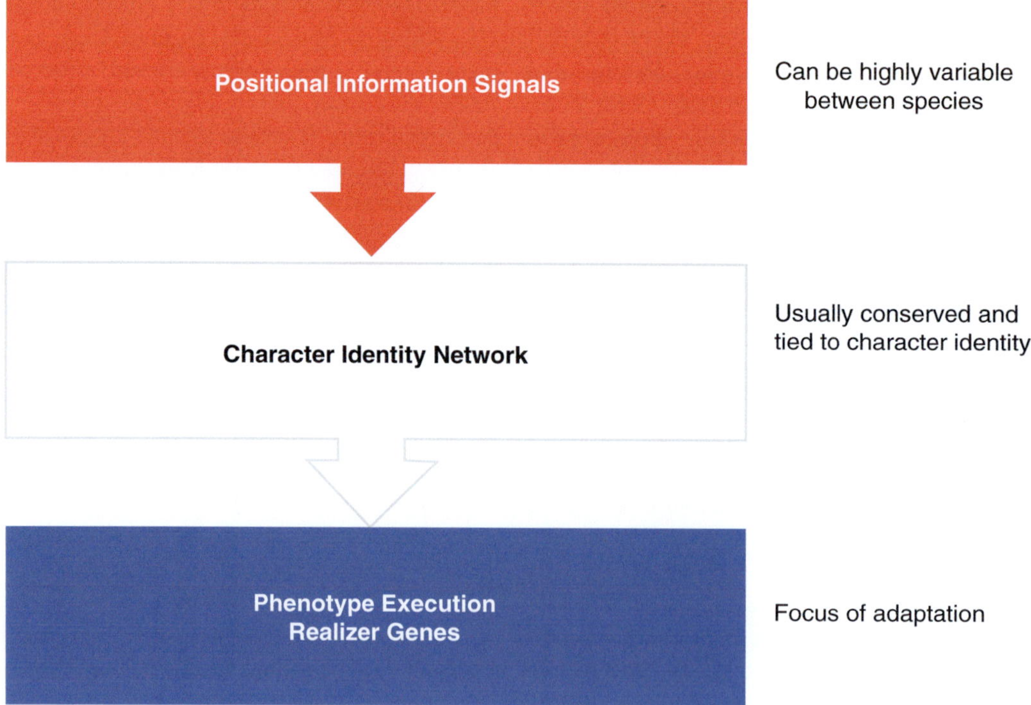

Fig. 2.2 The principal functional roles of genes in development: positional information, character identity determination, and execution of the phenotype (for more explanation, see text)

evolutionary identity of the body part. The members of the ChIN are usually transcription factor genes or transcriptional cofactor genes. The role of the ChIN then is to regulate the expression of a set of "realizer genes." These are genes that produce proteins that do physiological work such as enzymes, extracellular matrix proteins, and cytoskeletal proteins. The activity of these genes determines the phenotype that confronts the environment. For that reason, these genes and their regulation are the primary target of natural selection and thus are the focus of adaptive evolutionary change.

The notion that character identity is decoupled from character states or phenotype is now broadly accepted in the case of genes. Genes are homologous (or better orthologous) when they are derived from the same gene in a common ancestor. This definition applies regardless of how similar the gene sequences are. What matters is historical continuity of the piece of DNA that codes for the protein rather than the degree of sequence similarity or the biochemical function of the protein (Graur and Li 2000). It is true that sequence similarity is a fairly reliable indicator for orthology, but orthology is not *defined* via similarity. Orthology is discovered by similarity and even potentially applies when similarity is minimal. In the same way, homology, or character identity of morphological characters, is based on historical continuity and is not rigidly tied to similarity. Again the same logic applies. Morphological similarity is a way to discover potentially homologous characters, but homology can hold even among quite dissimilar body parts.

CHARACTER MODALITIES

Before we proceed to a more detailed discussion of character identity networks, I want to address another pattern of morphological variation that has been confused with character identity and needs to be discussed for clarity. In the previous sections, the emphasis was on the distinction between the identity of a body part and its character states. Character states can be minor differences in shape, size, or color, which may be easily reverted in evolution. However, there are differences in character states that are more radical and fundamental than mere shape and size differences and which need to be recognized, since they are subject to intense study and often mark important evolutionary transformations. I am thinking here of character transformations like the origin of the elytra in the stem lineage of coleopterans. Forewings and elytra are clearly corresponding body parts and thus are homologs. Nevertheless, there is something fundamentally different about wing blades and elytra that asks for recognition. I proposed to call classes of character states that mark radical transformations of the same body part as *character modalities* (Wagner 2014).

For instance, pectoral fins and forelimbs are examples of different character modalities: two sets of character states of the same organ, which differ so fundamentally that transitions between these character state sets are rare. Character modalities represent two different ways of being the same character. Lineages tend to remain in one or the other modality for a long time. Clearly, forelimbs have been derived from pectoral fins, but once they became limbs they never reverted back to fins, even when they reacquire their function as swimming organs (e.g., flippers in whales).

Recognizing character modalities implies that the character states representing these modalities differ in their developmental constraints. Each modality has certain character states that are easy to realize and others are not, separating the set of character states that represent the character modalities (Fig. 2.3). Often, character modalities are characterized by the presence or absence of character identities at a lower level of organization. For instance, tetrapod limbs have digits but lack fin rays. On the other hand, say, teleost fins have fin rays, but not digits. The developmental difficulty of "reinventing" fin rays after they were lost for a long time is likely one of the developmental constraints that separate limbs from most fins.

Character modalities do not need to be distinguished by the presence or absence of certain subsidiary character identities (e.g., fin rays within paired fins), however. An example is classes of flower symmetry, as documented by Ree and

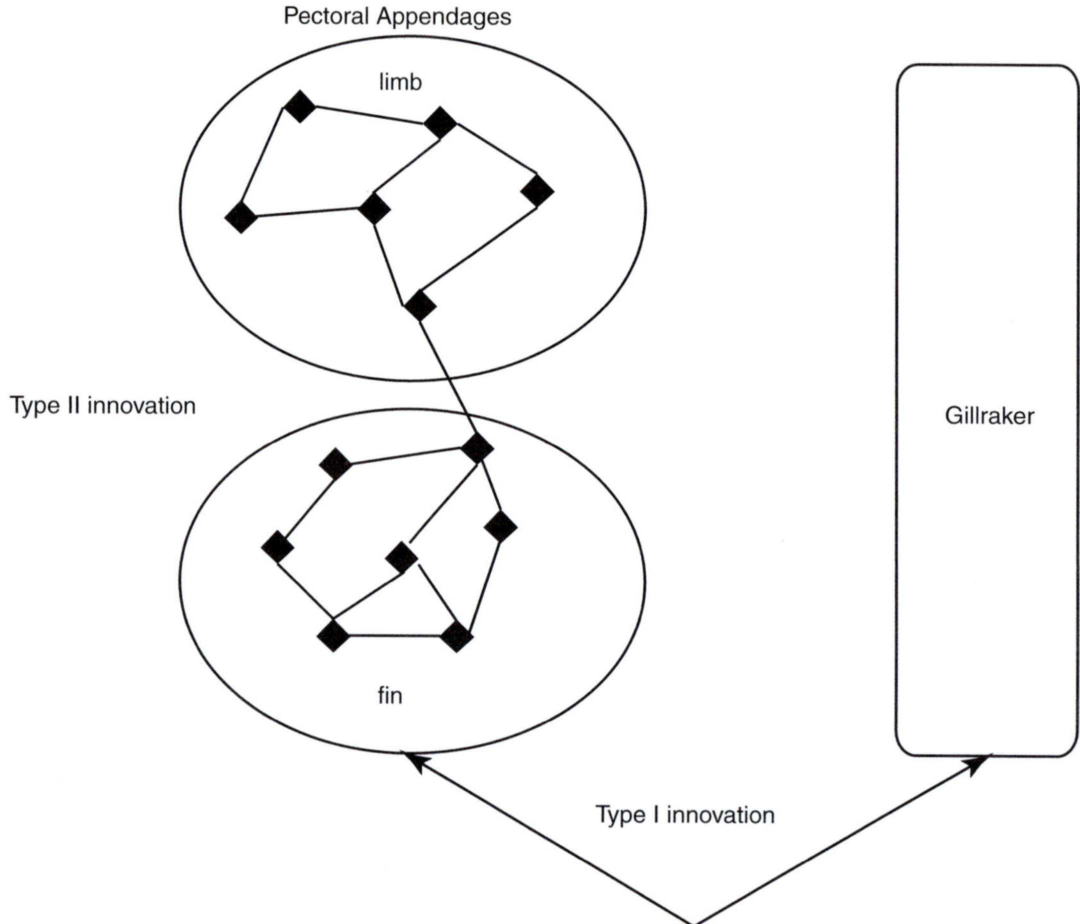

Fig. 2.3 Character identities, character modalities, and forms of innovation. Pectoral appendages and gill rakers are two character identities that may have been derived from a common ancestral structure according to Gegenbaur's theory (Gegenbaur 1876). Their origin would thus be a type I innovation, i.e., the origin of novel character identities. Among the pectoral appendages, there are two major forms: pectoral fins and forelimbs. Pectoral fins and forelimbs clearly represent the same character identity, i.e., forelimbs are derived from pectoral fins, but are radically different in their organization. These different modes of pectoral appendages are called character modalities (Wagner 2014). Each character modality represents a large number of possible character states, symbolized here as black diamonds, with many possibilities of transformation among them. But the transition between the two character modalities is rare because there are only a few (in the illustration shown as just one) mutational paths that lead from one character modality to the other. The transition from one character modality to another is called a type II innovation

Donoghue (1999). Symmetry classes are sets of flower shapes that easily transform into each other but rarely mutated to a state in another symmetry class (for details, see Ree and Donoghue 1999).

Clearly, the origin of a derived character modality, like the fin-limb transition, is a significant event in the history of life. Intuitively they also qualify as evolutionary novelties. Nevertheless, they represent a different kind of evolutionary event than the origin of a novel character identity. This suggests to distinguish between two kinds of novelties (Müller 2010; Wagner 2014): type I novelties can be called the origins of novel character identities like the origin of a novel cell type. Type II novelties are the origination of a novel character modality, i.e., a largely irreversible transformation of the character state (Fig. 2.3).

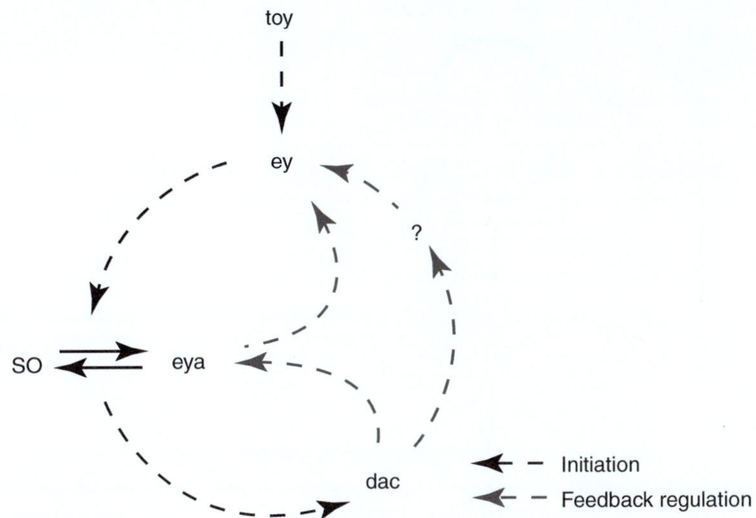

Fig. 2.4 The character identity network of the insect compound eye. Note that the induction of the ChIN genes by *toy* eventually leads to the establishment of a positive feedback which sustains the expression of the core ChIN member genes *eya* (*eyes absent*), *ey* (*eyeless*), *so* (*sine oculis*), and *dac* (*dachshund*). Loss of any one of these genes leads to the loss of eye identity (After Czerny et al. 1999)

CHARACTERIZATION OF CHARACTER IDENTITY NETWORKS

The comparison of a variety of character identity networks (ChINs) from certain cell types and some multicellular characters, discussed in greater detail in Wagner (2014), leads to a number of preliminary generalizations that can be used as a guide to experimentally identify ChINs and their constituent genes.

One broad generalization is that each character, be it a single cell type or a multicellular anatomical structure, has a core of regulatory genes that are jointly necessary for the development of the focal character. This is most obvious in the case of the core regulatory network of insect eyes (Fig. 2.4), where three of the participating genes have names that express the absence of eyes when the gene is mutated: *eyeless*, *eyes absent*, and *sine oculis*.

An explanation for the fact that some genes are jointly necessary for initiating the development of a character is that many of these networks are positive feedback circuits. Positive feedback locks the set of genes into what Eric Davidson calls a "cross-regulatory embrace" (Oliveri et al. 2008), where the genes sustain each other's expression (Fig. 2.4). Removing one of the members of the feedback circle cuts the circle open and interrupts the self-sustaining character of the network structure. The requirement for the positive feedback ensures co-expression of the members of the core gene regulatory network and also leads to a high degree of integration between the members of the core regulatory network (Pavlicev and Widder 2015).

While the positive feedback among core regulatory genes is widely acknowledged as a characteristic of core regulatory networks (e.g., Graf and Enver 2009), another equally important feature is receiving much less attention, although it is experimentally well documented. That is the fact that often some or all of the transcription factors coded for by genes in the core regulatory network are functionally cooperating with each other in regulating target genes. By functional cooperativity, I mean the situation where two or more transcription factor proteins have to physically interact to effect the expression of their target genes. Removing any one of them leaves the remaining transcription factor proteins unable to activate a certain target gene.

Joint necessity, positive feedback, and transcription factor cooperativity are the three most important characteristics of ChIN genes (Fig. 2.5). This is a preliminary generalization based on a small number of examples but nevertheless is specific enough to imply predictions for the identification of ChINs and for research into the origin of novel character identities.

In order to explain how an evolutionary novelty arises, it is necessary to explain the origin

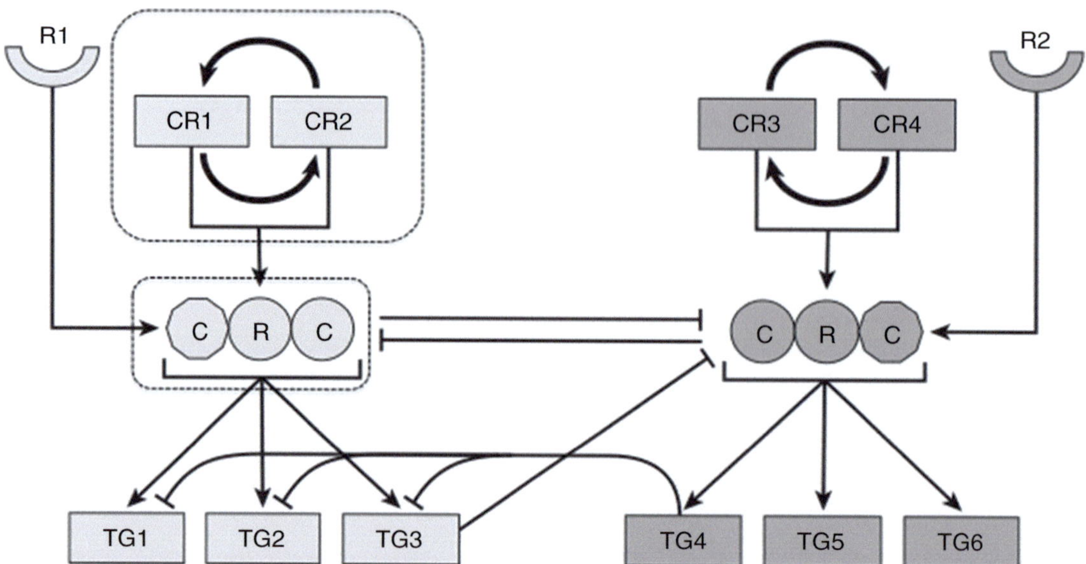

Fig. 2.5 Structure of a character identity network (*ChIN*). The ChIN consists of a set of core regulatory genes (*CRx*) that form a positive feedback to sustain each other's activity. The CR genes produce regulatory proteins, transcription factors, and transcriptional cofactors, some of them form a core regulatory complex (CRC). The CRC is the principal regulatory agent. The CRC regulates downstream target genes (*TGx*) which are the realizer genes of the model in Fig. 2.2. The activity of the CRC proteins is also influenced by some signals mediated by receptor proteins (*Rx*). Alternative character identities are realized by different ChINs. Alternative ChINs suppress each other's activity via a variety of direct and indirect pathways, as indicated by the blunt arrows connecting the two alternative ChINs. The explanation of a type I innovation, the origin of a derived character identity, requires us to explain how a novel core network evolved and how a novel core regulatory complex evolved (*dashed boxes*)

of a novel ChIN. One would expect that the novel ChIN has the characteristic of a gene regulatory network with positive feedback. One further predicts that the character-specific regulatory activities of the ChIN transcription factor genes would include functional cooperativity among the transcription factor proteins. Hence, it is likely that, during the origin of a novel character, the involved transcription factor genes will show signs of adaptive amino acid substitutions and derived cooperativity. The former, adaptive evolution of transcription factor proteins, can be tested by statistical sequence analysis with standard methods of sequence analysis, like the well-known dN/dS ratio methods (Nei 1987). The latter, derived functional cooperativity, however, requires experimental methods where the transcription factors from different species are tested for their regulatory activity (see, e.g., Lynch et al. 2008, 2011). Methods for the study of gene regulatory network structure and comparison will be discussed in the next section.

THE EXPERIMENTAL STUDY OF CHARACTER IDENTITY

Concepts are only as useful as they enhance our ability to learn from nature through further empirical investigation. For that reason this chapter cannot stop short of addressing issues of how the study of character identity and its evolution can take advantage of the technical opportunities that are presented to the evolutionary biologist by the advances of molecular biology. In this short section, I will thus explain how I see the ideas summarized above (as well as in Wagner 2014) can be put to work together with the techniques of functional genomics. First, I want to address to what degree the experimental study of character identity affects the recognition of homology in the comparative study of biodiversity. Then I want to discuss the natural connection between the idea of character trees as developed by Geeta (2003) and Oakley (2003, 2007) and the comparative study of cell and tissue transcriptomes. The phylogenetic ideas expressed in character trees naturally lead to the identification of candidate genes and thus lead us into the experimental study of the mechanism of character individuation. Finally, I add a short section on a few technical issues on the comparative analysis of transcriptomes.

Even though models of gene regulatory networks play a central role in explaining the nature of character identity, it is clearly neither necessary nor reasonable to expect that every single homology hypothesis is backed up with experimental genetic data. The classical indirect methods for supporting homology hypotheses are adequate in all but a few cases (Remane 1952; Riedl 1978; Patterson 1982; Rieppel 1988). Where experimental characterization of ChINs is necessary is for research programs that aim at investigating the origin of evolutionary novelties, i.e., derived character identities (type I novelties; see above). Investigating the genomic underpinnings of novel characters is an increasingly important area of evolutionary biology, and for that reason I will provide a brief discussion of experimental methods used to identify the genes involved in character identity determination and how to uncover the structure of the gene regulatory network underlying character identity.

Homology implies that there are developmentally individualized body parts that exhibit evolutionary continuity. Very often novel body parts are the result of differentiation of repeated, serially homologous body parts or characters (Riedl 1978; Weiss 1990). This is, for instance, implied in the so-called sister cell type model of cell type origination. This model assumes that new cell types arise by sub-functionalization of an ancestral multifunctional cell type (Arendt 2008). The same logic applies to many multicellular organs, like the variety of arthropod eyes (Oakley 2003; Oakley et al. 2007), parts of angiosperm plants (Geeta 2003), digit identity (Wang et al. 2011) or epidermal appendages (Musser et al. 2015). In

either case the consequence is that characters or cell types can be thought of as forming a tree of descent, where the bifurcation events are either speciation events or type I innovations (Fig. 2.6), i.e., events in which an ancestral structure or cell type is replaced by two individualized parts or cell types in the descendant lineage. This logic suggests that the evolutionary history of many characters can be represented as a "character tree" (Geeta 2003; Oakley 2003; Oakley et al. 2007).

Character trees can be reconstructed using transcriptome data and phylogenetic methods such as parsimony and maximum likelihood. In the case of cell types and even multicellular organs, transcriptomes have proven a powerful tool to infer character relatedness and also can be used to identify candidate gene recruitment events responsible for the origin of the new cell or character origination events (see, for instance, Oakley et al. 2007; Arendt 2008). Transcriptomes can be used as such to calculate a measure of phenetic similarity or can be transformed into a table of 0/1 values representing genes that are either not expressed (=0) or expressed (=1). The justification for one or the other approach will be discussed below.

The reconstruction of the character tree from transcriptome data only requires that we use some phylogenetic reconstruction methods, like parsimony or maximum likelihood. Once we have a tree that is well supported and biologically meaningful, we can use the character tree to identify genes that likely have been recruited at the time of origin of certain character identities. That means that the phylogenetic analysis of character or cell type gene expression data can directly lead to hypotheses about the genetic basis for the origin of the novel character (Kin et al. 2015). These candidate genes can then be tested for their importance for the development of the derived character identity by using any of the standard knockout or knockdown experimental techniques like morpholinos ® or RNAi and iCrisper. Hence, reconstructing character trees is both an exercise in the reconstruction of evolutionary history and a step toward the identification of the underlying genetic events responsible for the origin of a derived character identity.

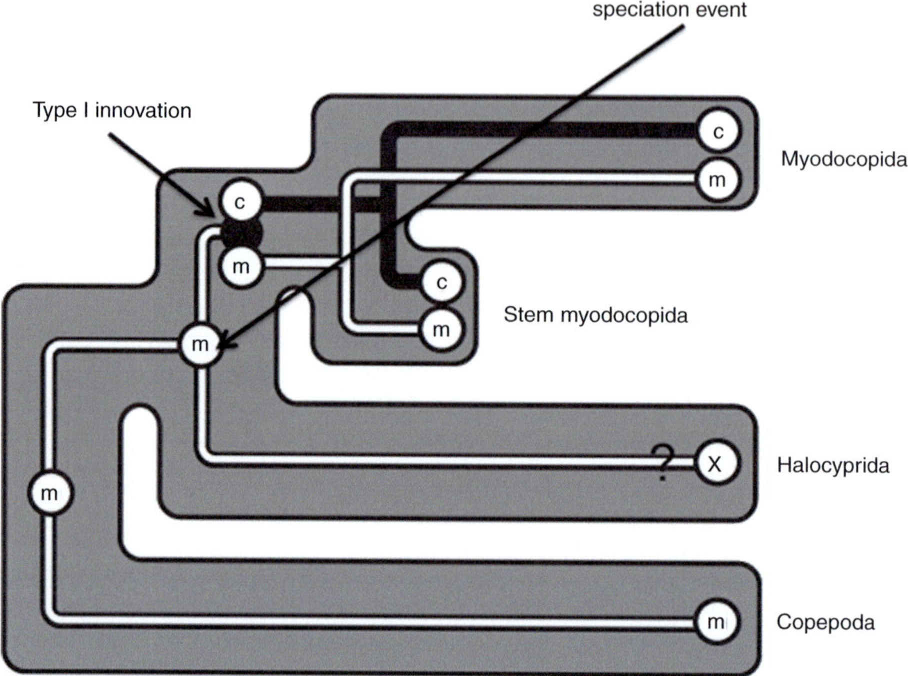

Fig. 2.6 Structure tree of the eyes of myodocopid ostracods according to Oakley et al. (2007). There are two eye types, a median (*m*) and a compound eye (*c*). The median eye is the ancestral form in this group and the compound eyes evolved in the stem lineage of Myodocopida by developmental field splitting. Hence, the origin of the new eye is a type I innovation represented by a node in the structure tree. In addition, the tree contains nodes that represent speciation events, like the one that gave rise to the myodocopid and the halocyprid lineages (Modified after Oakley et al. (2007))

Technical Issues in the Use of Transcriptomes for the Study of Innovation

This section focuses on the study of cell type or tissue innovation. The development of larger multicellular characters has a stronger spatial component that needs to be addressed in ways that we have limited experience with. The reconstruction of sister cell type trees can entirely be done by using information about the transcriptomes of cells from one species. The idea is that novel cell types arise from phylogenetic precursor cells that then differentiate into two closely related cell types. These new cell types will share much of their ontogenetic history and thus are expected to be also more similar to each other than any other cell type in this species to each of them (Arendt 2008). This is the idea of molecular fingerprinting to identify closely related cell types. The cell type tree then is a hypothesis of the cell type origination events in evolution.

Using different cells from the same species in a sister cell type analysis is the least challenging way to analyze the relationships among cell types, since all the transcriptomes are mapped to the same genome. This avoids a number of issues that arise when cells from different species are compared. One of them is related to differences in the quality of genome annotations. Differences in genome annotation can lead to a variety of artifacts in the normalization of RNA abundance measures. One source of artifacts is that the gene models used in different genome annotations could be of different size which leads to systematic differences in the normalization of RNA abundance measures. This problem can be avoided if one uses gene models that only include orthologous regions of the genomes, i.e., regions that are annotated in all species/genomes compared. Finally, the set of genes included in the RNA quantification needs to be restricted to those that are 1-1 orthologs among all species compared and are all present in all the species. RNA abundance measures are normalized to the total number of transcripts identified. If different sets of transcripts are annotated in the genomes of different species, the RNA abundance measures will be systematically different between species (Wagner et al. 2012).

In our work we have often used discretized data, i.e., data that indicate whether a gene is expressed in a particular cell or not. The rationale is that quantitative measures of RNA abundance are prone to experimental artifacts either due to cell culture conditions or the requirements for tissue harvesting. On the other hand, the majority of the phylogenetic signal seems to be residing in differences between expressed and non-expressed genes rather than in quantitative differences in gene expression levels.

Discretization of RNA abundance data requires a biological justification for the threshold used to categorize genes as either expressed or non-expressed. There are several methods available to justify such a classification threshold. One is a statistical method, which assumes that the distribution of RNA abundance across species consists of two components. One is a broad distribution of expression levels of genes that are actively transcribed. The other is a distribution of RNA abundances that comes from non-expressed genes, i.e., RNA transcripts due to leaky transcription (Wagner et al. 2013). This model is compared to the observed RNA abundance distribution, and a nonlinear regression is performed to determine the parameters of the mixed distribution model. The threshold is then placed at the expression level that corresponds to a small probability of expression due to leaky transcription. For a variety of samples from cells, this is about 2 tpm (transcripts per million, Wagner et al. 2012) which corresponds to about 1 RPKM, aka "reads per kilobase and million reads" (Mortazavi et al. 2008).

An alternative method comes from comparing the histone modifications specific for active promoters (H3K4me1) with the associated RNA

expression levels. Fortunately, for mammalian cells the biochemical and the statistical methods are highly congruent and suggest a robust threshold that seems to be independent of cell type (Hebenstreit et al. 2011; Kin et al. 2015). It is 1 RPKM or 2–3 tpm for pure cell samples, which corresponds to about one transcript per cell (Hebenstreit et al. 2011).

Discretized gene expression data from the same species can then be analyzed like any other categorical data, most easily by parsimony. Analyzing discretized expression data leads to easily interpreted ancestral state reconstructions. These ancestral state reconstructions lead to hypotheses about the gene expression profile of the ancestral cell type and can also be used to infer candidate gene recruitment events associated with the origin of novel cell identities (Kin et al. 2015). Among the candidate gene reconstructions, it is useful to eliminate genes that have expression levels in the gray zone around the threshold of 2 tpm. For that reason it might be useful to discretize using a threshold interval, so that genes are considered non-expressed if they have <1 tpm or expressed if they have, say, >6 tpm and ignore all the genes that are in between. These upper and lower thresholds, which should be chosen to bracket the biologically meaningful threshold of 2 tpm, are arbitrary and can be tuned to limit the number of candidate genes considered.

While cell type and tissue type trees can be reconstructed from data from a single species, these data do not allow to place a particular innovation event on the species phylogeny, since the bifurcation events on the cell type tree are not attached to a particular phylogenetic branch. There are two principal ways to proceed to a more specific evolutionary scenario. One could think of combining transcriptome data from different species on the same tree and thereby also establish the homology of corresponding cell types across species. The problem, however, is that the transcriptomes of homologous cell types and tissues from different species not only carry the signal of their homology but also a signature of species-specific modifications. In other words, when we compare transcriptomes from different species, we are dealing with both a cell type-specific and a species-specific signal. The resulting tree often then sorts the samples first by species and then by character relatedness. Only if the divergence of gene expression among the cell types is very strong compared to that of the species can the tree reflect both the character relatedness and the homology among cells from different species. In these cases the relatedness structure is often not very interesting, i.e., it is not surprising to find that muscle cells are related to muscle cells and less so to nerve cells (Brawand et al. 2011). In most cases where there is an interesting biological question to be answered, the species signal tends to overwhelm the signal for the homology among the cell types. There are a variety of ways to proceed from this point.

The simplest way to address the issue of strong species signal is to analyze the cell or tissue type data from each species independently and see whether the resulting trees are congruent. Congruent cell type trees suggest both robustness of the result and also correspondence of the cell types included. This interpretation requires that the two species represent lineages that bifurcated after all the most recent cell type innovation events, i.e., when both species have only strictly homologous cell types.

The other direction to pursue is to try to separate the species signal from the character-specific signal and then analyze the data that only contains the cell type signal. Ideally, then homologous cell types would fall out as independent clades. To my knowledge there is no reliable method to eliminate species signals. This is an important problem in comparative transcriptomics.

Gene Regulatory Network Reconstruction

The comparative analysis of transcriptomes is a powerful tool to reconstruct cell type history and gene recruitment events. This method, however, is limited in its ability to uncover the functional relationships among the regulatory genes that mediate their role in character identity determination. Ultimately, one would like to reconstruct the evolutionary history of the gene regulatory networks that led to the origin of novel cell type and character identities rather than just the gene sets themselves.

Experimental gene regulatory network reconstruction has been a laborious enterprise requiring many person-years and large amounts of money to reconstruct the gene regulatory network of a single cell in a single species (Davidson 2001). More recently, methods have been developed that have the potential to be scalable, i.e., can be used on more than one cell type and more than one species with a reasonable amount of effort. As many other methods, they only rely on the availability of the genome sequence of the species and living cells. These methods all are a form of foot printing of accessible chromatin regions (Neph et al. 2012a, b; Buenrostro et al. 2013). They mainly differ in the biochemical method of probing for accessible chromatin regions, but their basic idea is similar.

The logic of these methods is that expressed genes and the cis-regulatory regions that regulate them are characterized by "open" chromatin, i.e., parts of the chromatin where the DNA is less protected from degradation. A sample of chromatin extracted from a cell population is subjected to a degradation agent (DNAse I or hyperactive transposase), and the so-produced DNA fragments are extracted and sequenced. The sequences are then mapped back onto the genome, which gives a map of where in the genome the cells have open chromatin. In addition, one can map the frequency of cut sites, i.e., the exact location where the DNAse cut the DNA or where transposable elements were inserted. The expectation is that at sites where the DNA is associated with a transcription factor, the DNA is protected from cutting. Hence, when one finds a "valley" of cut site frequency in a region of open chromatin, this small region is likely bound by a transcription factor in the cells analyzed. Which transcription factor binds at such a "footprint" cannot be directly observed with these methods. In many cases, however, the transcription factor can be inferred from the DNA sequence under the footprint. Classes of transcription factors tend to bind to characteristic DNA sequence motifs, and many of the motifs are known. Any transcription factor binding upstream of a gene is a potential upstream regulator of this gene.

The footprinting method outlined above can be used to infer the structure of a gene regulatory network. If we restrict our attention only to the upstream regions of all known transcription factor genes, then this method can be used to reconstruct the structure of the core regulatory network. The footprints upstream of transcription factor genes lead to a network of transcription factor genes that regulate other transcription factor genes (Neph et al. 2012a). Among those regulatory connections must also be the ones that constitute the cell type identity network.

The attraction of the methods outlined above is that in principle each reconstruction of a gene regulatory network is based on a single experiment (as well as a lot of sequencing and computation). Thus, this method of reconstructing gene regulatory networks is scalable, i.e., can be performed on many cell types and species in one study. This amount of data is required to study the dynamics of gene regulatory network evolution and thus the evolution of character identity networks. It will be exciting to see how to overcome the inevitable practical problems associated with this class of methods.

CODA

The study of body plan diversity and evolution has its roots in the beginnings of modern biology in the eighteenth and nineteenth century and at times seemed to have been superseded by the study of molecular and cellular mechanisms. In recent years, however, the molecular methods have matured to a degree that enables the return to questions that seemed inert to a mechanistic understanding. These are questions of character identity, cell type identity, body plan innovations, and others. These techniques, with all their power, however, will do us not much good if the community interested in these questions does not afford the intellectual discipline and rigorous

standards of evidence required to take advantage of these new opportunities. With intellectual discipline, I mean clarity of what the questions are that we try to address and what the alternative models are we want to discriminate between. Too often one finds in the morphological literature statements about the presence or absence of a cell or a character without clarity what the criteria are that the author used to come to a particular conclusion. Also, it is necessary to distinguish between what we observe and what we infer. For instance, it was necessary to distinguish the relative positions of digits in the hand of birds or skinks and to clearly distinguish these designations from the inferred character identities (e.g., Young et al. 2009; Wang et al. 2011).

Intellectual discipline is also necessary for making sure we are explicit about distinctions that reflect different biological realities. The origin of paired appendages, i.e., the origin of novel body parts, is a different kind of evolutionary process than the transformation of paired fins into limbs. To call both of them novelties is not useful, since these are different kinds of processes, just as it is necessary to distinguish between adaptation (due to natural selection) and speciation (which may or may not be related to natural selection). Also, it will be critical to develop widely accepted standards of evidence for this field of research. This will be a painful and controversial process but one we cannot do without.

Acknowledgments I thank Professor Andreas Wanninger for the invitation to participate in this important project as well as for comments and corrections to a previous version of this paper. I am also grateful to Jake Musser for suggestions and edits of the manuscript. Finally, I thank all current and former members of my lab for intellectual companionship during the long time these ideas were developed.

References

Amundson R (2005) The changing role of the embryo in evolutionary thought: roots of Evo-Devo. Cambridge University Press, Cambridge, xii, 280pp

Arendt D (2008) The evolution of cell types in animals: emerging principles from molecular studies. Nat Rev Genet 9:868–882

Brawand D, Soumillon M, Necsulea A, Julien P, Csardi G, Harrigan P et al (2011) The evolution of gene expression levels in mammalian organs. Nature 478:343–348

Buenrostro JD, Giresi PG, Zaba LC, Chang HY, Greenleaf WJ (2013) Transposition of native chromatin for fast and sensitive epigenomic profiling of open chromatin, DNA-binding proteins and nucleosome position. Nat Methods 10:1213–1218

Carroll SB (2008) Evo-Devo and an expanding evolutionary synthesis: a genetic theory of morphological evolution. Cell 134:25–34

Czerny T, Halder G, Kloter U, Souabni A, Gehring WJ, Busslinger M (1999) Twin of eyeless, a second Pax-6 gene of Drosophila, acts upstream of eyeless in the control of eye development. Mol Cell 3:297–307

Damen WG (2007) Evolutionary conservation and divergence of the segmentation process in arthropods. Dev Dyn 236:1379–1391

Davidson E (2001) Genomic regulatory systems. Academic, San Diego, xii, 261

de Beer GR (1971) Homology, an unsolved problem. Oxford University Press, London, 16pp

Deutsch J (2005) Hox and wings. Bioessays 27:673–675

Dobzhansky T (1937) Genetics of the evolutionary process. Columbia University Press, New York, xiii, 505

Donoghue MJ (1989) Phylogenies and the analysis of evolutionary sequences, with examples from seed plants. Evolution 43:1137–1156

Donoghue MJ (1992) Homology. In: Keller EF, Lloyd EA (eds) Keywords in evolutionary biology. Harvard University Press, Cambridge, MA, pp 171–179

Felsenstein J (2003) Inferring phylogenies. Sinauer, Sunderland, xx, 664

Friedrich M (2006) Ancient mechanisms of visual sense organ development based on comparison of the gene networks controlling larval eye, ocellus, and compound eye specification in Drosophila. Arthropod Struct Dev 35:357–378

Geeta R (2003) Structure trees and species trees: what they say about morphological development and evolution. Evol Dev 5:609–621

Gegenbaur C (1876) Zur Morphologie der Gliedmassen der Wirbeltiere. Morphologisches Jahrb 2:396–420

Ghiselin MT (2005) Homology as a relation of correspondence between parts of individuals. Theory Biosci 124:91–103

Graf T, Enver T (2009) Forcing cells to change lineages. Nature 462:587–594

Graur D, Li W-H (2000) Fundamentals of molecular evolution, 2nd edn. Sinauer Ass. Inc, Sunderland, xiv, 481

Hall BK (2003) Descent with modification: the unity underlying homology and homoplasy as seen through an analysis of development and evolution. Biol Rev Camb Philos Soc 78:409–433

Hallgrimsson B, Jamniczky H, Young NM, Rolian C, Parsons TE, Boughner JC et al (2009) Deciphering the palimpsest: studying the relationship between morphological integration and phenotypic covariation. Evol Biol 36:355

Hebenstreit D, Fang M, Gu M, Charoensawan V, van Oudenaarden A, Teichmann SA (2011) RNA sequencing reveals two major classes of gene expression levels in metazoan cells. Mol Syst Biol 7:497

Hennig W (1966) Phylogenetic systematics. University of Illinois Press, Urbana, 263

Hobert O (2011) Regulation of terminal differentiation programs in the nervous system. Annu Rev Cell Dev Biol 27:681–696

Kin K, Nnamani MC, Lynch VJ, Michaelides E, Wagner GP (2015) Cell type phylogenetics and the origin of endometrial stromal cells. Cell Reports 10:1398–1409

Lankester RE (1870) On the use of the term homology in modern zoology, and the distinction between homogenetic and homoplastic agreement. Annu Mag Nat Hist 6:34–43

Lynch VJ, Tanzer A, Wang Y, Leung FC, Gellersen B, Emera D et al (2008) Adaptive changes in the transcription factor HoxA-11 are essential for the evolution of pregnancy in mammals. Proc Natl Acad Sci U S A 105:14928–14933

Lynch VJ, May G, Wagner GP (2011) Regulatory evolution through divergence of a phosphoswitch in the transcription factor CEBPB. Nature 480:383–386

Maddison WP, Donoghue MJ, Maddison DR (1984) Outgroup analysis and parsimony. Syst Zool 33:83–103

Mayr E (1942) Systematics and the origin of species. Columbia University Press, New York, xiv, 334

Mayr E (1982) The growth of biological thought. The Belknap Press, Cambridge, ix, 974

Mortazavi A, Williams BA, McCue K, Schaeffer L, Wold B (2008) Mapping and quantifying mammalian transcriptomes by RNA-Seq. Nat Methods 5:621–628

Müller GB (2010) Epigenetic innovation. In: Pigliucci M, Müller GB (eds) Evolution – the extended synthesis. MIT Press, Boston, pp 307–332

Müller GB, Newman SA (1999) Generation, integration, autonomy: three steps in the evolution of homology. In: Bock GR, Cardew G (eds) Homology. Wiley, New York, pp 65–79

Müller GB, Wagner GP (1991) Novelty in evolution: restructuring the concept. Ann Rev Ecol Syst 22:229–256

Musser J, Wagner GP, Prum RO (2015) Nuclear β-catenin expression supports homology of feather, avian scutate scale, and alligator scale early development J Exp Zoo (Mol Dev Evol) [in press]

Nei M (1987) Molecular evolutionary genetics. Columbia University Press, New York, x, 512

Neph S, Stergachis AB, Reynolds A, Sandstrom R, Borenstein E, Stamatoyannopoulos JA (2012a) Circuitry and dynamics of human transcription factor regulatory networks. Cell 150:1274–1286

Neph S, Vierstra J, Stergachis AB, Reynolds AP, Haugen E, Vernot B et al (2012b) An expansive human regulatory lexicon encoded in transcription factor footprints. Nature 489:83–90

Oakley TH (2003) The eye as a replicating and diverging, modular developmental unit. Trends Ecol Evol 18:623–627

Oakley TH, Plachetzki DC, Rivera AS (2007) Furcation, field-splitting, and the evolutionary origins of novelty in arthropod photoreceptors. Arthropod Struct Dev 36:386–400

Oliveri P, Tu Q, Davidson EH (2008) Global regulatory logic for specification of an embryonic cell lineage. Proc Natl Acad Sci U S A 105:5955–5962

Owen R (1848) On the archetype and homologies of the vertebrate skeleton. Voorst, London, viii, 203

Patterson C (1982) Morphological characters and homology. In: Joysey KA, Friday AE (eds) Problems of phylogenetic reconstruction. Academic, London, pp 21–74

Pavlicev M, Widder S (2015) Wiring for independence: positive feedback motifs facilitate individuation of traits in development and evolution. J Exp Zool (Mol Dev Evol) 324B:104–113

Raff R (1996) The shape of life. Chicago University Press, Chicago, xxiii, 520

Ree RH, Donoghue MJ (1999) Inferring rates of change in flower symmetry in asterid angiosperms. Syst Biol 48:633–641

Remane A (1952) Die Grundlagen des natürlichen systems, der vergleichenden Anatomie und der Phylogenetik. Akademische Verlagsgesellschaft Geest & Portig, Leipzig, vi, 364

Riedl R (1978) Order in living organisms: a systems analysis of evolution. Wiley, New York, xx, 313

Rieppel OC (1988) Fundamentals of comparative biology. Birkhäuser, Basel

Roth VL (1988) The biological basis of homology. In: Humphries CJ (ed) Ontogeny and systematics. Columbia University Press, New York, pp 1–26

Simpson GG (1961) Principles of animal taxonomy. Columbia University Press, New York, 247pp

Sommer RJ, Sternberg PW (1994) Changes of induction and competence during the evolution of vulva development in nematodes. Science 265:114–118

Spemann H (1915) Zur Geschichte und Kritik des Begriffs der Homologie. In: Chun C, Johannsen W (eds) Allgemeine Biologie, vol 3. Teubner, Leipzig, pp 63–86

Tomoyasu Y, Wheeler SR, Denell RE (2005) Ultrabithorax is required for membranous wing identity in the beetle Tribolium castaneum. Nature 433:643–647

Wagner GP (1994) Homology and the mechanisms of development. In: Hall BK (ed) Homology: the hierarchical basis of comparative biology. Academic, San Diego, xiii, 478pp

Wagner GP (2007) The developmental genetics of homology. Nat Rev Genet 8:473–479

Wagner GP (2014) Homology, genes and evolutionary Innovation. Princeton University Press, Princeton

Wagner GP, Lynch VJ (2010) Evolutionary novelties. Curr Biol 20:R48–R52

Wagner GP, Misof BY (1993) How can a character be developmentally constrained despite variation in developmental pathways? J Evol Biol 6:449–455

Wagner GP, Kin K, Lynch VJ (2012) Measurement of mRNA abundance using RNA-seq data: RPKM measure is inconsistent among samples. Theory Biosci 131:281–285

Wagner GP, Kin K, Lynch VJ (2013) A model based criterion for gene expression calls using RNA-seq data. Theory Biosci 132:159–164

Wake DB (2003) Homology and homoplasy. In: Hall BK, Olson WM (eds) Keywords and concepts in evolutionary developmental biology, vol 191–201. Harvard University Press, Cambridge, MA

Wang Z, Young RL, Xue H, Wagner GP (2011) Transcriptomic analysis of avian digits reveals conserved and derived digit identities in birds. Nature 477:583–586

Weatherbee SD, Halder G, Kim J, Hudson A, Carroll S (1998) Ultrabithorax regulates genes at several levels of the wing-patterning hierarchy to shape the development of the Drosophila haltere. Genes Dev 12:1474–1482

Weatherbee SD, Frederik Nijhout H, Grunert LW, Halder G, Galant R, Selegue J et al (1999) Ultrabithorax function in butterfly wings and the evolution of insect wing patterns. Curr Biol 9:109–115

Weiss KM (1990) Duplication with variation: metameric logic in evolution from genes to morphology. Yearb Phys Anthropol 33:1–23

Wilkins AS (2002) The evolution of developmental pathways. Sinauer Assoc, Sunderland, xvii, 603pp

Wray GA, Abouheif E (1998) When is homology not homology? Curr Opin Genet Dev 8:675–680

Young RL, Caputo V, Giovannotti M, Kohlsdorf T, Vargas AO, May GE et al (2009) Evolution of digit identity in the three-toed skink Chalcides chalcides: a new case of digit identity frame shift. Evol Dev 11:647–658

Embryology in Deep Time

Philip C.J. Donoghue, John A. Cunningham, Xi-Ping Dong, and Stefan Bengtson

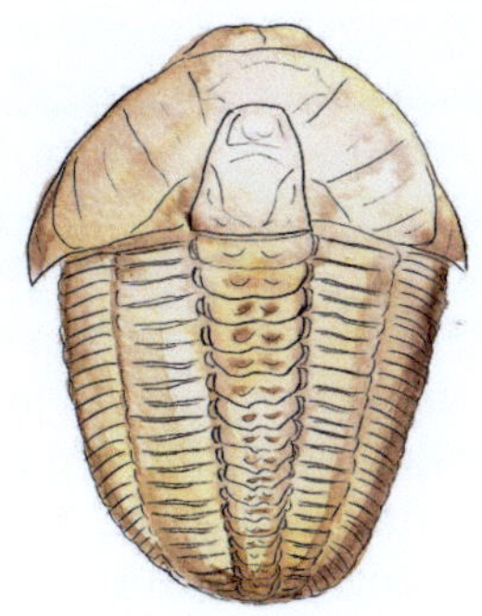

Chapter vignette artwork by Brigitte Baldrian.
© Brigitte Baldrian and Andreas Wanninger.

P.C.J. Donoghue (✉) • J.A. Cunningham
School of Earth Sciences, Life Sciences Building,
University of Bristol, Bristol BS8 1TQ, UK
e-mail: phil.donoghue@bristol.ac.uk

X.-P. Dong
School of Earth and Space Sciences,
Peking University, Beijing, China

S. Bengtson
Department of Palaeobiology,
Nordic Center for Earth Evolution,
Swedish Museum of Natural History,
Stockholm, Sweden

A. Wanninger (ed.), *Evolutionary Developmental Biology of Invertebrates 1:
Introduction, Non-Bilateria, Acoelomorpha, Xenoturbellida, Chaetognatha*
DOI 10.1007/978-3-7091-1862-7_3, © Springer-Verlag Wien 2015

INTRODUCTION

For anyone who has cared for animal embryos, it beggars belief that these squishy cellular aggregates could be fossilised. Hence, with hindsight, it is possible to empathise with palaeontologists who found such fossils and, in their naming of *Olivooides*, *Pseudooides*, etc., drew attention to their likeness to animal eggs and embryos but without going so far as to propose such an interpretation. However, in 1994, Zhang Xi-guang and Brian Pratt described microscopic balls of calcium phosphate from Cambrian rocks of China, one or two of which preserved polygonal borders that resembled blastomeres on the surface of an early cleaving animal embryo (Zhang and Pratt 1994). In retrospect, these fossils are far from remarkable, some of them may not be fossils at all, and it is not as if anyone ever conceived Cambrian animals as having lacked an embryology. But Zhang Xi-guang and Brian Pratt dared the scientific world, not least their fellow palaeontologists, to believe that the fragile embryonic stages of invertebrate animals could be fossilised, that there was a fossil record of animal embryology, that this record hailed from the interval of time in which animal body plans were first established, and that it had been awaiting discovery in the rocks, for want of looking. The proof of this concept came a few years later, when phosphatised Cambrian fossils from China and Siberia were shown to display indisputable features of animal embryonic morphologies (Bengtson and Yue 1997). In the case of *Olivooides*, a series of developmental stages from cleavage to morphogenesis through hatching and juvenile growth could be tentatively identified; in *Markuelia*, the coiled-up body of an annulated worm-like animal could be clearly seen within its fertilisation envelope.

It is not as if palaeontologists had been sitting on their hands until then. There has long been a strong tradition of assaying rocks of all ages, including these, for microscopic phosphatic fossils, principally conodonts (Donoghue et al. 2000) and elements of the enigmatic small shelly faunas (Bengtson 2005), driven principally by attempts to establish a global stratigraphy as a basis for establishing a relative timescale for Earth history. Indeed, the majority of discoveries of fossil embryos made subsequently have been based on the redescription, reinterpretation, and augmentation of knowledge of fossils that had been described long before or the discovery of new fossils from the deposits that had, on re-examination, previously yielded embryonic remains. There was palpable excitement in these early days that an extra dimension to the fossil record had been revealed and evolutionary biologists would soon be integrating the embryology of trilobites, ammonites, and anomalocaridids, with that of their living kin, effecting tests of developmental evolution that would be as direct as possible without the aid of a time machine, settling centuries-old debates over the plesiomorphy of gastrulation modes and the like (Donoghue and Dong 2005). Indeed, embryos and larvae of a great diversity of animals have been reported, including stem-metazoans (Hagadorn et al. 2006), sponges (Chen et al. 2000, 2009a), cnidarians (Bengtson and Yue 1997; Kouchinsky et al. 1999; Yue and Bengtson 1999; Chen et al. 2000, 2002, 2009a; Chen and Chi 2005; Dong et al. 2013), ctenophores (Chen et al. 2007), bilaterians generally (Chen et al. 2000, 2006, 2009a, b), or, more specifically, arthropods (Chen et al. 2004) and scalidophorans (Dong et al. 2004, 2005, 2010; Donoghue et al. 2006a; Steiner et al. 2014), the majority of which are from the Ediacaran Doushantuo Formation and the Early Cambrian Kuanchuanpu Formation, both of South China. Not all of these interpretations have withstood scrutiny, principally because palaeobiologists and embryologists have been unprepared in interpreting these most remarkable of fossil remains.

DISINTERRING THE BIOLOGY OF FOSSIL EMBRYOS

Given that the preservation of purported Ediacaran and Cambrian fossil embryos extends beyond the cellular to the subcellular and organelle level (Hagadorn et al. 2006; Huldtgren et al. 2011), it seems that there might be a compelling case to make direct comparisons to the embryos of living animals. However, fossils are not merely the decayed remains of once living organisms, and, somewhat ironically, exceptionally preserved

Fig. 3.1 Biological features and diagenetic artefacts in the Ediacaran Doushantuo biota. (**A–D**) are scanning electron micrographs, (**E–F**) are synchrotron radiation X-ray tomographic microscopy-based reconstructions. (**A**) An alga from Doushantuo with algal anatomy preserved in a low atomic number phase (*black arrowhead*); high atomic number material encrusts the algal cells and fills spaces between cells (*white arrowhead*). (**B**) Detail of the same specimen as (**A**) showing that the high atomic number phase consists of elongate crystals with their long axes normal to the surface of the alga (*arrowhead*). (**C**) An embryo-like fossil with structures interpreted as lipid vesicles or yolk droplets within the cells. (**D**) Detail of the same specimen as (**C**) showing that the spaces between these structures are filled by layered diagenetic cements. (**E**) Embryo-like specimen that preserves subcellular anatomy including possible nuclei (*arrowheads*). (**F**) Embryo-like specimen that preserves only surface anatomy. Parts (**A–D**) also figured by Cunningham et al. (2012a); parts (**E–F**) also figured by Cunningham et al. (2012b). Relative scale bar: (**A**) 50 μm, (**B**) 30 μm, (**C**) 145 μm, (**D**) 27 μm, (**E**) 200 μm, (**F**) 125 μm

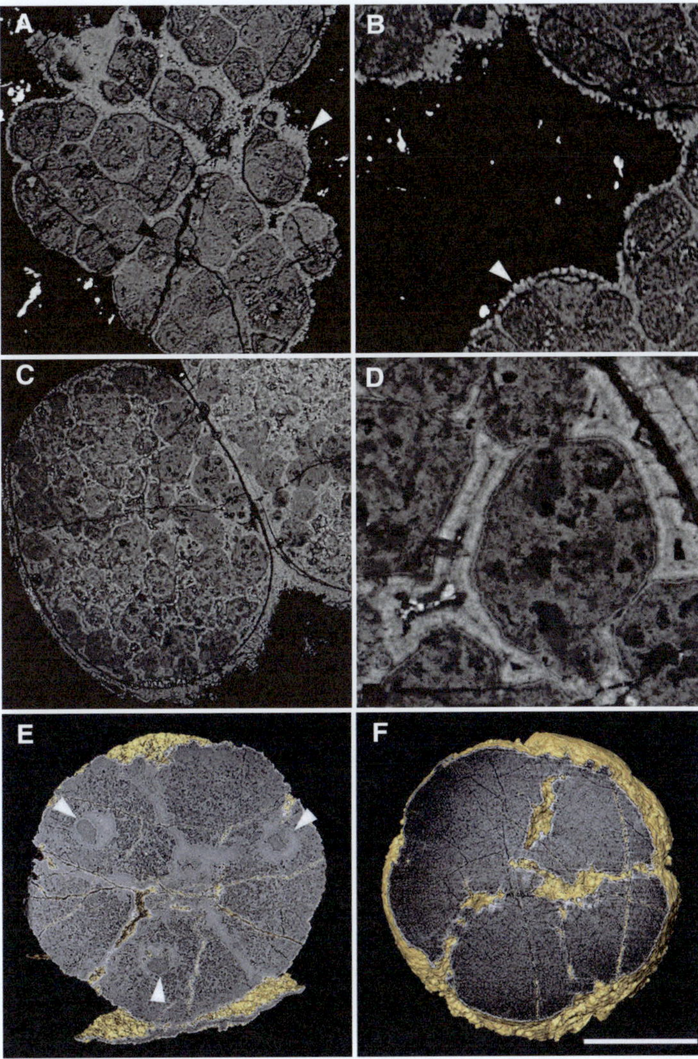

fossils are among the most difficult to interpret. This is because decay is an essential prerequisite to the mineralisation of labile biological tissues, which is invariably mediated microbially (Briggs 2003). Thus, the biological substrates that are available for mineralisation will not reflect perfectly the in vivo condition, which will have been defiled by the heinous processes of death, autolysis, and microbial decay, at the very least. What is more, organic structures decay at different rates and may be more or less predisposed to mineral replication by fossilisation. Hence, most exceptionally preserved fossils constitute a mineralogical mélange of crystal growth on or within original biological structures that will have undergone a spectrum of decay across different structures

(Fig. 3.1A–F), both within and between individual carcasses. While some biological structures are preserved by mineral impregnation or templating, residual structures decay away to unrecognisable clumps of organic matter that serve as templates for mineralisation or leave voids that are filled much later by percolating fluids rich in mineral ions during the process of sedimentary diagenesis. The resulting complex geode-like diagenetic mineralisation patterns can be readily mistaken for original biological structure (Bengtson and Budd 2004; Donoghue and Purnell 2009). Fossilisation history can be further complicated by later phases of mineral growth that obliterate original biological and intervening diagenetic structure. All of this may be confused further by the laboratory processes that palaeontologists use to recover the fossils from the rock, which invariably employ acids that exploit differences in the solubility of the mineral comprising the fossil versus the mineral cement that binds the sedimentary rock, such that the fossils may be recovered from the disaggregated matrix. However, it can be difficult to control the pH and chemistry of these experiments, leading to artefacts introduced into the fossils by etching or through removal or one or more of the phases of mineralisation introduced during their fossilisation history (Jeppsson et al. 1985).

Fossil embryos are far from immune from the introduction of artefacts as a consequence of these processes of fossilisation and fossil recovery (Xiao and Knoll 2000; Cunningham et al. 2012a). Indeed, fossils interpreted to reflect the earliest stages of embryonic development are simple geometric arrangements of spheroids that can themselves be difficult to discriminate from inorganic structures. However, in interpreting these fossils, it can be difficult to move beyond gainsaying and to obtain an objective approach to discriminating mineral phases that preserve biological structure versus later phases associated with the mineralisation of decayed remains, void filling, or still later phases of mineral growth.

The interpretation of purported embryo fossils from the Ediacaran Doushantuo Formation have proven particularly contentious, with claims of exceptional preservation of labile structures matched by counterclaims that the critical structures on which these interpretation are based merely represent void-filling cement fabrics (Xiao et al. 2000; Bengtson 2003; Cunningham et al. 2012a). Cunningham et al. (2012a) discriminate these phases on the basis of their mineralogical fabric and elemental chemistry. Generally, the earliest mineral phases that preserve biological structure grow within organic substrates, and so the crystals are irregularly arranged and comparatively small (typically at most only a few micrometres in length; Fig. 3.1A–F). Void-filling phases of mineralisation nucleate on the existing mineral substrates and grow approximately perpendicularly to these substrates, yielding an aligned and centrifugally and centripetally layered mineral fabric charac-

terised by comparatively large crystals (typically tens of microns in length), and commonly a botryoidal texture (Fig. 3.1A–F). These two phases of mineralisation are also invariably correlated to differences in elemental chemistry. By demonstrating these characteristics in fossils or features from the same deposit whose biology can be interpreted uncontroversially, it is possible to discriminate mineral phases that preserve original biological structure in more controversial fossils. In this manner, it has been possible to reject many claims for the presence of derived embryonic animals in the Ediacaran Doushantuo Formation (Cunningham et al. 2012a).

ONTOGENY AND TAPHONOMY

Discriminating the biology of preserved fossil embryos is just the beginning of the process of obtaining material insights into the embryology of fossil organisms. The embryology of living animals is difficult enough to study in itself, but at least it is possible to observe the development of a single organism within a Petri dish. The study of fossil embryos requires that developmental stages are correctly identified as such within a fossil assemblage, and there is no guarantee that all stages are preserved. The only insights we have into the relative preservation potential of different developmental stages is based on experimental studies of the decay of *Artemia salina*, which showed quite surprisingly that the rate of decay increases with development, from the encysted diapause stages through postembryonic larval stages through to adults (Gostling et al. 2009). When maintained under reducing conditions, the dead encysted embryos remained stable as physical substrates available for mineral replication, for a period of more than a year (Gostling et al. 2009), a timescale that is readily compatible with the establishment of conditions required for microbially mediated mineralisation of those substrates (Briggs et al. 1993). Indeed, the long-term physical stability of embryonic structure post-mortem appears to be a general phenomenon for animal embryos in reducing conditions (Raff et al. 2006; Gostling et al. 2008),

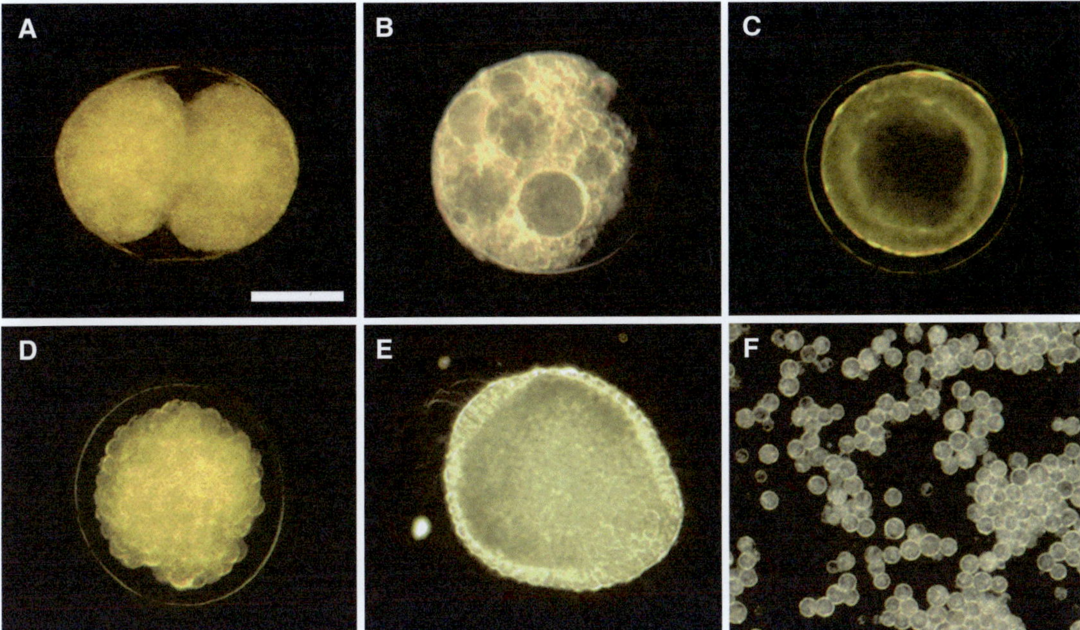

Fig. 3.2 Experimental taphonomy of embryos of the echinoid *Lytechinus pictus*. (**A**) The physical structure of the 2-cell embryo is intact after 26 days post-mortem in a medium of seawater and beta-mercapto ethanol to simulate the reducing conditions necessary for authigenic mineralisation. (**B**) The physical structure of the 2-cell embryo has deteriorated as a consequence of autolysis and consequent lipid coalescence after just 5 h post-mortem in normal seawater. (**C**) Live unhatched blastula showing the columnar cells of the wall of blastula and the blastocoel within. (**D**) Unhatched blastula as in (**C**) but euthanised in seawater containing beta-mercapto ethanol; the component cells are intact, but they have lost adhesion and the blastocoel has collapsed. (**E**) Live hatched blastula. (**F**) Hatched blastula as in (**E**) but euthanised in seawater containing beta-mercapto ethanol; the component cells are intact but they have lost adhesion and, in the absence of a fertilisation envelope, they have disaggregated (From Raff et al. (2006)) Scale bar: (**A–F**) 48 μm

perhaps in large part a consequence of the consumption and replication of the original tissue, cellular, and/or subcellular structure by microbial biofilms (Raff et al. 2008). Regardless, the insights afforded by *Artemia* into the relative preservation potential of developmental stages may go a long way to explain why deposits such as the Doushantuo and Kuanchuanpu Formations preserve some but not all developmental stages (Gostling et al. 2009).

Even accepting that assemblages of fossil embryos preserve only snapshots into the development of the component organisms, it is important to consider whether or not the fossilised developmental stages faithfully reflect the organisation of the embryo in vivo. Raff et al. (2006, 2013) have shown that through the process of autolysis, cytological structure is disrupted through the condensation of lipids (Fig. 3.2A, B). This process is halted or diminished under the reducing conditions required for fossilisation via authigenic mineralisation, when the gross physical integrity of early cleavage stage embryos can be maintained for weeks to months (Fig. 3.2A; Raff et al. 2006). However, within experiments, the component cells in later embryonic and larval developmental stages can lose adhesion and reorganise relative to their original in vivo arrangement (Fig. 3.2C, F). Thus, though the integrity of component cells is maintained, much of their biological context is lost such that evidence of a blastocoel, gastrulation, an archenteron, etc., can be lost as a consequence of the loss of cell adhesion (Fig. 3.2C, F; Raff et al. 2006). Furthermore, while during embryonic stages the component cells remain associated because they are enclosed within the fertilisation envelope (Fig. 3.2C, D), postembryonic stages are readily disaggregated, and evidence of the origin of the component cells (Fig. 3.2E, F), if fossilised, is lost entirely (Raff et al. 2006).

Evidently, fossil remains of animal embryos must be interpreted with great caution. The fidelity of their preservation can be beguiling, but careful analysis of their mineralisation history, discriminating biology from geology and interpreting that biology in light of knowledge of biases in the preservation of developmental stages and the faithfulness with which such fossils reflect their original embryology, can yield material insights into developmental evolution. Since some of the fossils are among the very oldest fossil evidence for the existence of animals in evolutionary history, the stakes could not be higher in our aim of uncovering the role of developmental evolution in effecting the origin and early diversification of animal biodiversity. We now cast a critical eye upon fossil embryos themselves and evaluate competing interpretations of their biological affinity and, consequently, their evolutionary significance.

FOSSIL INVERTEBRATE EMBRYOS

The sum total of fossil remains of embryonic stages of animal development does not extend far beyond the initial deposits from which they were reported, the Early Cambrian Kuanchuanpu and Pestrotsvet Formations, Middle Cambrian Gaotai Formation, and the Ediacaran Doushantuo Formation, though possible eggs and embryos have been recovered from a small number of other deposits in the Middle Cambrian through Early Ordovician. Some of these reports are tenuous and constitute little more than spheroids comprised of calcium phosphate or silica that are more or less filled with diagenetic cement (Lin et al. 2006; Pyle et al. 2006; Broce et al. 2014; Mathur et al. 2014). Whether or not these fossils represent embryos and some are more convincing than others (Broce et al. 2014), they constitute little more than curios of fossilisation until they can be joined with other developmental stages and their phylogenetic affinity constrained (Donoghue and Dong 2005). These criteria are,

as yet, met by precious few species known from fossilised embryonic remains, described below.

Markuelia

The first fossilised embryos to be described as such are attributable to *Markuelia* (Fig. 3.3A–C), though they were then known only from cleavage stages and were interpreted as arthropod embryos (Zhang and Pratt 1994). Recovery of further material from the original site revealed the cleavage embryos to be associated with *Markuelia*, an annulated vermiform organism, coiled into an approximation of a sphere, enclosed within a fertilisation envelope (Fig. 3.3A–D; Zhang et al. 2011). However, *Markuelia* was originally described much earlier as a globular fossil (<1 mm diameter) of unknown affinity with parallel double-walled septa, from the Early Cambrian of Siberia (Val'kov 1983, 1987). It was later shown that these were fossilised embryos, with spines associated with their transverse annulae and a series of paired and variably recurved spines associated with their posterior end (Fig. 3.3A, C, E, F; Bengtson and Yue 1997). *Markuelia hunanensis*, *M. qianensis*, and *M. spinulifera* are known species from the Middle and Late Cambrian of South China, *M. secunda* from the Lower Cambrian of Siberia, *M. lauriei* and *M. waloszeki* from the Middle Cambrian of Australia, as well as undetermined species from the Lower Ordovician of the USA (Donoghue et al. 2006b; Dong et al. 2010; Zhang et al. 2011). Affinities to lobopods (extinct onychophoran-like panarthropods; Bengtson and Yue 1997), annelids (Bengtson and Yue 1997), and halkieriids (armoured worms currently interpreted as stem-mollusks; Conway Morris 1998) were considered until discovery of specimens preserving the anterior end of the organism revealed a terminal mouth surrounded by rings of teeth-like scalids. This narrowed phylogenetic debate to the clade Scalidophora, which is comprised of the phyla Kinorhyncha, Loricifera, and Priapulida (see Vol. 3, Chapter 1), all characterised by the presence of circumoral rings of scalids associated with their introvert (Dong et al. 2004). A more precise affinity can be established for *Markuelia* based principally on details of the number of

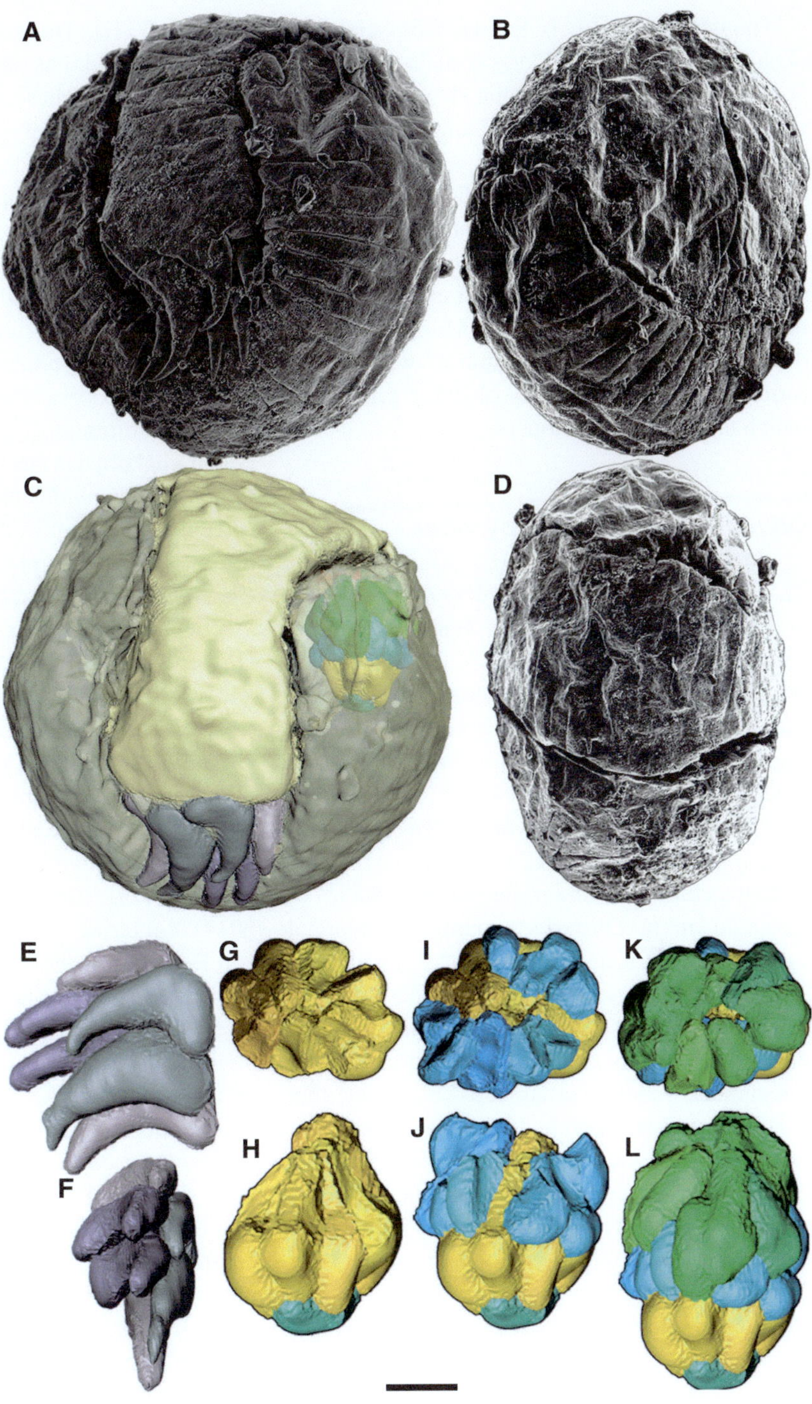

Fig. 3.3 *Markuelia*, a scalidophoran from the Cambrian of Australia, China, and Siberia and the Ordovician of the USA, known only from embryonic stages of development. (**A–L**) *Markuelia waloszeki* (Dong et al. 2010) from the Cambrian of Australia. (**A, B, D**) are scanning electron micrographs; (**C, E–L**) are synchrotron radiation X-ray tomographic microscopy-based reconstructions. (**A**) Embryo with tail (*centre*) and head (*upper right*) juxtaposed. (**B, D**) Same embryo rotated to show the opposing sides, revealing the annulated trunk coiled in a broad S-shaped loop. (**C**) Virtual model of the same embryo (in broadly the same orientation as in (**A**)) derived from synchrotron tomography characterisation of the fossil, showing the tooth-like scalids within the head. (**E, F**) Three pairs of tail spines, recurved ventrally, oriented about anal opening. (**G–L**) The assemblage of scalids that comprise the introvert, viewed from rostrum (**G, I, K**) and lateral (**H, J, L**). These specimens were figured by Dong et al. (2010). Relative scale bar: (**A–B**) 47 μm, (**C–D**) 50 μm, (**E–F**) 24 μm, (**G–L**) 25 μm

scalids arranged in the first three rings around the mouth cone (8-8-9), comprising 25 longitudinal rows (Fig. 3.3G–L). Such characters, along with a terminal anus surrounded by three pairs of bilaterally arranged spines (Fig. 3.3A, C, E, F) and the apparent absence of an armoured pharynx, resolve *Markuelia* as a stem-group scalidophoran (Dong et al. 2004, 2005, 2010; Donoghue et al. 2006a; Harvey et al. 2010; Duan et al. 2012). It is not known whether the extensive annulation of the trunk of *Markuelia* reflects segmentation, but it suggests that annulation may be a shared primitive feature of scalidophorans and, indeed, cycloneuralians.

Though *Markuelia* is known from cleavage (Zhang and Pratt 1994; Dong et al. 2004; Zhang et al. 2011) and late embryonic stages (Bengtson and Yue 1997; Dong 2007; Dong et al. 2004, 2005, 2010; Donoghue et al. 2006a, b; Haug et al. 2009; Zhang et al. 2011), little can be gleaned concerning its development, save that because the late embryonic stages are readily reconcilable with adult scalidophorans, and cycloneuralians more generally, that *Markuelia* was a direct developer. This contrasts with living loriciferans and the majority of living priapulids, which are indirect developers.

Olivooides and *Quadrapyrgites*

The olivooids, *Olivooides* and *Quadrapyrgites*, are known only from the Early Cambrian Kuanchuanpu Formation of South China (Fig. 3.4A–H). Two principal components of the life cycle were originally described independently, viz., the cone-shaped postembryonic theca stage hitherto named 'Punctatus' (Fig. 3.4F–H) and the embryonic stage *Olivooides* (Fig. 3.4D, E), which has taxonomic seniority. Both the embryonic and postembryonic stages of development in *Olivooides* exhibit pentaradial symmetry, manifested through the apex and the single terminal orifice which are folded in five principal rays (Fig. 3.4D, F). The anatomy of the embryo is known mainly from features of its integument, which is ornamented by stellae that resemble twisted bundles of fibres, approximately 10 μm in length (Fig. 3.4D, E). The embryo increases in size through the episodic release of striated integument from the orifice, ultimately developing the tube-shaped characteristic of the postembryonic theca (Fig. 3.4G, H). Indeed, the principal evidence supporting the link between the embryonic and postembryonic stages of development is the stellate and striate integument that envelops the embryo and the base of the hatched theca (Fig. 3.4D, E, G, H; Bengtson and Yue 1997; Yue and Bengtson 1999). The theca expanded in length throughout life through the episodic release of striate integument from the otherwise closed aperture, reflected in a series of circumferential growth rings

(Fig. 3.4G, H). The nature of the internal anatomy of the adult has only been inferred based on an assumed hypothesis of affinity.

A small number of specimens have shown that the external pentaradial symmetry is imposed more fundamentally upon the internal anatomy which has been preserved in only the most exceptional of circumstances, revealing a series of circumference parallel walls with pentaradial divisions and diverticulations, leading to an open adapertural space otherwise occupied by an axial pentaradial process (Dong et al. 2013; Han et al. 2013). Additionally, a set of two juxtaposed pentaradial objects has been described in association with *Olivooides* and interpreted as ephyrae (juvenile medusae) in the process of strobilation (budding in scyphozoan cnidarians; see Chapter 6) (Dong et al. 2013).

Quadrapyrgites occurs in association with *Olivooides*, and in general terms its embryonic and postembryonic growth and developmental stages are identical to those described for *Olivooides*, with the principal distinction that *Quadrapyrgites* is tetraradial (Steiner et al. 2014).

Debate over the affinity of the olivooids has been wide-ranging, including echinoderms (Chen 2004), scalidophorans (Steiner et al. 2014), cnidarians (Bengtson and Yue 1997; Yue and Bengtson 1999; Han et al. 2013), and diploblastic stem-eumetazoans (Yasui et al. 2013). The hypothesis of echinoderm affinity is based on little more than pentaradial symmetry and can be rejected since the olivooids lack key echinoderm apomorphies, not least a mineralised skeleton comprised of calcite stereom (Dong et al. 2013; Han et al. 2013; Steiner et al. 2014). The scalidophoran interpretation is based principally on the general similarity between the theca of olivooids and the lorica of loriciferans and larval priapulids, together with similarities in general symmetry and the requirement of the aperture to open and close akin to a scalidophoran introvert (Steiner et al. 2014). However, these similarities are vague, and the demonstrable absence of a through gut and a scalid-bearing introvert are incompatible with a scalidophoran interpretation of the olivooids. The diploblast stem-eumetazoan interpretation is based principally on the assumption that a single specimen of theca,

Fig. 3.4 *Olivooides multisulcatus* from the Early Cambrian Kuanchuanpu Formation of South China, known from embryonic and postembryonic stages of development. (**A–C**) Associated cleavage and gastrulation stages; (**A, D–H**) are scanning electron micrographs, and (**B, C**) are synchrotron radiation X-ray tomographic microscopy-based reconstructions. (**A**) Cleavage embryo. (**B, C**) Surface model (*blue*) of putative gastrula from synchrotron tomography characterisation of the fossil, showing the interpreted blastopore as a deep sulcus (*arrowed*; **B**) and cells (*orange, yellow*) within, some of which have been reconstructed in 3D (**C**). (**D, E**) Embryonic stages of *Olivooides* with the characteristic stellate integument and remains of the fertilisation envelope obscuring the pentaradial aperture. (**F–H**) Postembryonic developmental stages with the adapertural stellate ornament retained from the embryo and the characteristic pentaradial aperture through which the additional striate integumentary tissue is released to increase the length of the theca. Specimens figured by Dong et al. (2013) except for B and C (figured by Donoghue et al. 2006a, b). Relative scale bar: (**A**) 98.5 μm, (**B–C**) 145 μm, (**D**) 119 μm, (**E**) 144 μm, (**F**) 197 μm, (**G**) 282 μm, (**H**) 254 μm

which bears a diagenetic mineral plug beneath the aperture, reflects a miniature gut and infers that the remaining theca constituted a vast body cavity (Yasui et al. 2013). However, this is a misinterpretation of mineralised decayed remains as reflecting in vivo anatomy.

The cnidarian interpretation of the olivooids was based originally on comparison that had been drawn between the theca and conulariids, an extinct group of cnidarians that have been interpreted as coronate scyphozoans (Conway Morris and Chen 1992; Yue and Bengtson 1999), with the theca interpreted as the sclerotised integument of the adult polyp (Chapter 6). The pattern of direct development, from embryo to adult theca, is unusual for scyphozoans, specifically, and cnidarians generally. The pattern of pentaradial symmetry seen in *Olivooides* is also unusual for cnidarians, as is the aperture which is difficult to rationalise with the presence of a polyp. Many of these concerns are diminished if not dismissed based on recent description of internal anatomy, which Han et al. (2013) interpret (and likely grossly over-interpret) in the image of a cubozoan polyp. Details of this comparison are problematic, not least the evident fact that the embryo of *Olivooides* develops into a sessile polyp (if it is indeed a cnidarian), not a medusa, and what is more, the polyp stage of living cubozoans is grossly reduced. Indeed, the principal points of similarity between the olivooids and cubozoans, as interpreted by Han and colleagues (2013), are cnidarian or at least medusozoan symplesiomorphies (cf. Chapter 6). Thus, the internal anatomy of olivooids, as evidenced by *Olivooides*, supports a cnidarian affinity at the least. Concerns over differences in the pentaradial symmetry of *Olivooides* and the generally (but far from exclusively) tetraradiality of cnidarians may certainly be dismissed on the description of *Quadrapyrgites*, which differs materially from *Olivooides* only in terms of its tetraradial symmetry. Finally, the description of minute pentaradial strobilating medusae in association with *Olivooides* would appear to settle debate over its affinity (though this is disputed by Steiner et al. 2014). In sum, the available evidence supports the interpretation of the olivooids as medusozoan cnidarians, and their similarity to scyphozoans must represent either shared derived or shared primitive characteristics; only a better understanding of the interrelationships of extant cnidarians and morphological character evolution among them will aid a more precise classification for the olivooids. Either way, the olivooids evidence the fact that although indirect development is the norm among extant cnidarians, known lineages in the Cambrian underwent direct development.

Pseudooides

Pseudooides prima has also been described from the Early Cambrian Kuanchuanpu Formation from a number of localities in South China (Fig. 3.5A–C; Qian 1977; Steiner et al. 2004a, b; Donoghue et al. 2006a). After the initial description as a globular microfossil of unknown affinity (Qian 1977), Steiner and colleagues recognised it as an embryo, typically 250–500 μm in diameter, characterised by a segmented 'germband' that can extend around the majority of the diameter of the fossil (Fig. 3.5A–C). The remainder of the surface is undifferentiated, and no internal anatomy appears to have been preserved in any of the material described to date (Donoghue et al. 2006a). The segmented band pinches out at its extremity (Fig. 3.5A) and is divided longitudinally along the midline and transversely into up to twelve compartments (Fig. 3.5C). The centre of the band may also be pinched before the central compartments develop (Fig. 3.5B), leading to the inference that the compartments are added from the centre (Donoghue et al. 2006a). Steiner et al. (2004b) present specimens that indicate that the compartments develop through progressive division of a band that is initially undifferentiated save for the longitudinal furrow, without which it would be difficult to attribute the embryos to *Pseudooides*. A number of cleavage and gastrulation stage embryos have been attributed to *Pseudooides* as opposed to co-occurring olivooids (Fig. 3.4A–C), on the basis of their size (Donoghue et al. 2006a; Steiner et al. 2004b, 2014). None of these data are particularly phylogenetically informative, although Steiner et al. (2004b) associated these embryos with fragmentary remains of an arthropod or arthropod-like organism. The apparent pattern of germband development, if that is what it represents, is extremely unusual for an arthropod or, indeed, any bilaterian (Donoghue et al. 2006a). *Pseudooides* requires further study before its embryology, phylogenetic affinity, and evolutionary significance can be determined.

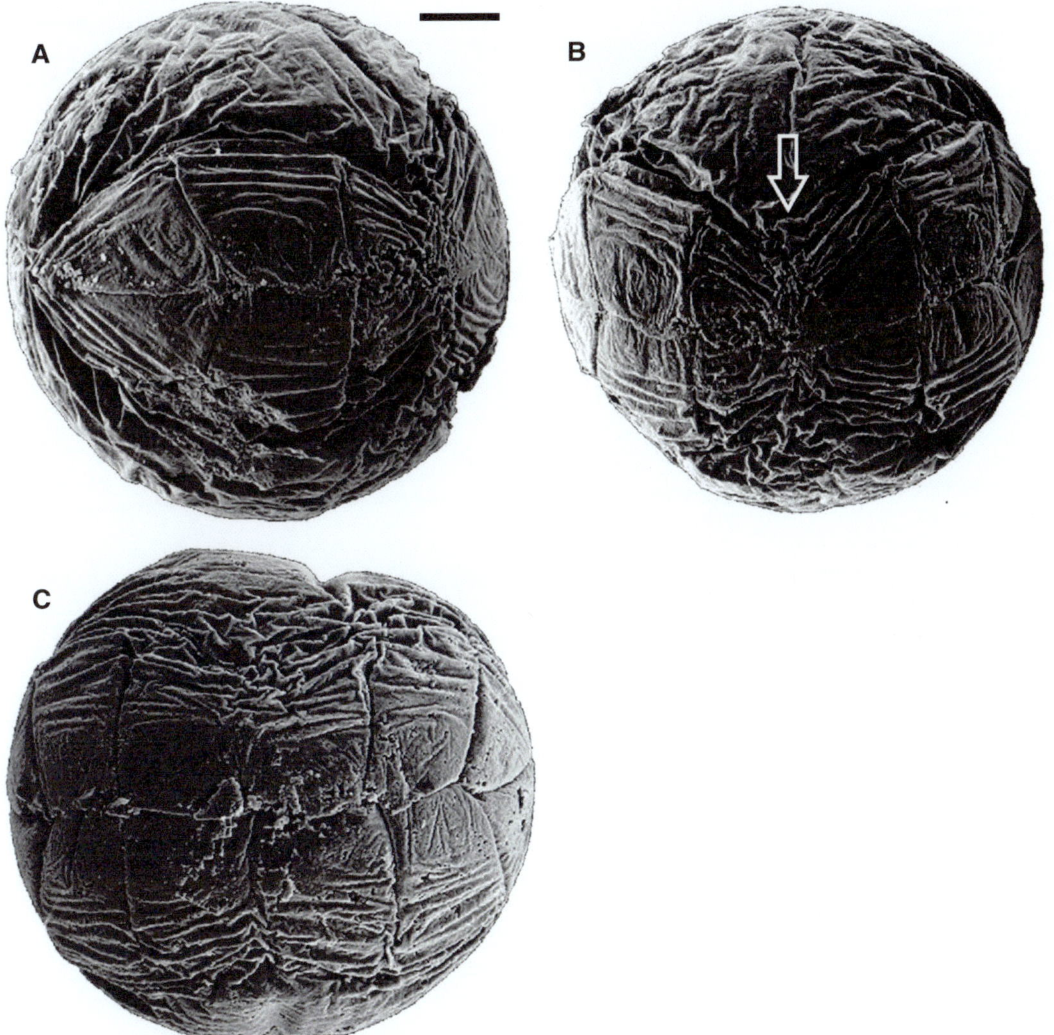

Fig. 3.5 Scanning electron micrographs of *Pseudooides prima* from the Early Cambrian Kuanchuanpu Formation of South China, known only from embryonic stages of development. (**A, B**) The segmented 'germband' with a central pinch (*arrowed*), interpreted to the point from which new segments are developed. (**C**) Twelve-segment germband without central pinch (These figures are reproduced with permission from the publishers from Steiner et al. (2004b)). Relative scale bar: (**A, B**) 53 μm, (**C**) 56 μm

Tianzhushania

In addition to the well-accepted embryos from the Cambrian and lowermost Ordovician, there have also been more contentious reports of animal cleavage embryos (Fig. 3.6A–D; Xiao et al. 1998) from the Ediacaran (i.e., the final period of the Precambrian). The fossils in question are from the Doushantuo biota of southern China and are approximately 570 million years old

(Xiao et al. 1998). The Doushantuo Formation varies in its composition throughout its broad occurrence in South China, ranging from black shales through cherts and phosphorites, although the purported fossils of animal embryos are preserved in calcium phosphate in both chert and phosphorite. In the phosphorite at least, the fossilisation occurred elsewhere, and the fossils were resedimented and size sorted (Xiao et al. 2007).

Fig. 3.6 Embryo-like fossils from the Ediacaran Doushantuo biota assigned to the genera *Tianzhushania* (**A–D**), *Spiralicellula* (**E**), and *Helicoforamina* (**F**); (**A–D**, **F**) are synchrotron radiation X-ray tomographic microscopy-based reconstructions, while (**E**) is a scanning electron micrograph. (**A**) *Tianzhushania* specimen with an ornamented outer envelope. (**B**) Early cleavage stage of *Tianzhushania*. (**C**) Early cleavage stage of *Tianzhushania* showing possible nuclei. (**D**) Later cleavage stage of *Tianzhushania*. (**E**) *Spiralicellula* specimen showing cells coiled helicospirally. (**F**) *Helicoforamina* specimen showing a single helicospiral body (Parts **A**, **B**, **D**, and **F** also figured by Cunningham (2012); part **C** was also figured by Cunningham et al. (2014); part **E** was previously figured by Tang et al. (2008)) Relative scale bar: (**A**) 140 μm, (**B**, **C**) 105 μm, (**D**) 115 μm, (**E**) 200 μm, (**F**) 120 μm

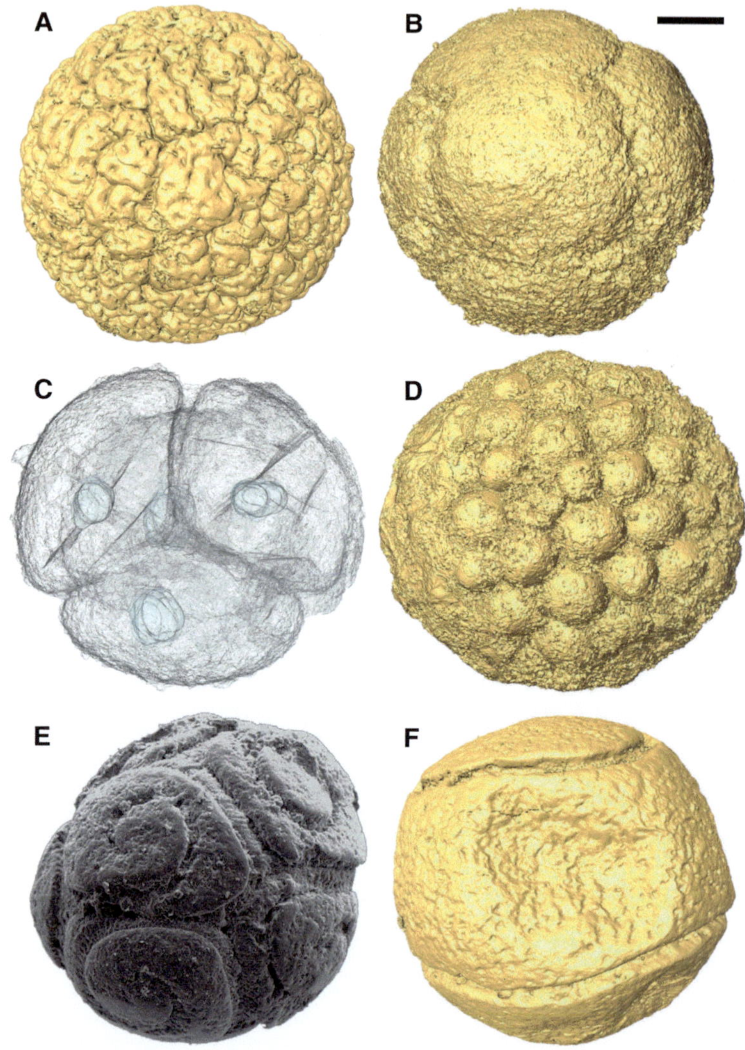

The vast majority of putative embryos can be assigned to the genus *Tianzhushania* (senior synonym of *Megasphaera*, *Parapandorina*, and *Megaclonophycus*; see Yin et al. 2004). They would originally have been encased within a spinose cyst, but this has been lost in most specimens recovered through acid dissolution of the surrounding matrix. Specimens range from single cells through early cleavage stages (Fig. 3.6B, C) to those with thousands of cells. *Tianzhushania* shares a number of features with animal embryos including reductive division, Y-shaped junctions between cells, and a multi-layered ornate envelope surrounding the cells (Fig. 3.6A; Xiao 2002). These microfossils are preserved with remarkable fidelity with cellular and even subcellular details including possible nuclei having been fossilised (Figs. 3.1E and 3.6C; Hagadorn et al. 2006). They have attracted much attention, not only because they could represent the oldest animal fossils in the entire record but also because they might potentially allow palaeontologists to study embryology at the time when animal body plans were first starting to become established. However, other workers have challenged the animal affinities of these fossils. Firstly, Bailey et al. (2007a, b) suggested that the embryo-like fossils might be better interpreted as giant sulphur bacteria comparable to the extant *Thiomargarita*, which can reach 750 μm in diameter and undergoes reductive division. This provided an explanation for the

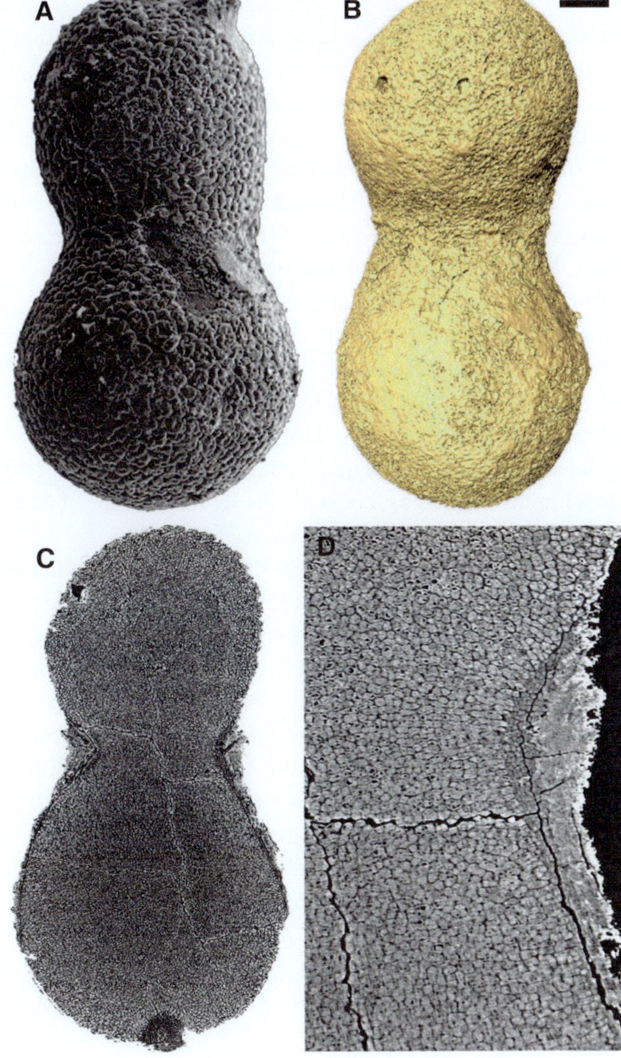

Fig. 3.7 A peanut-shaped fossil from the Ediacaran Doushantuo biota. (**A**) Scanning electron micrograph of a peanut-shaped specimen preserving the ornamented outer envelope. (**B**) Synchrotron radiation X-ray tomographic microscopy-based surface rendering of a peanut-shaped specimen. (**C**) SRXTM section through the specimen in (**B**). (**D**) Detail of an SRXTM section through the specimen in (**B**) showing many cells. These specimens were figured by Huldtgren et al. (2011). Relative scale bar: (**A–C**) 100 µm, (**D**) 36 µm

lack of associated later developmental stages. Subsequent work has revealed a number of problems with this model. Bacteria do not produce structures with complexity of the ornamented envelope that surrounds the fossils (Xiao et al. 2007), they do not have nuclei (Huldtgren et al. 2011), and they can only achieve this giant size by means of a vacuole that takes up around 98 % of the volume, which is absent in the fossils (Donoghue 2007). Moreover, decay experiments suggest that, unlike animal embryos, giant sulphur bacteria would be unlikely to be preserved three-dimensionally because of collapse of the vacuole (Cunningham et al. 2012b). Secondly, Huldtgren et al. (2011, 2012) argued that the

embryo-like fossils were in fact cyst-forming protists. They showed that none of the characters that had been used to identify the fossils as animals are metazoan synapomorphies. Although these features are compatible with an animal interpretation, they are at best animal symplesiomorphies, found in more universal clades. Thus, the characters identified in the fossils and used to evidence an animal interpretation may be necessary to identify *Tianzhushania* as an animal, but they are not sufficient. Huldtgren et al. (2011) also reported fossils interpreted as later stages in the life cycle of *Tianzhushania*, on the basis that they possess an identical ornamented envelope (Fig. 3.7A–D). These fossils have features that cannot be reconciled with metazoan development. In particular, the specimens have hundreds of thousands of cells and yet show no sign of tissue differentiation – something that is present in all extant animals by this stage of development. While the authors could not definitively rule out a placement within the stem group of animals, they found that there was no evidence to place them here either. Others have suggested that the new fossils are not part of the life cycle of the same organism (Xiao et al. 2012) and instead present the same metazoan symplesiomorphies as evidence of animal affinity. Additional work to establish the variability of forms in the deposit is needed to assess whether or not intermediate developmental stages between the embryo-like forms and the new specimens exist.

A third possibility is that the fossils are multicellular green algae. Before being interpreted as animals, the fossils were initially compared to the alga *Pandorina* by Xue et al. (1995), and as a result the cleavage stages were named *Parapandorina*. This interpretation has been resurrected in a recent comment by Butterfield (2011). This possibility requires further investigation but seems unlikely given that extant green algae like *Pandorina* maintain cell adhesion in the cleavage stages by means of cytoplasmic bridges that are absent from the fossils.

Spiralicellula and *Helicoforamina*

Associated with *Tianzhushania*, but much rarer, are similar forms that differ in having each cell coiled into a spiral. These enigmatic forms are assigned to the genus *Spiralicellula* (Fig. 3.6E) and also contain nucleus-like structures (Huldtgren et al. 2011) and have also been considered to be embryos. In addition, *Helicoforamina* (Fig. 3.6F), a form with a helical groove running around a spherical body, is also known from the Doushantuo biota (Xiao et al. 2007). One suggestion is that *Helicoforamina* is an elusive later developmental stage of *Tianzhushania*, perhaps representing a coiled embryo of a vermiform or tubular organism (Xiao et al. 2007). On the other hand, *Spiralicellula* and *Helicoforamina*

have been associated together by various authors (Tang et al. 2008; Huldtgren et al. 2011; Zhang and Pratt 2014). These have been considered as embryos (with *Helicoforamina* being the single-celled stage), possibly representing the embryonic stages of the enigmatic ctenophore-like fossil *Eoandromeda*, which has eight spiral arms (Tang et al. 2008). Alternatively, they have also been interpreted as cyst (*Helicoforamina*) and dividing stages (*Spiralicellula*) of protists (Huldtgren et al. 2011) or green algae (Zhang and Pratt 2014).

Other Candidate Embryos from the Doushantuo Biota

There are different perspectives on the diversity of the organisms represented by the embryonic and larval stages from the Doushantuo Formation, with some arguing for a diverse assemblage of animals, including derived bilaterians (Chen et al. 2000, 2002, 2004, 2006, 2009b; Chen and Chi 2005), while at another extreme, others rationalise the majority of remains as representing one or a few species that may represent only stem-metazoans (Hagadorn et al. 2006; Xiao et al. 2012), or else that all or a majority of such fossils may not represent animals or embryos at all (Bailey et al. 2007a, b; Huldtgren et al. 2011, 2012; Zhang and Pratt 2014). Much of this equivocation will be resolved with debate over the phylogenetic affinity of *Tianzhushania*; however, the interpretation of a diverse biota is based principally on the spurious interpretation of diagenetic mineral fabrics as preserving original biological structures (Bengtson 2003; Xiao and Knoll 2000; Xiao et al. 2000; Cunningham et al. 2012a). For instance, Li et al. (1998) described sponge embryos and larvae of sponges based on effectively two-dimensional thin sections of rock; however, the critical structures interpreted as amoebocytes, blastomeres, flagellae, mesohyl, a plasma membrane, porocytes, sclerocytes, spongocoel, and spicules are indistinguishable from layered and clotted void-filling diagenetic mineralisation, unrelated to the replication of biological structure, that is common in phosphatised Doushantuo fossils (Xiao and Knoll 2000; Hagadorn et al. 2006; Cunningham et al. 2012a). Similarly, Chen et al. (2000, 2002) describe anthozoan gastrulae, larvae and polyps, as well as hydrozoan gastrulae, although these fossils preserve no more biological structure than what may constitute a fertilisation envelope infilled with an anastomosing diagenetic mineral cement. None of these records withstand scrutiny (Xiao and Knoll 2000; Xiao et al. 2000; Bengtson 2003; Cunningham et al. 2012a).

Other Records of Fossil Embryos

Phosphatised embryos have been described from the Early Cambrian of Yakutia, Siberia (Fig. 3.8A–C), that preserve a cross-like structure (Fig. 3.8A, B) that has been compared to the appearance of micromeres on spiralian blastula-stage embryos and to incipient tentacles in cnidarian actinula larvae, substantiating a tentative link to co-occurring anabaritids, long considered cnidarians (Kouchinsky et al. 1999). Silicified eggs and embryos have been described from the Middle Cambrian Kaili Biota of Ghizou, China (Lin et al. 2006); however, their phylogenetic affinity is unknown, and they provide no material insights into embryology.

THE NATURE OF THE FOSSIL RECORD OF EMBRYOS

The fossil record of embryos could hardly be considered representative. Precious few organisms are represented by fossilised embryonic stages, certainly not even those organisms known from fossil remains in the same deposits, and those that are preserved represent only a small proportion of their embryological development. It is not clear what, if anything, unites these organisms to justify the preservation of their embryological stages. Indeed, it may merely be a conspiracy of environmental circumstances rather than anything more intrinsically biological. Ultimately, however, there appear to be two principal classes of structures preserved: (i) dividing cells in early stages of palintomy (Doushantuo embryo-like

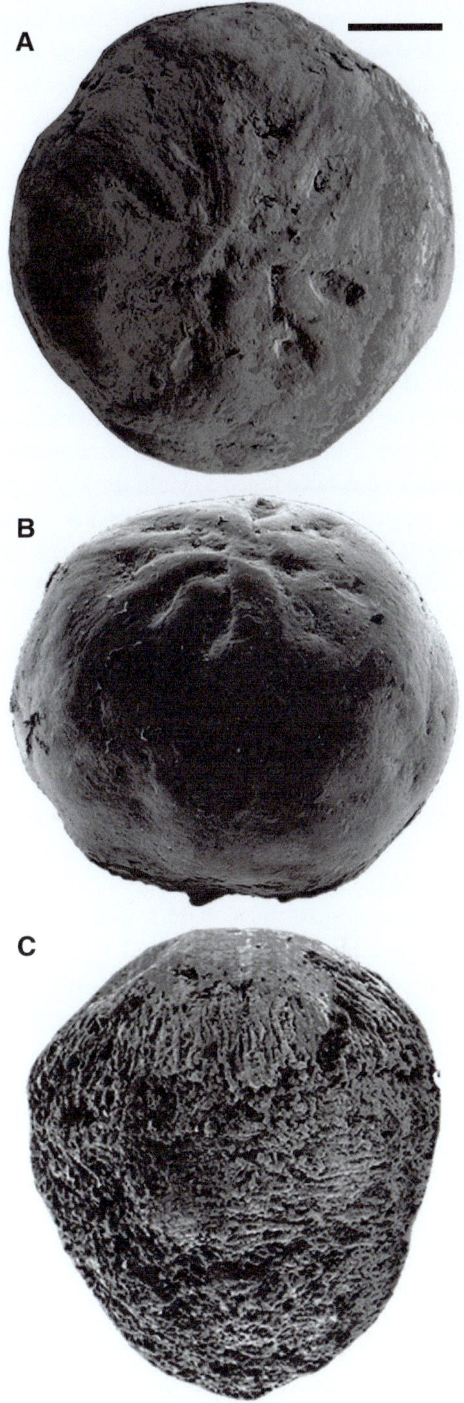

Fig. 3.8 Scanning electron micrographs of putative cnidarian embryos from the Early Cambrian of Siberia, known only from these embryonic stages of development. (**A, B**) Embryonic stage with cross-like structure. (**C**) Embryonic stage with spicule-like structures on its surface. These specimens were figured by Kouchinsky et al. (1999). Relative scale bar: (**A**) 100 μm, (**B**) 103 μm, (**C**) 89 μm

fossils), cleavage or gastrulation (Kuanchuanpu and *Markuelia* embryos), and (ii) cuticle and/or integument, as in the case of *Markuelia*, *Olivooides*, *Quadrapyrgites*, and possibly also *Pseudooides*. Primary larvae (as opposed the arthropod larvae which dominate the Orsten biota) are rarely preserved, and the record is dominated by large and, therefore, presumably yolky embryos. These taxonomic and developmental biases may be explained by the inherent biases in the pattern of decay seen in taphonomy experiments (Raff et al. 2006; Gostling et al. 2008, 2009). The preponderance of large embryos may not be a sampling bias for size since attempts to control for this have failed to yield further discoveries in sites where embryo fossils are already known (Donoghue et al. 2006b). The absence of fossilised primary larvae in the embryo-bearing deposits may be a taphonomic artefact, and so it does not follow that their absence from fossil assemblages is evidence of their absence during life. However, the mere presence of large marine invertebrate embryos in the Cambrian suggests that direct development may have evolved early among animal lineages and may be a primitive feature of metazoan development (Donoghue and Dong 2005).

Palaeontologists have dared to believe that there was a hitherto undiscovered fossil record of embryos and, without doubt, there will be further discoveries. However, it appears that the broad extent of this fossil record has been plumbed. Thus, the fossil record of marine invertebrate embryos is very clearly biased to the Ediacaran and/or the earliest Phanerozoic interval. This may reflect a combination of factors (Donoghue et al. 2006b), including the widespread deposition of marine phosphates at this time. Another factor must be the paucity of deposit feeders which, in later times, more effectively recycled organic remains directly and, indirectly, served to expand the depth of sediment oxygenation and, with it, aerobic microbial activity, the principal vector of decay. Nevertheless, would we wish for a fossil record of embryology from any interval of Earth history, it would be this one, and so we should make the most of what we have. Thus, future research should focus on better resolving the biological nature of fossils known to preserve embryological stages and to prospect for new remains to better understand embryology in deep time.

Acknowledgments We would like to thank the TOMCAT beamline scientists at the Swiss Light Source, Paul Scherrer Institut, Marco Stampanoni, Federica Marone, and Julie Fife, and the researchers at Beijing, Bristol, and Stockholm who have assisted in our research into fossil embryos over the years: Chen Fang, Cheng Gong, Duan Baichuan, Peng Fan, Zhi-kun Gai, Sam Giles, Neil Gostling, Jenny Greenwood, Guo Wei, Sandy Hetherington, Therese Huldtgren, David Jones, Joe Keating, Liu Jie, Liu Zheng, Duncan Murdock, Maria Pawlowska, Martin Rücklin, Ceri-Wynn Thomas, Zhao Yue, Chongyu Yin, and Zhang Huaqiao. We would also like to thank our colleagues with whom we have enjoyed fruitful discussions on embryology, fossilisation, and embryo fossils over the years. Finally, we thank Andreas Wanninger for inviting us to contribute to this wonderful and insightful volume on comparative embryology of invertebrates.

References

Bailey JV, Joye SB, Kalanetra KM, Flood BE, Corsetti FA (2007a) Evidence of giant sulfur bacteria in Neoproterozoic phosphorites. Nature 445:198–201

Bailey JV, Joye SB, Kalanetra KM, Flood BE, Corsetti FA (2007b) Palaeontology – undressing and redressing ediacaran embryos – reply. Nature 446:E10–E11

Bengtson S (2003) Tracing metazoan roots in the fossil record. In: Legakis A, Sfenthourakis S, Polymeni R, Thessalou-Legaki M (eds) The new panorama of animal evolution. Pensoft, Moscow, pp 289–300

Bengtson S (2005) Mineralized skeletons and early animal evolution. In: Briggs DEG (ed) Evolving form and function: fossils and development. Yale Peabody Museum, New Haven, pp 101–124

Bengtson S, Budd G (2004) Comment on "Small Bilaterian Fossils from 40 to 55 Million Years Before the Cambrian". Science 306:1291a

Bengtson S, Yue Z (1997) Fossilized metazoan embryos from the earliest Cambrian. Science 277:1645–1648

Briggs DEG (2003) The role of decay and mineralization in the preservation of soft-bodied fossils. Annu Rev Earth Planet Sci 31:275–301

Briggs DEG, Kear AJ, Martill DM, Wilby PR (1993) Phosphatization of soft-tissue in experiments and fossils. J Geol Soc Lond 150:1035–1038

Broce J, Schiffbauer JD, Sharma KS, Wang G, Xiao S (2014) Possible animal embryos from the Lower Cambrian (stage 3) Shuijingtuo Formation, Hubei Province, South China. J Paleontol 88:385–394

Butterfield NJ (2011) Terminal developments in Ediacaran embryology. Science 334:1655–1656

Chen JY (2004) The dawn of animal world. Jiangsu Science and Technology Press, China

Chen JY, Chi HM (2005) Precambrian phosphatized embryos and larvae from the Doushantuo Formation and their affinities, Guizhou (SW China). Chin Sci Bull 50:2193–2200

Chen J, Oliveri P, Li C-W, Zhou G-Q, Gao F, Hagadorn JW, Peterson KJ, Davidson EH (2000) Precambrian animal diversity: putative phosphatised embryos from the Doushantuo Formation of China. Proc Natl Acad Sci U S A 97:4457–4462

Chen J, Oliveri P, Gao F, Dornbos SQ, Li CW, Bottjer DJ, Davidson EH (2002) Precambrian animal life: probable developmental and adult cnidarian forms from Southwest China. Dev Biol 248:182–196

Chen J, Braun A, Waloszek D, Peng Q-Q, Maas A (2004) Lower Cambrian yolk-pyramid embryos from southern Shaanxi, China. Prog Nat Sci 14:167–172

Chen JY, Bottjer DJ, Davidson EH, Dornbos SQ, Gao X, Yang YH, Li CW, Li G, Wang XQ, Xian DC, Wu HJ, Hwu YK, Tafforeau P (2006) Phosphatized polar lobe-forming embryos from the Precambrian of Southwest China. Science 312:1644–1646

Chen J-Y, Schopf JW, Bottjer DJ, Zhang C-Y, Kudryavtsev AB, Tripathi AB, Wang X-Q, Yang Y-H, Gao X, Yang Y (2007) Raman spectra of a Lower Cambrian ctenophore embryo from southwestern Shaanxi, China. Proc Natl Acad Sci 104:6289–6292

Chen J-Y, Bottjer DJ, Davidson EH, Li G, Gao F, Cameron RA, Hadfield MG, Xian D-C, Tafforeau P, Jia Q-J, Sugiyama H, Tang R (2009a) Phase contrast synchrotron X-ray microtomography of Ediacaran (Doushantuo) metazoan microfossils: phylogenetic diversity and evolutionary implications. Precambrian Res 173:191–200

Chen J-Y, Bottjer DJ, Li G, Hadfield MG, Gao F, Cameron AR, Zhang C-Y, Xian D-C, Tafforeau P, Liao X, Yin Z-J (2009b) Complex embryos displaying bilaterian characters from Precambrian Doushantuo phosphate deposits, Weng'an, Guizhou, China. Proc Natl Acad Sci U S A 106:19056–19060

Conway Morris S (1998) Eggs and embryos from the Cambrian. BioEssays 20:676–682

Conway Morris S, Chen M (1992) Carinachitids, Hexangulaconulariids, and *Punctatus*: problematic metazoans from the Early Cambrian of South China. J Paleontol 66:384–406

Cunningham JA (2012) Fossil focus: animal embryos. Palaeontology 2:1–4 (Online)

Cunningham JA, Thomas C-W, Bengtson S, Kearns SL, Xiao S, Marone F, Stampanoni M, Donoghue PCJ (2012a) Distinguishing geology from biology in the Ediacaran Doushantuo biota relaxes constraints on the timing of the origin of bilaterians. Proc R Soc B Biol Sci 279:2369–2376

Cunningham JA, Thomas C-W, Bengtson S, Marone F, Stampanoni M, Turner FR, Bailey JV, Raff RA, Raff EC, Donoghue PCJ (2012b) Experimental taphonomy of giant sulphur bacteria: implications for the interpretation of the embryo-like Ediacaran Doushantuo fossils. Proc R Soc B Biol Sci 279:1857–1864

Cunningham JA, Donoghue PCJ, Bengtson S (2014) Distinguishing biology from geology in soft tissue preservation. In: Laflamme M, Schiffbauer JD, Darroch SAF (eds.), Reading and writing of the fossil record: preservational pathways to exceptional fossilization. Paleontol Soc 20: 275–287.

Dong X-P (2007) Developmental sequence of the Cambrian embryo *Markuelia*. Chin Sci Bull 52:929–935

Dong X-P, Donoghue PCJ, Cheng H, Liu J (2004) Fossil embryos from the Middle and Late Cambrian period of Hunan, south China. Nature 427:237–240

Dong X-P, Donoghue PCJ, Cunningham J, Liu J, Cheng H (2005) The anatomy, affinity and phylogenetic significance of *Markuelia*. Evol Dev 7:468–482

Dong X-P, Bengtson S, Gostling NJ, Cunningham JA, Harvey THP, Kouchinsky A, Val'kov AK, Repetski JE, Stampanoni M, Donoghue PCJ (2010) The anatomy, taphonomy, taxonomy and systematic affinity of *Markuelia*: early Cambrian to early Ordovician scalidophorans. Palaeontology 53:1291–1314

Dong X-P, Cunningham JA, Bengtson S, Thomas CW, Liu J, Stampanoni M, Donoghue PCJ (2013) Embryos, polyps and medusae of the Early Cambrian scyphozoan *Olivooides*. Proc Biol Sci Royal Soc 280: 20130071

Donoghue PCJ (2007) Palaeontology: embryonic identity crisis. Nature 445:155–156

Donoghue PCJ, Dong X-P (2005) Embryos and ancestors. In: Briggs DEG (ed) Evolving form and function: fossils and development. Yale Peabody Museum of Natural History, Yale University, New Haven, pp 81–99

Donoghue PCJ, Purnell MA (2009) Distinguishing heat from light in debate over controversial fossils. BioEssays 31:178–189

Donoghue PCJ, Forey PL, Aldridge RJ (2000) Conodont affinity and chordate phylogeny. Biol Rev 75: 191–251

Donoghue PCJ, Bengtson S, Dong X-P, Gostling NJ, Huldtgren T, Cunningham JA, Yin C, Yue Z, Peng F, Stampanoni M (2006a) Synchrotron X-ray tomographic microscopy of fossil embryos. Nature 442:680–683

Donoghue PCJ, Kouchinsky A, Waloszek D, Bengtson S, Dong X-P, Val'kov AK, Cunningham JA, Repetski JE (2006b) Fossilized embryos are widespread but the record is temporally and taxonomically biased. Evol Dev 8:232–238

Duan B, Dong X-P, Donoghue PCJ (2012) New palaeoscolecid worms from the Furongian (upper Cambrian) of Hunan, South China: is *Markuelia* an embryonic palaeoscolecid? Palaeontology 55: 613–622

Gostling NJ, Thomas C-W, Greenwood JM, Dong X-P, Bengtson S, Raff EC, Raff RA, Degnan BM, Stampanoni M, Donoghue PCJ (2008) Deciphering the fossil record of early bilaterian embryonic development in light of experimental taphonomy. Evol Dev 10:339–349

Gostling NJ, Dong X-P, Donoghue PCJ (2009) Ontogeny and taphonomy: and experimental taphonomy study of the development of the brine shrimp *Artemia salina*. Palaeontology 52:169–186

Hagadorn JW, Xiao S, Donoghue PCJ, Bengtson S, Gostling NJ, Pawlowska M, Raff EC, Raff RA, Turner FR, Yin C, Zhou C, Yuan X, McFeely MB, Stampanoni M, Nealson KH (2006) Cellular and subcellular structure of Neoproterozoic animal embryos. Science 314:291–294

Han J, Kubota S, Li G, Kubota S, Yao X, Yang X, Shu D, Li Y, Kinoshita S, Sasaki O, Komiya T, Yan G (2013) Early Cambrian pentamerous cubozoan embryos from South China. PLoS ONE 8:e70741

Harvey THP, Dong X, Donoghue PCJ (2010) Are palaeoscolecids ancestral ecdysozoans? Evol Dev 12:177–200

Haug JT, Maas A, Waloszek D, Donoghue PCJ, Bengtson S (2009) A new species of *Markuelia* from the Middle Cambrian of Australia. Mem Assoc Australas Palaeontol 37:303–313

Huldtgren T, Cunningham JA, Yin C, Stampanoni M, Marone F, Donoghue PCJ, Bengtson S (2011) Fossilized nuclei and germination structures identify Ediacaran "animal embryos" as encysting protists. Science 334:1696–1699

Huldtgren T, Cunningham JA, Yin C, Stampanoni M, Marone F, Donoghue PCJ, Bengtson S (2012) Response to comment on "fossilized nuclei and germination structures identify Ediacaran 'animal embryos' as encysting protists". Science 335:1169

Jeppsson L, Fredholm D, Mattiasson B (1985) Acetic acid and phosphatic fossils – a warning. J Paleontol 59:952–956

Kouchinsky A, Bengtson S, Gershwin L (1999) Cnidarian-like embryos associated with the first shelly fossils in Siberia. Geology 27:609–612

Li C, Chen JY, Hua TE (1998) Precambrian sponges with cellular structures. Science 279:879–882

Lin J-P, Scott AC, Li C-W, Wu H-J, Ausich WI, Zhao Y-L, Hwu Y-K (2006) Silicified egg clusters from a Middle Cambrian Burgess Shale-type deposit, Guizhou, south China. Geology 34:1037–1040

Mathur VK, Shome S, Nath S, Babu R (2014) First record of metazoan eggs and embryos from early Cambrian Chert Member of Deo ka Tibba Formation, Tal Group, Uttarakhand Lesser Himalaya. J Geol Soc India 83:191–197

Pyle LJ, Narbonne GM, Nowlan GS, Xiao SH, James NP (2006) Early Cambrian metazoan eggs, embryos, and phosphatic microfossils from northwestern Canada. J Paleontol 80:811–825

Qian Y (1977) *Hyolitha* and some problematica from the Lower Cambrian Meischucunian Stage in central and southwestern China. Acta Palaeontol Sin 16:255–275

Raff EC, Villinski JT, Turner FR, Donoghue PCJ, Raff RA (2006) Experimental taphonomy shows the feasibility of fossil embryos. Proc Natl Acad Sci U S A 103:5846–5851

Raff EC, Schollaert KL, Nelson DE, Donoghue PCJ, Thomas C-W, Turner FR, Stein BD, Dong X-P, Bengtson S, Huldtgren T, Stampanoni M, Yin C, Raff RA (2008) Embryo fossilization is a biological process mediated by microbial biofilms. Proc Natl Acad Sci 105:19359–19364

Raff EC, Andrews ME, Turner FR, Toh E, Nelson DE, Raff RA (2013) Contingent interactions among biofilm-forming bacteria determine preservation or decay in the first steps toward fossilization of marine embryos. Evol Dev 15:243–256

Steiner M, Li G, Zhu M (2004a) Lower Cambrian small shelly fossils of northern Sichuan and southern Shaanxi (China), and their biostratigraphic significance. Geobios 37:259–275

Steiner M, Zhu M, Li G, Qian Y, Erdtmann B-D (2004b) New early Cambrian bilaterian embryos and larvae from China. Geology 32:833–836

Steiner M, Qian Y, Li G, Hagadorn JW, Zhu M (2014) The developmental cycles of early Cambrian Olivooidae fam. nov. (?Cycloneuralia) from the Yangtze Platform (China). Palaeogeogr Palaeoclimatol Palaeoecol 398:97–124

Tang F, Yin C, Bengtson S, Liu P, Wang Z, Gao L (2008) Octoradiate spiral organisms in the Ediacaran of South China. Acta Geologica Sinica 82:27–34

Val'kov AK (1983) Rasprostranenie drevnejshikh skeletnykh organizmov i korrelyatsiya nizhnej granitsy kembriya v yugo-vostochnoj chasti Sibirskoj platformy [Distribution of the oldest skeletal organisms and correlation of the lower boundary of the Cambrian in the southeastern part of the Siberian Platform]. In: Khomentovsky VV, Yakshin MS, Karlova GA (eds) Pozdnij dokembrij i rannij paleozoj Sibiri, Vendskie otlozheniya. Inst Geol Geofiz SO AN SSSR, Novosibirsk, pp 37–48

Val'kov AK (1987) Biostratigrafiya nizhnego kembriya vostoka Sibirskoj Platformy (Yudoma-Olenyokskij region). [Lower Cambrian biostratigraphy of the eastern part of the Siberian Platform (Yudoma-Olenyok region)]. Nauka, Moscow

Xiao S (2002) Mitotic topologies and mechanics of Neoproterozoic algae and animal embryos. Paleobiology 28:244–250

Xiao S, Knoll AH (2000) Fossil preservation in the Neoproterozoic Doushantuo phosphorite Lagerstätte, South China. Lethaia 32:219–240

Xiao S, Zhang Y, Knoll AH (1998) Three-dimensional preservation of algae and animal embryos in a Neoproterozoic phosphorite. Nature 391:553–558

Xiao S, Yuan X, Knoll AH (2000) Eumetazoan fossils in terminal Proterozoic phosphorites? Proc Natl Acad Sci U S A 97:13684–13689

Xiao S, Zhou CM, Yuan XL (2007) Undressing and redressing Ediacaran embryos. Nature 446(7136):E9–E10

Xiao S, Knoll AH, Schiffbauer JD, Zhou C, Yuan X (2012) Comment on "Fossilized nuclei and germination structures identify Ediacaran 'animal embryos' as ecysting protists". Science 335:1169

Xue Y-S, Tang T-F, Yu C-l, Z C-M (1995) Large spheroidal Chlorophyta fossils from Doushantuo Formation phosphoric sequence (late Sinian), central Guizhou, South China. Acta Palaeontol Sin 34:688–706, In Chinese

Yasui K, Reimer JD, Liu Y, Yao X, Kubo D, Shu D, Li Y (2013) A diploblastic radiate animal at the dawn of Cambrian diversification with a simple body plan: distinct from Cnidaria? PLoS ONE 8:e65890

Yin C, Bengtson S, Zhao Y (2004) Silicified and phosphatized *Tianzhushania*, spheroidal microfossils of possible animal origin from the Neoproterozoic of South China. Acta Palaeontol Pol 49:1–12

Yue Z, Bengtson S (1999) Embryonic and post-embryonic development of the Early Cambrian cnidarian *Olivooides*. Lethaia 32:181–195

Zhang XG, Pratt BR (1994) Middle Cambrian arthropod embryos with blastomeres. Science 266:637–639

Zhang X-G, Pratt BR (2014) Possible algal origin and life cycle of Ediacaran Doushantuo microfossils with dextral spiral structure. J Palaeontol 88:92–98

Zhang X-G, Pratt BR, Shen C (2011) Embryonic development of a middle Cambrian (500 Myr old) scalidophoran worm. J Paleontol 85:898–903

Porifera

Bernard M. Degnan, Maja Adamska,
Gemma S. Richards, Claire Larroux, Sven Leininger,
Brith Bergum, Andrew Calcino, Karin Taylor,
Nagayasu Nakanishi, and Sandie M. Degnan

Chapter vignette artwork by Brigitte Baldrian.
© Brigitte Baldrian and Andreas Wanninger.

B.M. Degnan (✉) • C. Larroux • A. Calcino
K. Taylor • N. Nakanishi • S.M. Degnan
Centre for Marine Science, School of Biological
Sciences, University of Queensland,
Brisbane, QLD 4072, Australia
e-mail: b.degnan@uq.edu.au

M. Adamska • G.S. Richards
Centre for Marine Science, School of Biological
Sciences, University of Queensland,
Brisbane, QLD 4072, Australia

Sars International Centre for Marine Molecular
Biology, University of Bergen,
Thormøhlensgt. 55, Bergen 5008, Norway

S. Leininger
Sars International Centre for Marine Molecular
Biology, University of Bergen,
Thormøhlensgt. 55, Bergen 5008, Norway

Institute of Marine Research,
Nordnesgaten 50, Bergen 5005, Norway

B. Bergum
Sars International Centre for Marine Molecular
Biology, University of Bergen,
Thormøhlensgt. 55, Bergen 5008, Norway

Faculty of Medicine and Dentistry, University
of Bergen, P.O. Box 7804, Bergen 5020, Norway

A. Wanninger (ed.), *Evolutionary Developmental Biology of Invertebrates 1:
Introduction, Non-Bilateria, Acoelomorpha, Xenoturbellida, Chaetognatha*
DOI 10.1007/978-3-7091-1862-7_4, © Springer-Verlag Wien 2015

INTRODUCTION

Poriferans (sponges) are sessile aquatic (largely marine) animals that are found in almost all benthic habitats. There are an estimated 15,000 species living today, although many have not been described (reviewed in Hooper and Van Soest 2002). The sponge body plan is amongst the simplest in the animal kingdom and lacks nerve and muscle cells and a centralised gut (reviewed in Simpson 1984; Ereskovsky 2010; Leys and Hill 2012). Their body plan and ecology, and thus their evolution, appear to be intimately associated with the diversity of microbial symbionts they harbour (reviewed in Hentschel et al. 2012; Thacker and Freeman 2012), as is the case with other metazoans (McFall et al. 2013).

Sponges are separated from the external environment by an epithelial layer, the exopinacoderm. External pores in this outer boundary connect to an internal network of canals and chambers, which are lined by epithelial endopinacocytes and ciliated choanocytes, respectively. Choanocyte chambers pump water through this internal aquiferous canal system, drawing food into the sponge. This current also fulfils most of the sponge's physiological requirements, including respiration and excretion. Between the internal and external epithelial layers is the collagenous mesohyl, which is enriched with multiple cell types, often including pluripotent archaeocytes and skeletogenic sclerocytes that fabricate siliceous or calcareous spicules (Fig. 4.1). This juvenile/adult body plan is typically the outcome of the dramatic reorganisation of a radially symmetrical, bi- or trilayered larva at metamorphosis. Metamorphosis is deemed to be complete when the functional feeding juvenile is formed. This so-called rhagon or olynthus stage has been proposed to be the phylotypic stage for demosponges and calcisponges, respectively (reviewed in Ereskovsky 2010). It is often cited that poriferans lack true tissue-level organisation; however, there are numerous examples of tissue- and organ-like structures and functionalities in both larval and adult forms (e.g., the photosensitive pigment ring of many larvae).

Phylogenetic Position of Porifera

Porifera is traditionally regarded as the oldest surviving phyletic lineage of animals and in the past was often relegated into its own subkingdom, the Parazoa. Recent molecular phylogenetic analyses however have put forward a range of alternative proposals that either support

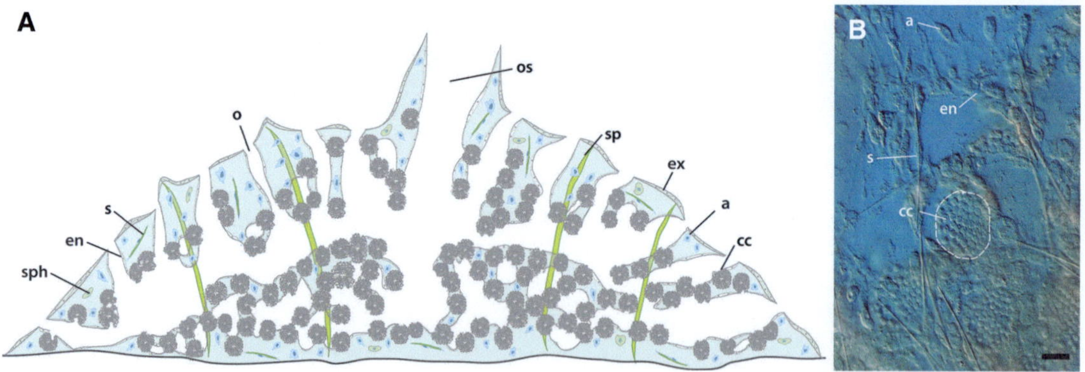

Fig. 4.1 Juvenile poriferan body plan. (**A**) Diagram of a sponge with leuconoid aquiferous system. Water flows into the internal aquiferous system via the ostium and out via the osculum. The mesohyl is shown in blue and populated by archaeocytes and other cell types, including sclerocytes and spherulous cells. (**B**) Optical section of a 3-day-old *Amphimedon queenslandica* juvenile showing internal morphology. Archeocyte (*a*), choanocyte chamber (*cc*), endopinacoderm (*en*), exopinacoderm (*ex*), ostium (*o*), osculum (*os*), sclerocyte (*s*), spicule (*sp*), and spherulous cell (*sph*). Scale bar: 10 μm

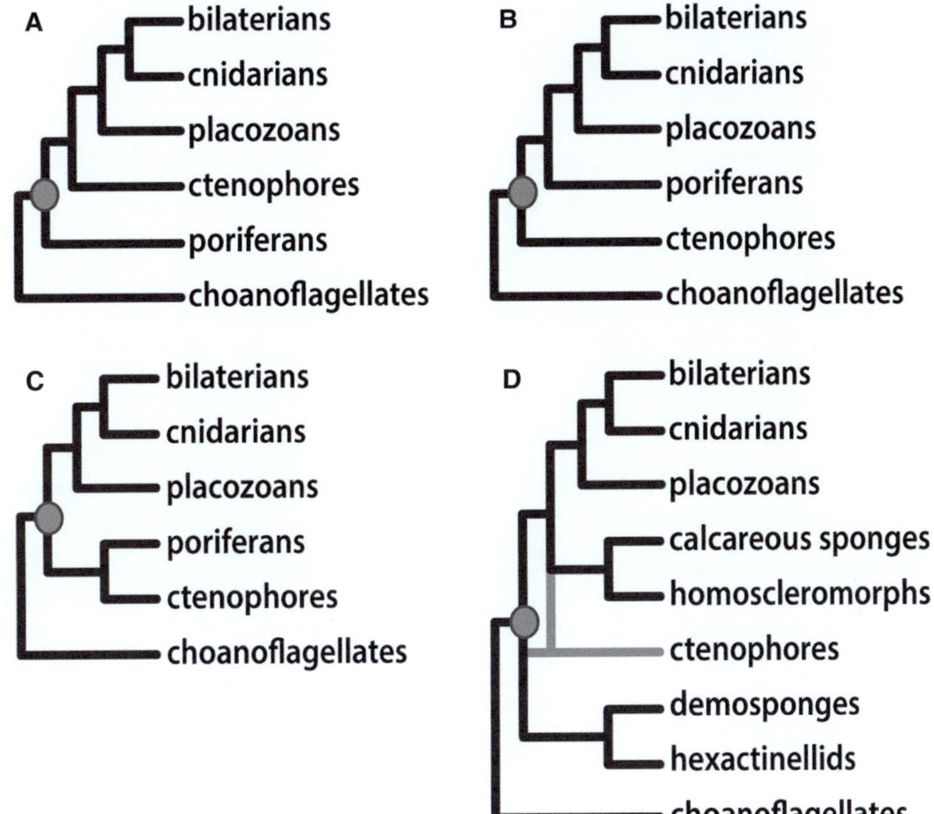

Fig. 4.2 Phylogenetic position of Porifera. Four hypotheses of the inter- and intrarelationship of sponges. These trees are largely derived from, and supported by, recent phylogenomic datasets. The grey node in each tree represents the last common ancestor to contemporary metazoans. (**A**) Porifera as the earliest branching metazoan phyletic lineage and sister to the Eumetazoa (Ctenophora, Placozoa, Cnidaria, and Bilateria); one possible tree topology is shown (Phillipe et al. 2009). (**B**) Ctenophora as the earliest branching metazoan phyletic lineage (Ryan et al. 2013; Moroz et al. 2014). (C) Porifera and Ctenophora form a clade separate from Placozoa, Cnidaria, and Bilateria. Recent detailed phylogenomic analyses raise the possibility that sponges and ctenophores may be sister taxa. This hypothesis is supported by the striking similarity of the developmental gene repertoires of the demosponge *Amphimedon queenslandica* and ctenophores. (**D**) Paraphyletic sponges with classes Calcarea and Homoscleromorpha more closely related to the Eumetazoa than to classes Demospongiae and Hexactinellida (Sperling et al. 2009); Ctenophora unresolved in this hypothesis. Note that hypotheses that place placozoans in the most basal position within the Metazoa or within a clade consisting of them, sponges, cnidarians, and ctenophores (e.g., Schierwater et al. 2009), are consistently rejected based on current genomic and transcriptomic data and thereby not included in this figure

(e.g., Philippe et al. 2009; Srivastava et al. 2010) or reject this traditional view (Fig. 4.2; e.g., Schierwater et al. 2009; Sperling et al. 2009; Ryan et al. 2013; Moroz et al. 2014). Specifically, current points of debate are whether poriferans or ctenophores are the sister group to all other animals and whether sponges are monophyletic.

Thus, interpretations of the sponge body plan in the context of metazoan evolution range from it representing a state similar to the last common ancestor (LCA) of contemporary metazoans to it being derived from a morphologically more complex LCA that possessed a gut, nerves, and muscles.

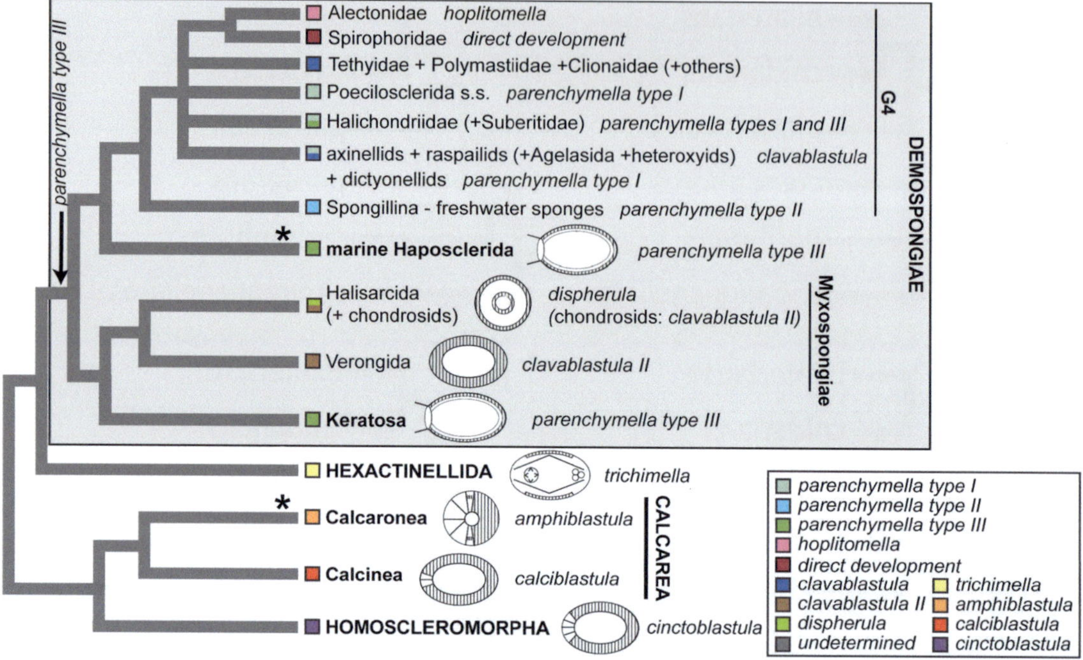

Fig. 4.3 Larval types mapped onto a recently proposed poriferan phylogeny. Stars mark lineages that include the demosponge *Amphimedon queenslandica* and calcareous sponge *Sycon ciliatum* (Figure adapted from Worheide et al. (2012))

Phylogenetic Relationships of Poriferan Classes

Classes Demospongiae, Calcarea, Homoscleromorpha, and Hexactinellida comprise phylum Porifera, with the Demospongiae being by far the most speciose. There is a growing consensus that demosponges and hexactinellids (sponges with a syncytial body organisation) and calcisponges and homoscleromorphs form pairs of sister classes (Fig. 4.3; Philippe et al. 2009; Gazave et al. 2012; Woerheide et al. 2012; Hill et al. 2013; Redmond et al. 2013; Thaker et al. 2013). However, there currently exist two broad views as to the exact relationship of these classes, one in which they form a monophyletic phylum (e.g., Philippe et al. 2009) and another where a clade comprised of demosponges and hexactinellids are separate from a clade comprised of calcisponges and homoscleromorphs + eumetazoans (Fig. 4.3; e.g., Sperling et al. 2009). Regardless, it appears that these classes diverged from each other over 600 million years ago, well before eumetazoan cladogenesis and the Cambrian explosion (Erwin et al. 2011).

Developmental Commonalities Within the Porifera

Poriferans exhibit a wide range of embryonic and larval types that are formed through a diversity of morphogenetic processes, many of which appear similar to those used during bilaterian development (Figs. 4.3 and 4.4). For instance, morphogenetic mechanisms, such as cell delamination (e.g., the hexactinellid *Oopsacas minuta*; Boury-Esnault et al. 1999), ingression (e.g., the halisarcid *Halisarca dujardini*; Gonobobleva and Ereskovsky 2004), egression (e.g., the homoscleromorph *Oscarella* sp.; Ereskovsky and Boury-Esnault 2002), and invagination (e.g., the halisarcid *Halisarca dujardini*; Gonobobleva and Ereskovsky 2004), are employed during sponge embryogenesis, albeit often in a taxon-restricted manner. For detailed descriptions of sponge

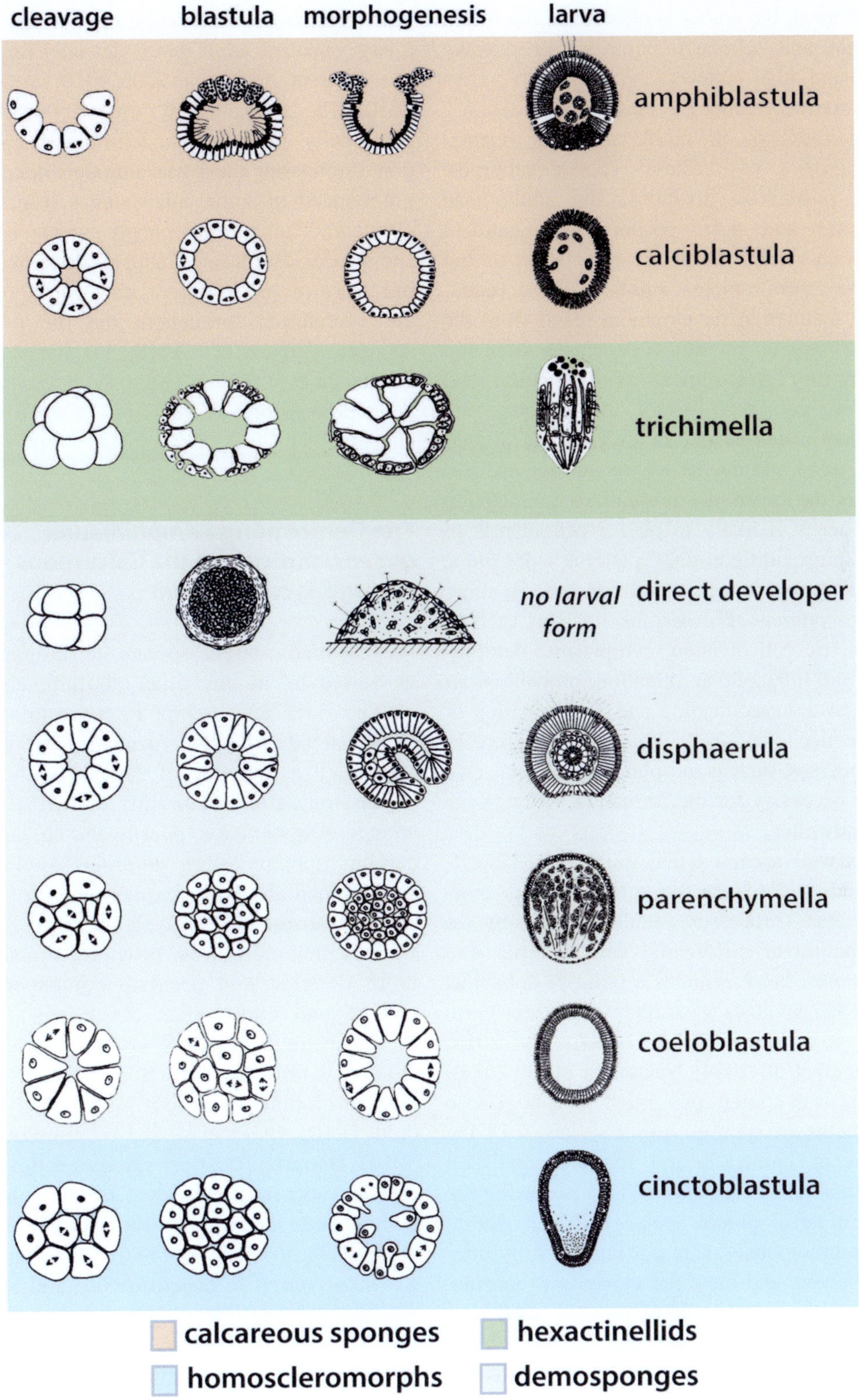

Fig. 4.4 Summary of the modes of poriferan development (Figure adapted from Ereskovsky (2010))

development, the reader is directed to the recent excellent and scholarly book by Ereskovsky (2010) and other reviews (e.g., Leys 2004; Leys and Ereskovsky 2006; Leys and Hill 2012).

Regardless of the differences in external characteristics of different sponge embryos, larvae, postlarvae, juveniles, and adults (see Figs. 4.3 and 4.4), poriferan development employs a similar morphogenetic toolkit to that used by more complex animals. These fundamental features of development result from the spatiotemporal regulation of gene expression and include the establishment of differential cell affinities, cell type-specific movements, and structural changes and the regulation of cell proliferation and death. As we will see in later sections, the localised expression of transcription factor genes is likely to play a central role in establishing differential patterns of gene expression in sponges, just as it does in more complex animals. Further, mechanisms such as asymmetric cell division, cytoplasmic determinants, and intracellular signalling probably contribute to the specification and determination of cell identity in sponges. Other symmetry-breaking processes, such as morphogen gradients, also appear necessary for the formation of the poriferan body plan.

It is well accepted that multilayered poriferan body plans form through a series of morphogenetic processes underpinned by a combination of differential cell affinities and movements, but it remains a point of debate as to whether sponges gastrulate or possess germ layers (e.g., Leys 2004; Ereskovsky 2010; Leininger et al. 2014; Nakanishi et al. 2014). Perhaps less contentious, most morphogenetic movements in sponges can be viewed in the context of epithelial and mesenchymal cell behaviours and interactions. That is, during the course of development, sponge cells can operate semi-autonomously or in partially or fully integrated layers and have the capacity to migrate into or from a cell layer as an individual or a group (reviewed in Ereskovsky 2010).

With few exceptions, sponges have a biphasic pelagobenthic life cycle with a tiny, planktonic ciliated larva that metamorphoses and grows into a large, benthic adult that is sexually reproductive (Degnan and Degnan 2006, 2010; Ereskovsky 2010). The body plans of a majority of sponges continually undergo remodelling and regeneration throughout their life although this is less pronounced in some calcisponges (e.g., *Sycon ciliatum*). Thus, the morphogenetic mechanisms and processes outlined above, along with the processes of transdifferentiation and apoptosis, are operational throughout the life of most sponges (Ereskovsky 2010; Funayama 2012; Nakanishi et al. 2014). This also applies to asexual reproduction, which is not covered in this chapter.

The Demosponge *Amphimedon queenslandica* and the Calcareous Sponge *Sycon ciliatum*

As outlined above, sponge development is as varied as in any other phylum, and this chapter does not attempt to cover this diversity. Instead, here we focus on two species for which a majority of developmental gene expression patterns currently exist, the demosponge *Amphimedon queenslandica* and calcareous sponge *Sycon ciliatum*. Analysis of developmental gene expression, combined with experimental analysis of embryogenesis and metamorphosis, provides a means for more detailed and accurate comparisons of sponge and eumetazoan development. Both *A. queenslandica* and *S. ciliatum* have well-annotated draft genomes supported by extensive developmental transcriptomes (Srivastava et al. 2010; Anavy et al. 2014; Leininger et al. 2014). Importantly, these sponges differ markedly in their mechanisms of development. As demosponge and calcareous sponge lineages most likely diverged well before the Cambrian, a comparison of *A. queenslandica* and *S. ciliatum* genomes and development has the potential to provide insights into their common ancestor, which existed over 600 million years ago (Erwin et al. 2011).

The Demosponge *Amphimedon queenslandica*
Amphimedon queenslandica (class Demospongiae, order Haplosclerida, family Niphatidae) is currently the sponge with the most extensive developmental gene expression data. Its draft genome was published in 2010, further enhancing its utility for understanding demosponge development and metazoan evolution. A number of studies have been published about the evolution of metazoan cell types and gene families using *A. queenslandica* (Table 4.3). Akin to other sponges and indeed other animals, development in *Amphimedon queenslandica* progresses through a series of recognisable phases. It begins with the subdivision of the fertilised oocyte into progressively smaller blastomeres, followed by the acquisition of embryonic polarity and the sorting of cells into layers via broad-scale cell migrations. Activity then centres on the patterning and differentiation of diverse cell types in defined localities throughout the embryo and the morphogenesis of larval structures (Fig. 4.6). Differing from typical eumetazoan development in which patterning processes occur before terminal differentiation, embryonic pigment cells, ciliated epithelial cells, and sclerocytes in *A. queenslandica* express terminal differentiation characters, pigment granules, cilia, and spicules, respectively, directly following cleavage. Early differentiation of cells during embryogenesis appears to be a shared feature amongst many sponges.

Amphimedon queenslandica. (**A**) Adult in situ on the southern Great Barrier Reef. (**B**) Larva with posterior pigment ring (*pp*) to the *left*. (**C**) A 3-day-old juvenile settled on crustose coralline alga (*ca*) with osculum (*os*) pointed upwards

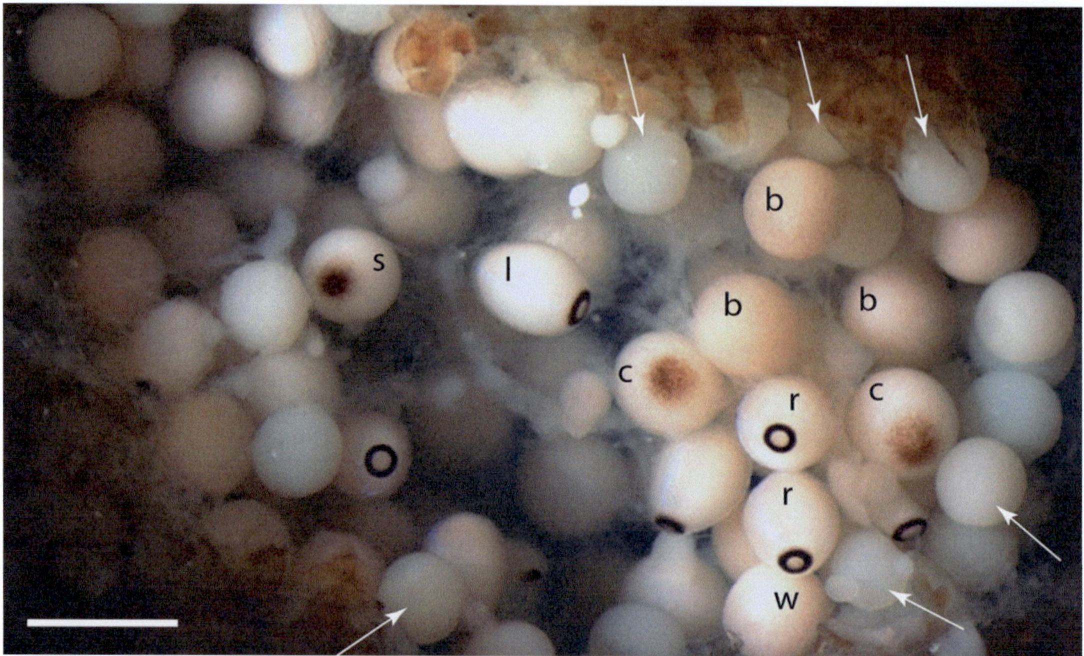

Fig. 4.5 An *Amphimedon queenslandica* brood chamber. Small white and translucent oocytes are located on the outer edge of the brood chamber (*white arrows*), which is surrounded by brown somatic cells. White (*w*), brown (*b*), cloud (*c*), and ring (*r*) stage embryos and unhatched larvae (*l*) are mixed within the chamber. Scale bar: 1 mm

THE *AMPHIMEDON QUEENSLANDICA* BROOD CHAMBER: GAMETOGENESIS AND FERTILISATION

Amphimedon queenslandica is viviparous, and its embryos are concentrated into brood chambers, several of which can occur in a single adult all year round (Fig. 4.5). However, the developmental origins of its gametes are unknown. In other viviparous sponges, oocytes appear to differentiate from either archaeocytes or choanocytes (reviewed in Simpson 1984; Kaye 1990; Ereskovsky 2010). The engulfment of nutrient-laden nurse cells (also called trophocytes/spherulous cells; Simpson 1984; Ereskovsky 2010) is considered an important feature of oogenesis (e.g., Fell 1969; Saller and Weissenfels 1985). Pre-cleavage stages are known to reside at the edges of the brood chambers in *A. queenslandica*

and can be identified by their translucent, smaller appearance and flattened shape in comparison to the more spherical and opaque embryos (Fig. 4.5; Leys and Degnan 2001; Adamska et al. 2010). Embryos and unhatched larvae are mixed and located more centrally in the brood chamber, with later stages tending to be towards the middle of the chamber. Developmental stages are identifiable and named by the presence and pattern of pigment cells, which first appear during cleavage: white stage embryos comprise a range of early cleavage stages; brown embryos have pigment cells distributed throughout the embryo and mark the transition from cleavage to the two-layered embryo; cloud stage appears after the anterior-posterior (AP) axis is established, with the pigment cells concentrating towards the future posterior pole; and spot to ring stages are identified by pigment cells concentrated at the posterior pole either as a spot or

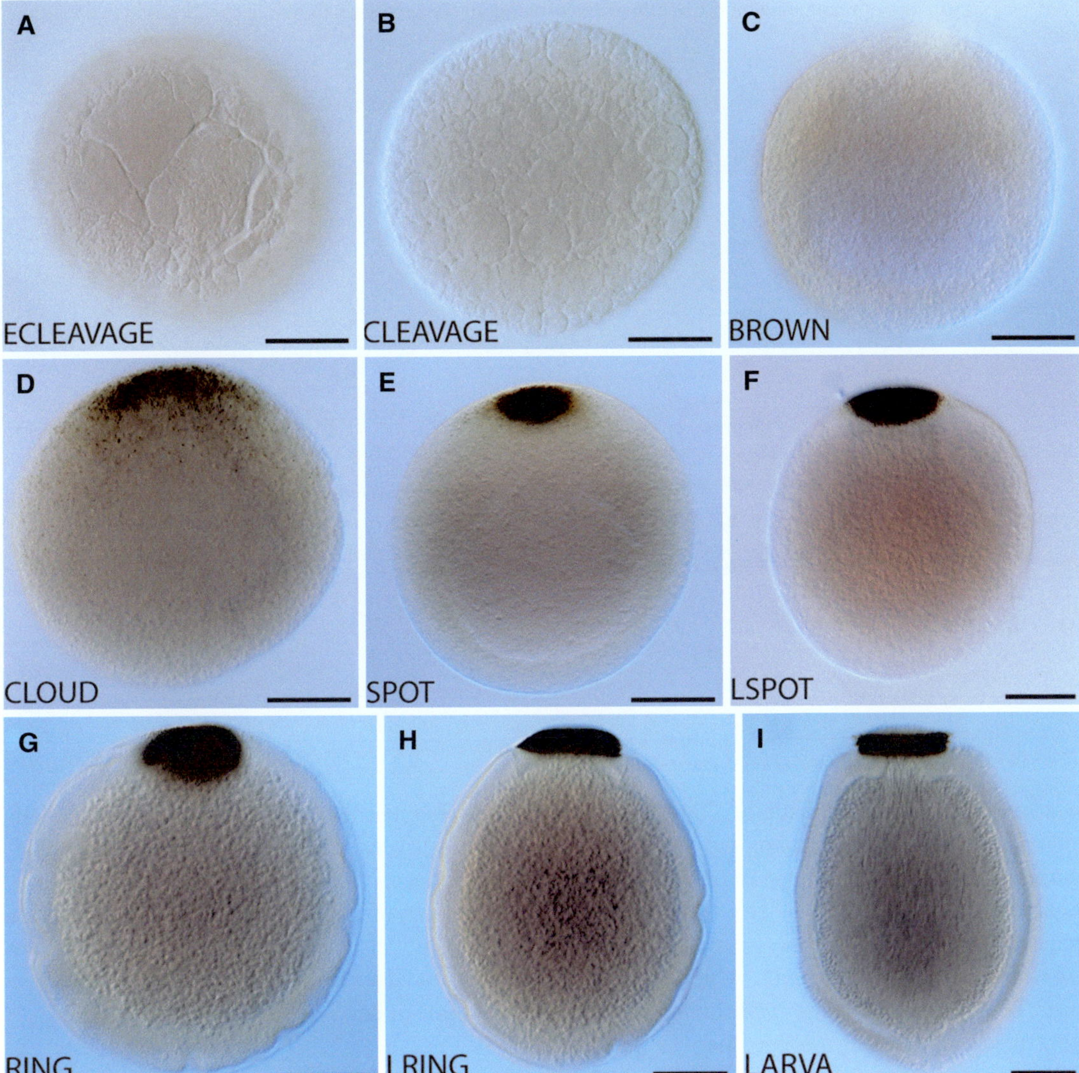

Fig. 4.6 Developmental stages of *Amphimedon queenslandica*. Whole-mount light micrographs of fixed and cleared *Amphimedon queenslandica* embryos and larva. Posterior is to the *top* in (**C–I**); orientation unknown in (**A, B**). *E* early, *L* late in (**A, F, H**). Scale bar: 100 μm

ring – ring stage follows spot stage – and characterised by morphogenesis of specific cell types and tissues (Fig. 4.6).

Like many sponges, *Amphimedon queenslandica* is a hermaphrodite that reproduces by spermcast spawning, followed by the apparent passive uptake by maternal adults of free spermatozoa from what is presumably a very dilute suspension in the water column. Genotyping of microsatellite loci with high allelic diversity (that together yield a combined paternal exclusion probability of 95 %) has revealed that up to 26 different paternal adults can be attributed to progeny being brooded by a single maternal adult at any particular time (Table 4.1). Near neighbours (within 4 m radius) can account for most of the fertilisations, but some progeny appear to be fathered by sperm

Table 4.1 Estimates of minimum number of fathers contributing to the genotypes of brooded embryos in two maternal adults

Maternal adult ID	Nos. embryos genotyped	Nos. potential fathers in 4 m radius in field	Microsatellite locus ID	# paternal alleles	Minimum nos. fathers per maternal adult
A	73	33	MS16	18	~9
			MS34	30	~15
			MS9	22	~11
			MS19	15	~8
			Mean[a]	*21.25 ± 3.25*	*~11*
B	315	17	MS16	37	~19
			MS34	51	~26
			MS9	32	~16
			MS19	22	~11
			Mean[a]	*35.50 ± 6.04*	*~18*

[a]Mean number of paternal alleles across four microsatellite loci ± standard error

Table 4.2 Proportion of embryos at different stages of development in four example maternal adults

Maternal adult ID	Proportion of each developmental stage						Total nos.
	White	Brown	Cloud	Spot	Ring	Larva	
A	0.40	0.18	0.04	0.14	0.16	0.08	73
C	0.24	0.16	0.07	0.13	0.23	0.17	315
B	0.28	0.14	0.07	0.21	0.14	0.16	159
D	0.51	0.11	0.04	0.18	0.16	0.00	107

sourced from a greater distance (Table 4.1). In any particular brood chamber, up to 30 % of white stage embryos appear to be of maternal origin only (i.e., unfertilised). However, as noted below, early cleaving embryos include a large number of maternal nurse cells, which eventually die. The import of these additional maternal genomes may prevent the detection of paternal microsatellite alleles in early embryos by increasing the ratio of maternal to paternal amplifiable DNA.

Although *Amphimedon queenslandica* fecundity is highest in the warmer months of the year, embryos of most developmental stages are present in most brood chambers at most times (Table 4.2). While overall there can be a large number of paternal contributors to fertilisation in a single maternal adult, only a subset of these fathers appear to contribute to any one brood chamber. There is no indication, however, that all embryos at a particular developmental stage within a brood chamber represent a single input of sperm from a single paternal source. Together, these observations suggest that *A. queenslandica* adults maintain a constant supply of all developmental stages, leading to a steady daily release of mature larvae (Maritz et al. 2010), by (i) passively accepting sperm that are trickle released continuously from neighbouring paternal adults and possibly also (ii) active regulation of fertilisation by stored sperm and/or regulation of the initiation or rate of development of fertilised eggs.

EARLY EMBRYONIC DEVELOPMENT IN *AMPHIMEDON QUEENSLANDICA*

Cleavage

Cleaving *Amphimedon queenslandica* embryos (Figs. 4.7, 4.8, 4.9, and 4.10) are similar to other demosponge embryos (Fig. 4.4; Fell 1969; Saller and Weissenfels 1985; De Vos et al. 1991; Kaye and Reiswig 1991; Leys and Ereskovsky 2006; reviewed in Ereskovsky 2010). The entire embryo is enveloped in a layer of squamous follicle cells with large nuclei, presumably of maternal origin (Fig. 4.8; Leys and Degnan 2002). In early cleaving stages, the blastomeres are unequal in size

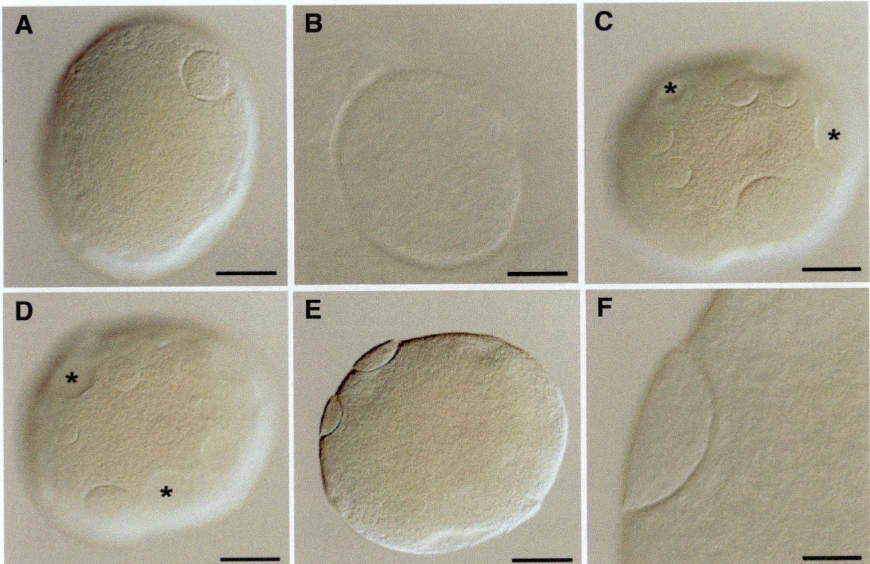

Fig. 4.7 Whole-mount micrographs of very early cleaving *Amphimedon queenslandica* embryos. Optical sections of two fixed and cleared embryos (**A**, **B** and **C–F**), from which follicle membranes have been removed. (**A**) A single large macromere has cleaved and is shown in higher magnification in (**B**). (**C**) Multiple large macromeres have cleaved in the second embryo, and several of them have fallen off during preparation of the sample (*asterisks* indicate 'prints' of some of the missing cells), demonstrating a lack of cell adhesion between the blastomeres in these early embryos. (**C–D**) are different optical sections; (**F**) is a higher magnification of one of the cells shown in (**E**). Scale bars: (**A**, **C–E**), 100 µm; (**B**, **F**) 20 µm

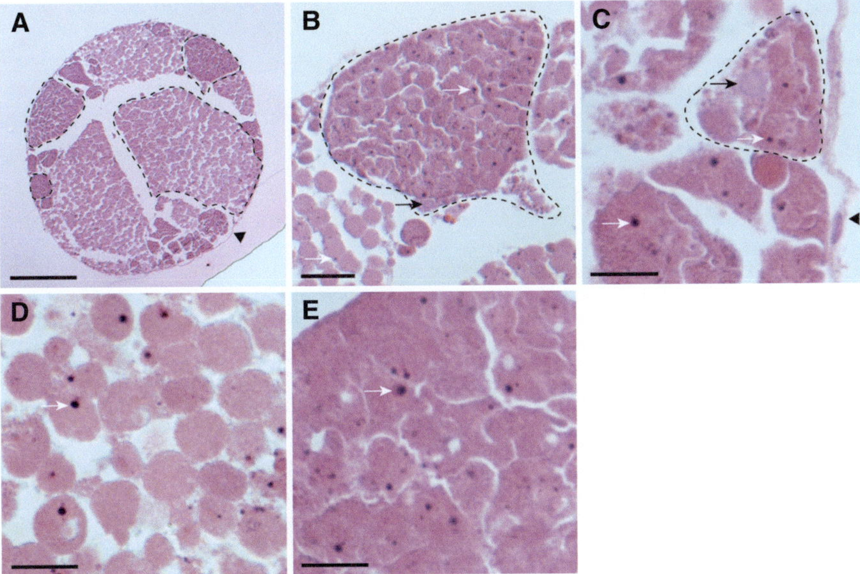

Fig. 4.8 Early cleavage in *Amphimedon queenslandica*. (**A**) Whole embryo. (**B**) Large blastomere with peripheral cytoplasm around nurse cell mass. (**C**) Smaller blastomere with more central nucleus. (**D**) *Light pink* nurse cells. (**E**) *Dark pink* nurse cells. All panels show H+E-stained median sections; *arrowhead*, maternal layer; *dashed lines*, boundary of blastomeres; *black arrows*, blastomere nuclei; *white arrows*, pyknotic nuclei of nurse cells. Scale bars: (**A**) 100 µm; (**B**) 20 µm; (**C–E**) 10 µm

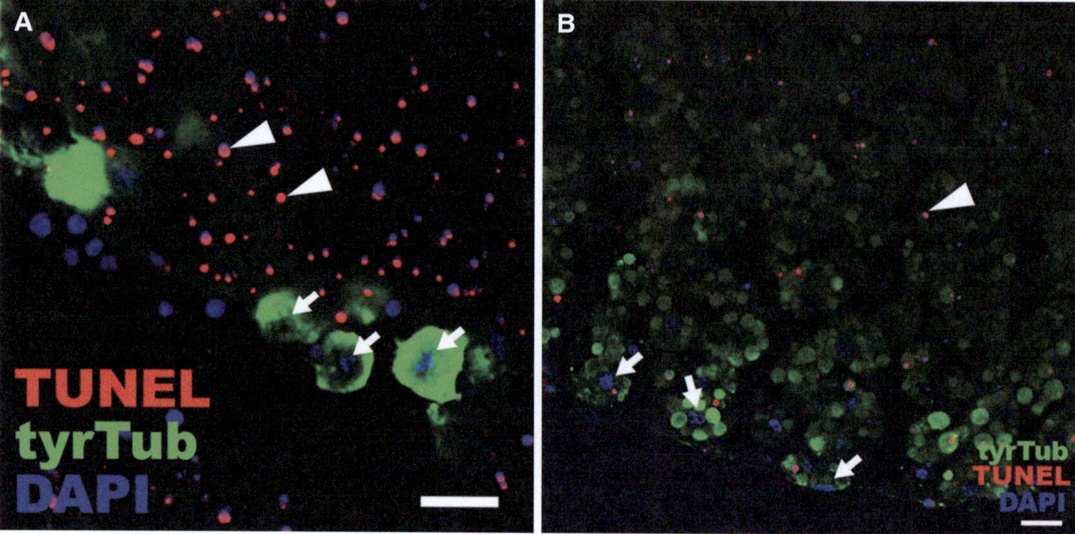

Fig. 4.9 TUNEL analysis of early *Amphimedon queenslandica* development. (**A**) Early cleavage. Note the periphery of the macromeres and the cytoplasm of the micromeres are enriched in tyrosine tubulin. (**B**) Solid blastula. Scale bars: 10 μm. Both are embryos stained with DAPI (*blue*), anti-tyrosine tubulin (*green*) and TUNEL (*red*); *arrowhead*, example TUNEL positive nuclei or nuclear fragments; *arrows*, embryonic nuclei

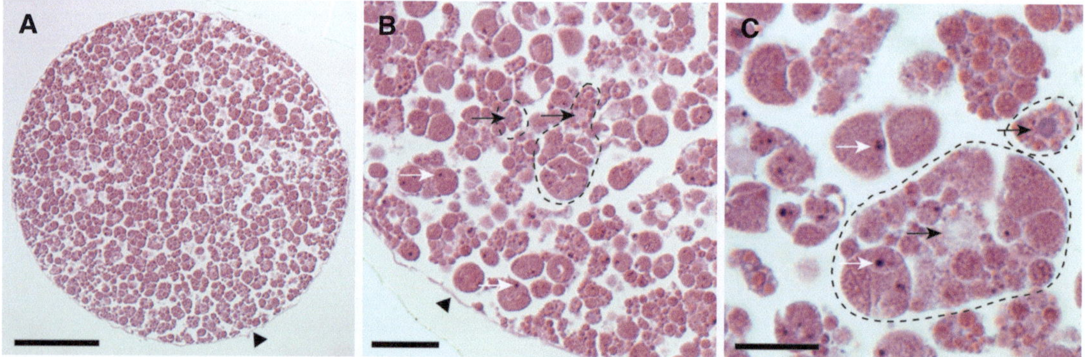

Fig. 4.10 Late cleavage in *Amphimedon queenslandica*. (**A**) Whole embryo. (**B**) Blastomeres in loose aggregation. (**C**) Variation in blastomere sizes. All panels show H+E-stained median sections; *arrowhead*, maternal layer; *dashed lines*, boundary of blastomeres; *black arrows*, blastomere nuclei; *white arrows*, pyknotic nuclei of nurse cells. Scale bars: (**A**) 100 μm; (**B**) 20 μm; (**C**) 10 μm

and loaded with yolk that appears to be derived from nutritive maternal nurse cells (e.g., Fell 1969). These large blastomeres contain large quantities of smaller eosinophilic (thus protein-rich) bodies that appear to be derived from nurse cells undergoing programmed cell death, based on the presence of compact pyknotic nuclei with intensely basophilic staining (Fig. 4.8). Terminal deoxynucleotidyl transferase dUTP nick end labeling (TUNEL) – a method that detects frag-mented DNA – of pyknotic nuclei in these bodies is consistent with these being derived from apoptosing nurse cells (Fig. 4.9A).

Blastomere cytoplasm appears as a thin layer around the mass of yolk-containing vesicles, with embryonic nuclei visible at the cell periphery (Fig. 4.8, black arrows). The extent of these yolk reserves means that neither blastomere cytoplasm nor cell boundaries are always apparent (Figs. 4.7 and 4.8; Leys and Degnan 2002).

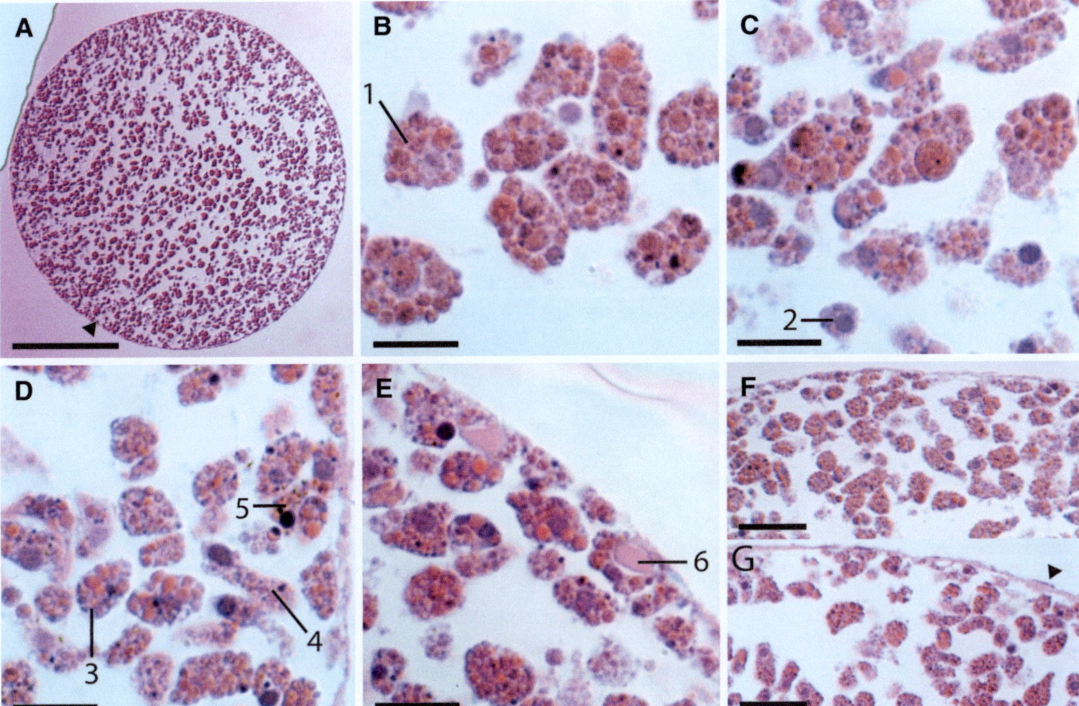

Fig. 4.11 Brown embryo stage. (**A**) Whole embryo. (**B**) Type 1, granular macromeres ('*1*'). (**C**) Granular macromeres and micromeres ('*2*'). (**D**) Type II, globular macromeres ('*3*'), amoeboid cells ('*4*'), and pigment cells ('*5*'). (**E**) Sclerocyte ('*6*'). (**F, G**), opposing sides of the embryo. All panels show H+E-stained median sections; *arrowhead*, maternal layer. Scale bars: (**A**) 100 μm; (**B–E**) 10 μm; (**F–G**) 20 μm

As cleavage progresses, the blastomeres reduce in size, and the number and concentration of pyknotic nuclei decreases (Figs. 4.9B and 4.10).

At the end of cleavage, a solid blastula forms, which is light brown in colour (Fig. 4.11). These 'brown' embryos contain a mixture of loosely aggregated cell types, of which six are readily identifiable: early sclerocytes around the outer margin of the embryo, two larger cell types with a variety of inclusions (designated here as type I and II macromeres), large amoeboid cells, pigment cells, and a minor micromere population (Fig. 4.11B–E). The type I and II macromeres are both spherulous and contain granular or homogenous inclusions, respectively. The amoeboid cells possess smaller and more lightly eosinophilic inclusions and larger nuclei with dense heterochromatin around their periphery. The early differentiation of some cell types, especially sclerocytes, occurs in a range of sponge embryos (Fell 1969; Maldonado and Berquist 2002; Leys 2003). Despite displaying characteristics of their final larval differentiated state (e.g., pigmentation, deposition of spicule matrix, ciliation), these cells remain to be patterned in the embryo and thus maintain the capacity to respond to positional signals and migrate appropriately.

Asymmetric Cell Division and Transcript Localisation

During cleavage an increasing number of small blastomeres are present on the periphery of the embryo and nestled between the macromeres. Following the fate of daughter cells originating from individual macromeres injected with high molecular weight tetramethylrhodamine dextran confirms that macromeres divide asymmetrically, giving rise to micromeres, typically 2–4 μm in diameter, and macromeres, initially most often >50 μm in diameter; symmetric cell divisions may occur in late cleavage (Fig. 4.12). The location of the daughter cells of individually labelled

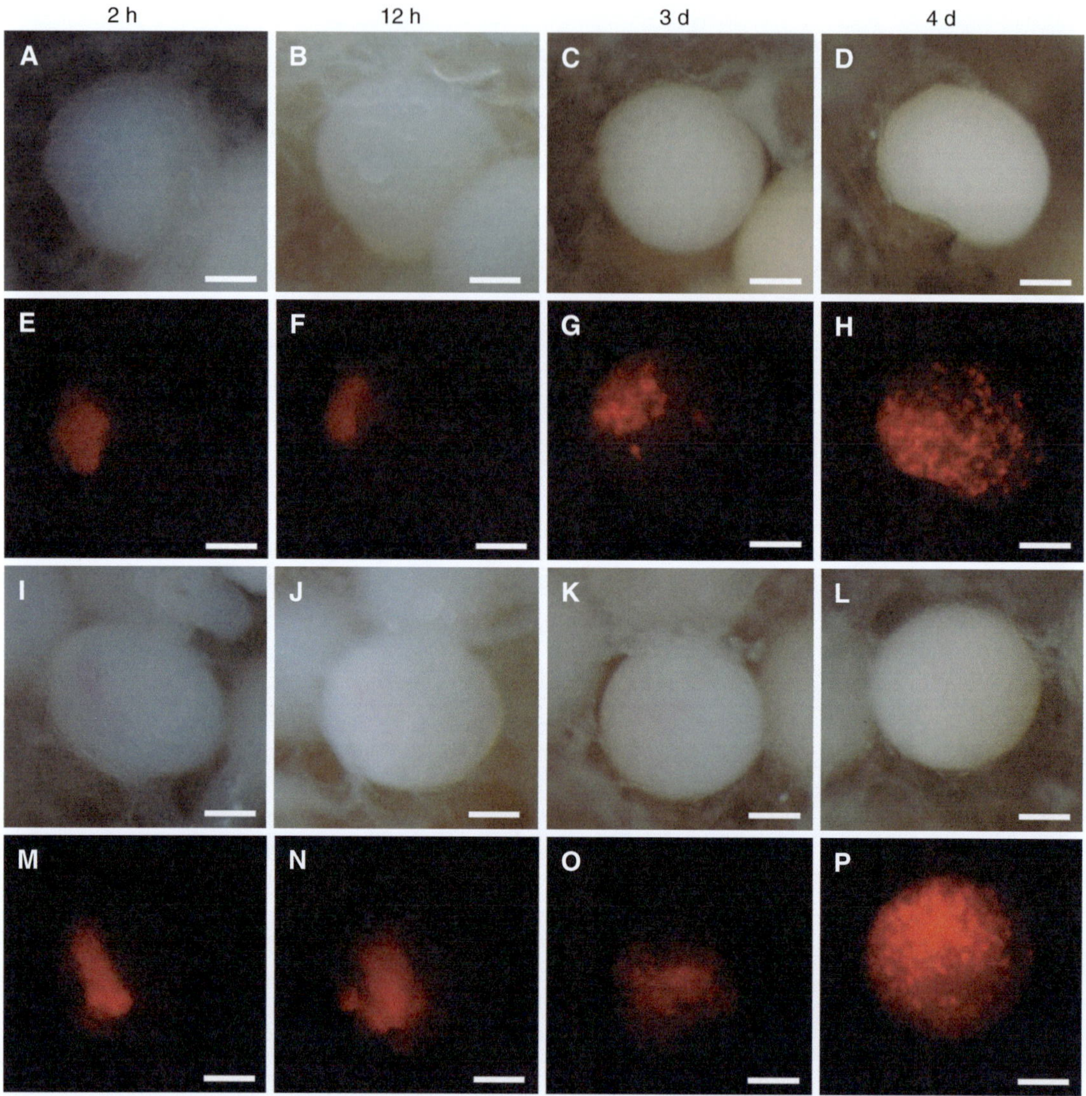

Fig. 4.12 Cell lineage analysis of randomly labelled *Amphimedon queenslandica* macromeres. Two experiments are shown (*top* (**A–H**) and *bottom* (**I–P**)). In both cases, a macromere is injected with a high molecular weight fluorescent dextran suspension, and the localisation of this fluorescent marker is traced over 4 days of development (time postinjection on *top*). Scale bars: 100 μm

macromeres provides no evidence of stereotypical or predictable cell lineages or cleavage patterns. Instead, cleavage appears to be chaotic.

Morphologically distinct pigment cells, sclerocytes, and ciliated cells are first detected during cleavage throughout the embryo (Fig. 4.11; Leys and Degnan 2002). Consistent with the early specification and determination of these cell types is the detection of transcripts encoding a number of conserved developmental transcription factors in subpopulations of micromeres at cleavage, including NK homeobox genes *Bsh* and *Tlx*, LIM homeobox gene *Lhx3/4*, and the nuclear receptor gene *NR1* (Fig. 4.13; Larroux et al. 2006, 2007; Fahey et al. 2008; Bridgham et al. 2010; Srivastava et al. 2010b).

Further analysis of *Lhx3/4* mRNA localisation at cleavage reveals transcripts are not only

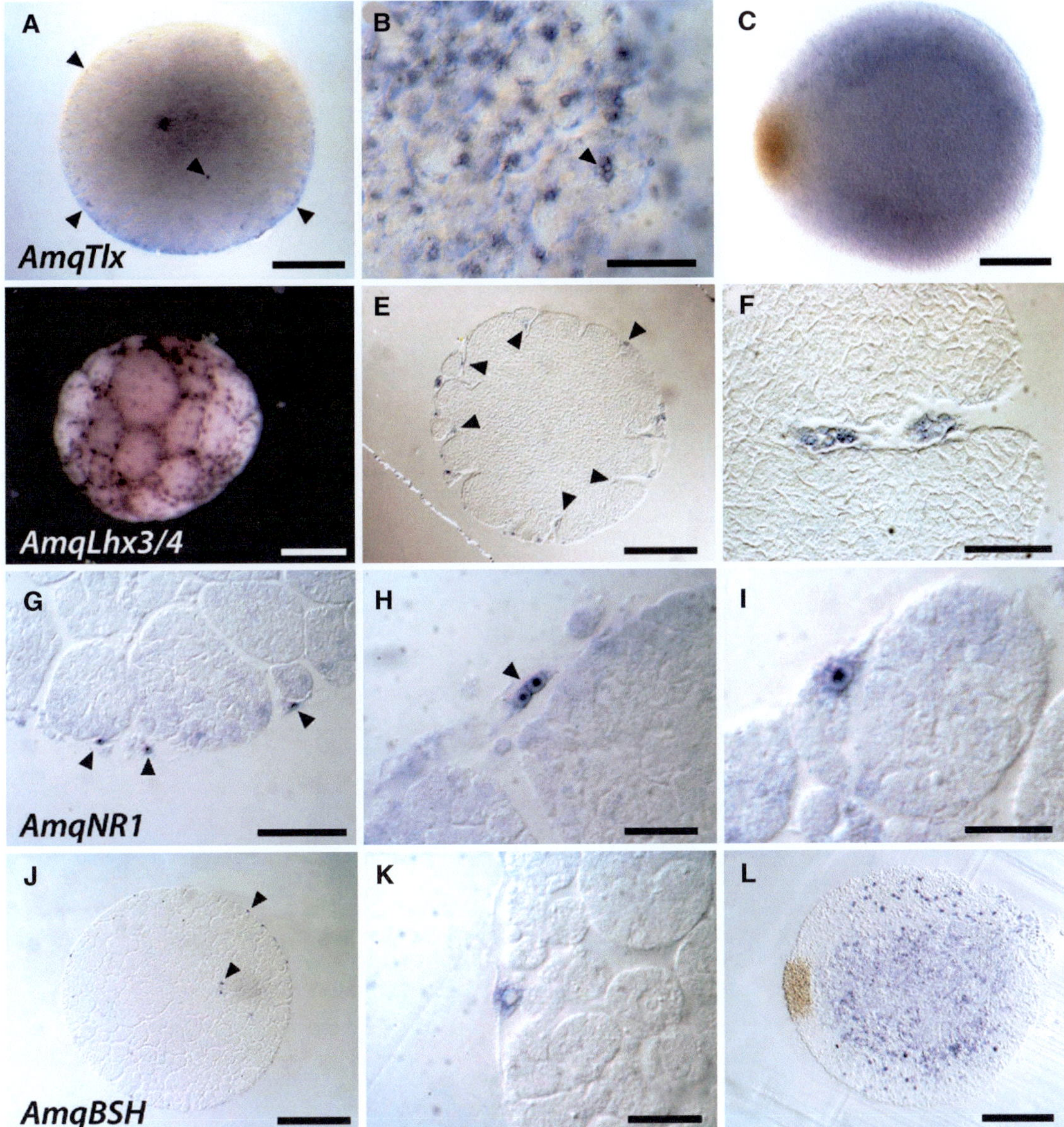

Fig. 4.13 Expression of transcription factor genes in micromeres at cleavage. (**A–C**) Localisation of *AmqTlx* in micromeres and sclerocytes during cleavage (**A, B**) and spot stage (**C**). (**D–F**) Localisation of *AmqLhx3/4* in micromeres during cleavage (**A, B**) and spot stage (**C**). (**G–I**) Localisation of *NR1* in micromeres during. (**J–L**) Localisation of *AmqBSH* in micromeres. *Arrowheads*, example cells enriched with the transcript. (**A, B, C, H, I, J, K,** and **L**), sections; (**D, E, F, G**), whole mounts. Scale bars: (**A, C, D, F, G, H**), 100 μm; (**B, I, K, L**), 10 μm; (**E, I**), 25 μm **L** (From Larroux et al. 2006)

present in individual micromeres but are also associated with the cortex of adjacent macromeres (Fig. 4.14). Macromere cortices and micromeres are enriched in microtubules (Fig. 4.15). Nuclei also localise to the macromere cortical region, consistent with these being regions of cell division. Asymmetric inheritance of cell fate determinants, in the form of localised mRNAs, is widespread in animal development (reviewed in Knoblich 2010; Medioni et al. 2012). This typically requires the localisation of particular mRNAs via cytoskeleton-mediated active transport to a defined cortical region of the cell. These results suggest that such a mechanism is operational during *Amphimedon queenslandica* cleavage and probably essential for the early specification and determination of micromere fate.

A more extensive survey of gene expression patterns in cleaving embryos indicates

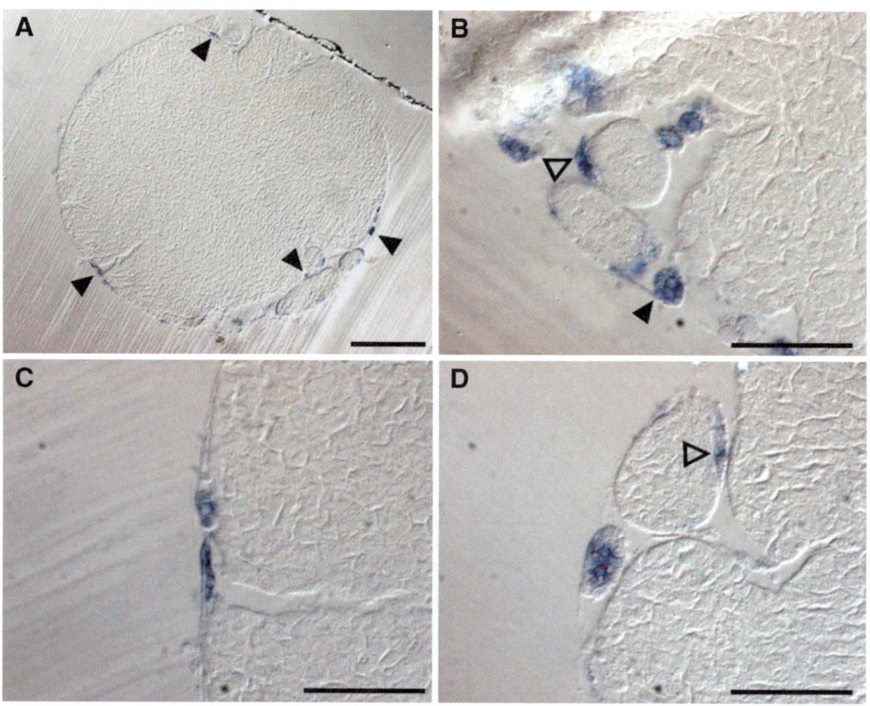

Fig. 4.14 Localisation of *AmqLhx3/4* mRNA in macromeres and micromeres. (**A**) Cleaving embryo, low magnification. (**B–D**) Higher magnification of cortex and micromeres. *Arrowheads*, example cells and cortical regions enriched with the transcript. Scale bars: (**A**) 100 μm; (**B–D**) 10 μm

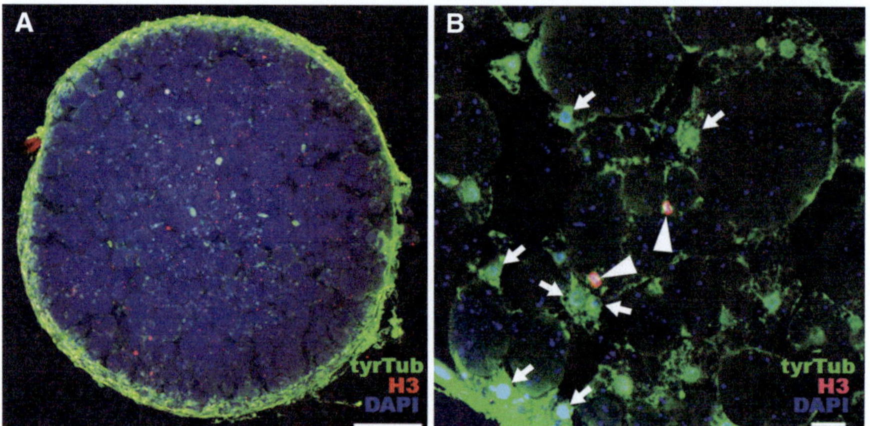

Fig. 4.15 Localisation of microtubules and mitotic figures to macromere cortices. (**A**) Cleaving embryo showing mitotic cell nuclei (*red*). (**B**) Higher magnification showing dividing nuclei are localised to macromere cortices, which are also enriched in microtubules. *Arrowheads*, phosphorylated histone 3 (H3) positive nuclei (*red*); *arrows*, example micromere nuclei (*blue*); anti-tyrosine tubulin immunoreactivity (*green*). Scale bars: (**A**) 100 μm; (**B**) 10 μm

that the localisation of transcripts to subsets of micromeres is widespread in *Amphimedon queenslandica* (Fig. 4.16). The lack of similarity in many of the in situ hybridisation patterns is consistent with different transcripts being localised to different sets of micromeres; this suggests that cell fate specification and determination starts early in *A. queenslandica* embryogenesis. This mode of cell specification may be an important feature of the early development of many sponges with similar external embryological characteristics (cf. Ereskovsky 2010).

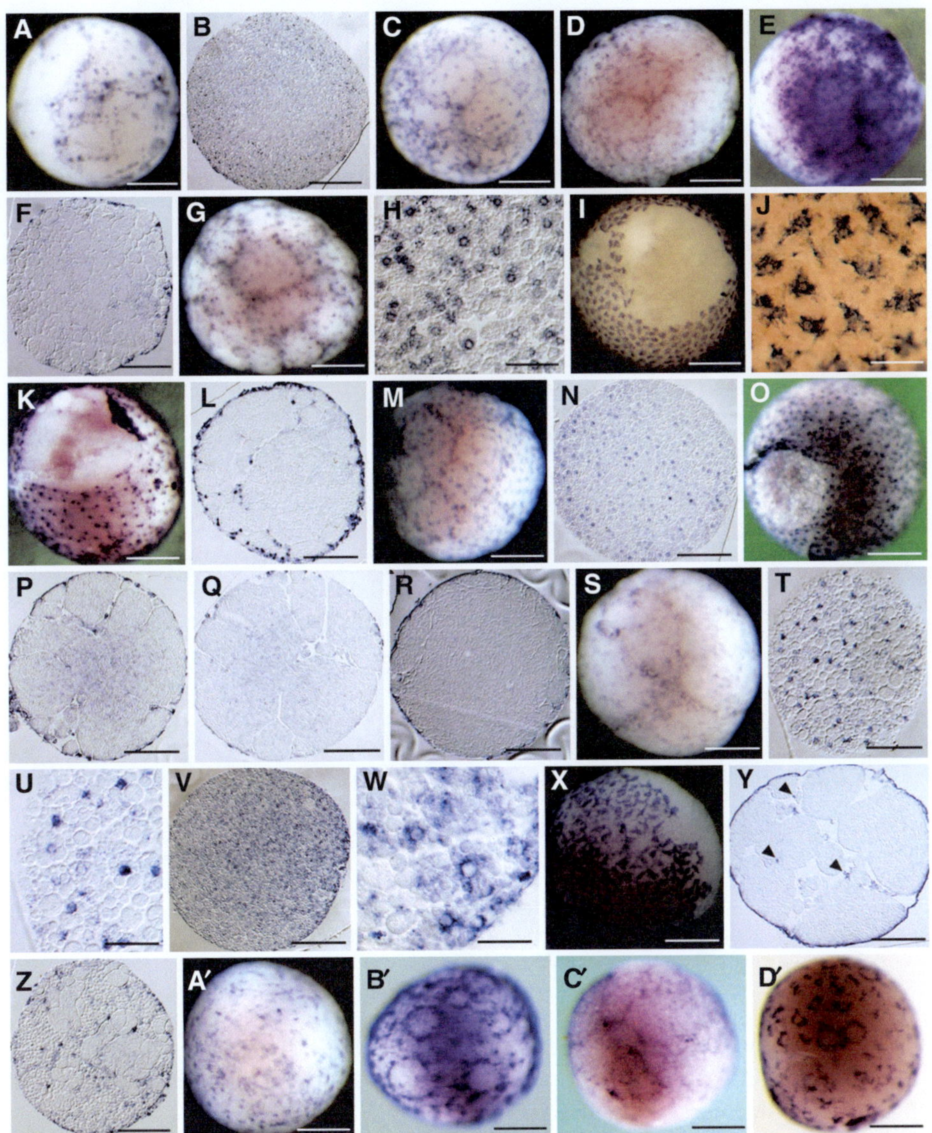

Fig. 4.16 Examples of localised transcripts in cleaving *Amphimedon queenslandica* embryos. (**A**) *AmqWntA*. (**B**) *AmqWntB*. (**C**) *AmqTGF-β*. (**D**) *AmqFzdA*. (**E, F**) *AmqFzdB*. (**G, H**) *Amqβ-catenin*. (**I, J**) *AmqSFRPD* in maternal follicle cell layer. (**K, L**) *AmqAxin* predominantly in external micromeres but also a smaller number of internal micromeres. (**M**) *AmqDsh*. (**N**) *Amq-Gro*. (**O**) *Amq-Lrp5/6* in maternal follicle cell layer. (**P**) *AmqGSK*. (**Q**) *AmqAPC*. (**R**) *AmqTcf*. (**S**) *AmqHedgling*. (**T, U**) *AmqPellino*. (**V, W**) *AmqMyD88*. (**X, Y**) *AmqSTmyhc* in maternal follicle cell layer and few internal micromeres. (**Z**) *AmqTollip*. (**A′**) *AmqCry1*. (**B′**) *PL10*. (**C′**) *Nanos*. (**D′**) *Vasa*. Scale bars: (**A–G, I, K–T, V, Y–D′**), 100 μm; (**H, U, W**), 25 μm; (**J**), 10 μm. (**A–R**) (From Adamska et al. 2010); (**T–W and Z**) (From Gauthier et al. 2010); (**X** and **Y**) (From Steinmetz et al. 2012); (**A′**) (From Rivera et al. 2012)

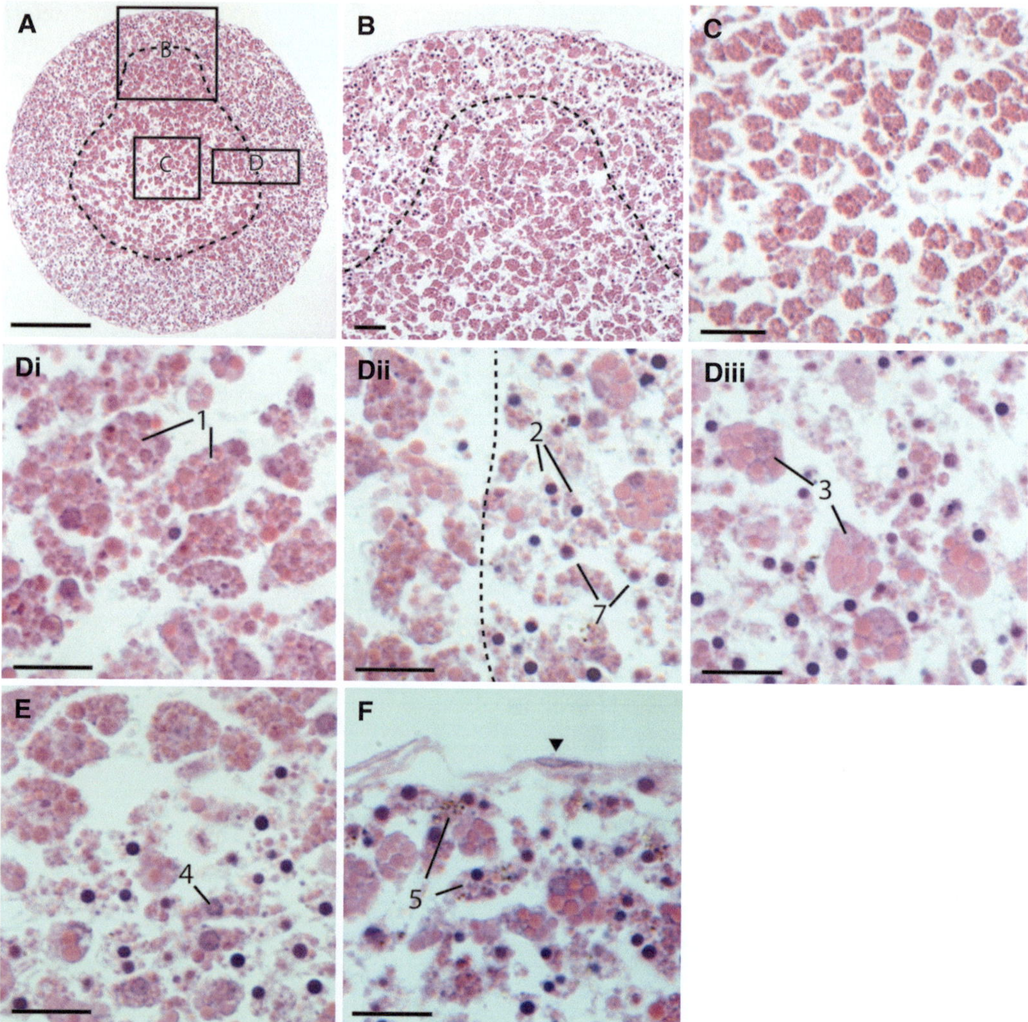

Fig. 4.17 Cloud stage embryos. (**A**) Whole embryo. (**B**) Posterior pole and underlying macromeres. (**C**) Inner layer. (**D**) Progression from inner (*i*) to outer (*iii*) layer showing locations of type 1 granular macromeres ('*1*'), type 1 micromeres ('*2*'), type II globular macromeres ('*3*'), and type II micromeres ('*7*'). (**E**) Amoeboid cells in the outer layer ('*4*'). (**F**) Pigment cells towards the posterior pole ('*5*'). All panels show H+E-stained median sections; *dashed line*, inner/outer cell layer boundary; *arrowhead*, maternal layer. Scale bars: (**A**), 100 µm; (**B–C**), 20 µm; (**D–F**), 10 µm

In addition to various transcription factor mRNAs, transcripts that localise to micromeres at cleavage include those encoding components of conserved signalling pathways, innate immunity factors, structural proteins, and RNA-binding proteins and presumptive germ line factors. It is worth noting that the last group, which includes *vasa*, *nanos*, and *PL10*, displays transcript enrichment around a subset of micromeres (Fig. 4.16B′–D′).

Cell Layer Formation and Establishment of Axial Polarity

Cleavage is followed by a period of differential cell movement that sorts these different cell types into inner and outer layers (Fig. 4.6C, D), with ciliated cells, sclerocytes, and pigment cells being enriched in the outer portion of the embryo (Figs. 4.11 and 4.17; Leys and Degnan 2002). The inner cell mass (ICM) is primarily composed of large granular cells (Fig. 4.17). At this stage, cells appear mesenchyme-like, lacking robust cell junctions and being surrounded by a collagenous extracellular matrix (ECM) (Leys and Degnan 2002). After this initial sorting, cells continue migrating to become patterned along the anterior-posterior (AP) axis (Fig. 4.1; Leys and Degnan 2002; Degnan et al. 2005; Adamska et al. 2007a, 2010).

Tracing cells on the surface of embryos, by labelling with the fluorescent lipophilic dye DiI, confirms that early *Amphimedon queenslandica* embryos undergo extensive cellular rearrangements between late cleavage (brown stage) and the establishment of the AP axis ('cloud' and 'spot' stages) (Fig. 4.18; Adamska et al. 2010). Accordingly, the spacing between cells in cleavage and early brown stages probably reflects a lack of robust intercellular adhesion as well as a lack of extensive extracellular material. A similar event has been inferred to occur in embryos of *Ephydatia* prior to the differentiation of cell layers (De Vos et al. 1991). These broad cell movements represent morphogenesis via 'differential centrifugal migration' or 'multipolar migration/delamination' and commonly rely on cell sorting via the relative adhesive properties of each cell type (Leys and Ereskovsky 2006).

Just before the cloud stage, *Amphimedon queenslandica* embryos undergo compaction (Adamska et al. 2010), which may mark the culmination of the mass cell migratory events. This also coincides with stronger cell adhesion in the embryo and the increased density of ECM (Leys and Degnan 2002; Adamska et al. 2010). The increased density of cells and the appearance of ECM in the inner layer of the cloud stage (Fig. 4.17; Leys and Degnan 2002) support this timing and interpretation of events. External cells labelled with DiI at the cloud stage and later in development do not migrate to the same extent as cells labelled during cleavage and brown stages (Fig. 4.18), suggesting that the majority of the cells in the outer layer of the embryo reach their final embryonic territory by the end of the cloud stage (i.e., spot stage).

Localisation of Wnt and TGF-β Transcripts in the Early Embryo

The establishment of the bilayered embryo with AP axial polarity is preceded by the localisation of *Wnt*-expressing cells in the future posterior half of the embryo (Adamska et al. 2007a, 2010). These cells initially appear to be evenly distributed in the cleaving embryo and then appear to migrate posteriorly (Fig. 4.19). The mechanisms underlying the coalescing of these *Wnt-expressing* cells towards the future posterior side of the embryo remain unknown. It is yet to be determined if the Wnt pathway or another mechanism, such as signalling from the maternal follicle layer, directs the formation of the embryonic axis. Nonetheless, many of the components of the Wnt pathway also are expressed in cleaving embryos, usually in subsets of micromeres, or the surrounding follicle layer (Fig. 4.16; Adamska et al. 2007a, 2010), suggesting that some embryonic cells are competent to respond to the Wnt ligand.

A gene encoding a TGF-β ligand is expressed initially in a fraction of small cells distributed throughout the outer layer of the cleaving *Amphimedon queenslandica* embryo (Adamska et al. 2007a). During the formation of the bilayered embryo, transcripts become differentially localised along the AP axis and enriched at the poles (Fig. 4.19). These results are consistent with Wnt and TGF-β pathways working together to pattern the embryonic AP axis, from which will form the radially symmetrical *A. queenslandica* larva. During the formation of the cell layers and the establishment of the primary (AP) axis, many genes are differentially expressed in the inner and outer cell layers (Fig. 4.20).

In most eumetazoans, Wnt and TGF-β pathways are responsible for patterning of the AP and dorsoventral body axes (reviewed in Martindale 2005; see also Hayward et al. 2002; Matus et al. 2006). The differential expression of Wnt and TGF-β along the demosponge AP axis suggests that these genes were used to pattern the body plans of the first (radially symmetrical) animals (Adamska et al. 2007a, 2010).

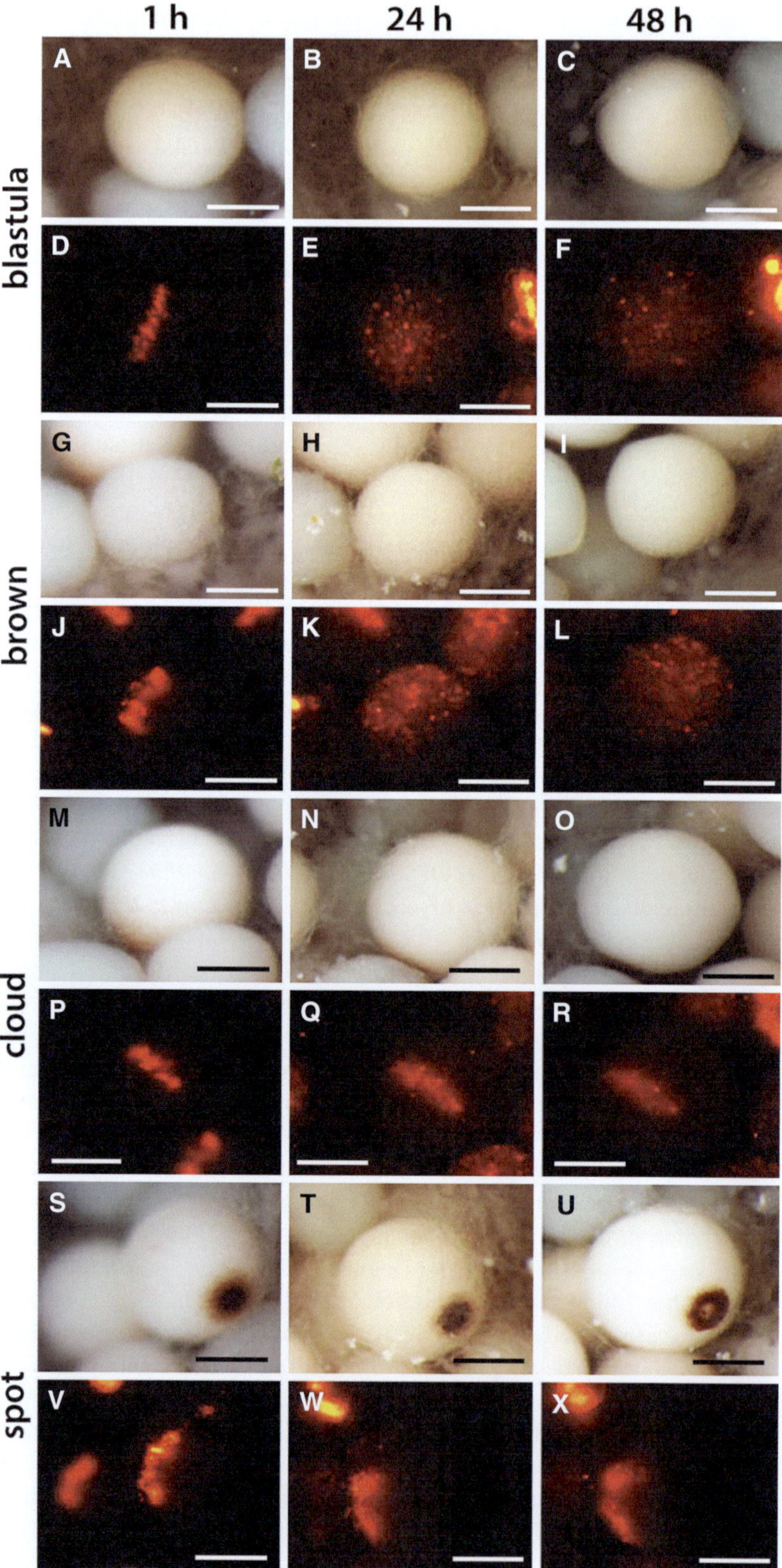

Fig. 4.18 Cell movement in *Amphimedon queenslandica* embryos. Embryos within brood chambers were 'tattooed' with a fluorescent lipophilic tracer, DiI, and photographed 1, 24, and 48 h after labelling. Embryos at four stages were labelled. (**A–F**) Blastula. (**G–L**) Brown stage. (**M–R**) Cloud stage. (**S–X**) Spot stage. *Top* panels (**A–C, G–I, M–O, S–U**): bright light images of the embryos; *bottom* panels (**D–F, I–L, P–R, V–X**): fluorescence images of the same embryos. Developmental stages are indicated to the *left*. Scale bars: 250 μm

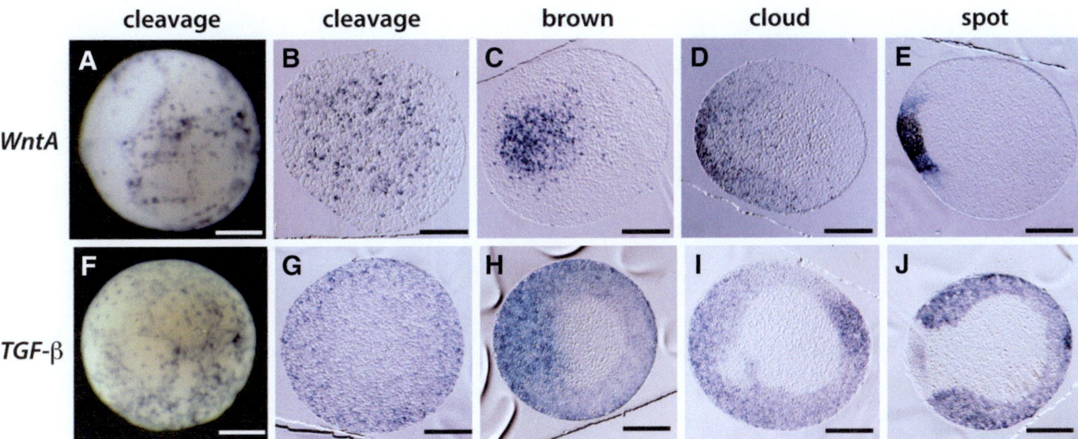

Fig. 4.19 *Wnt* and *TGF-β* expression in *Amphimedon queenslandica* early embryos. (**A–E**) *WntA* in situ hybridisations. (**F–J**) *TGF-β* in situ hybridisations, except (**H**), which is a double in situ hybridisation with *WntA* in *light blue* and *TGF-β* in *purple*. (**A, F**) Whole mounts. (**B–E, G–J**) Sections with posterior to the left. Scale bars: 100 μm (Modified from Adamska et al. (2007a) and (2010))

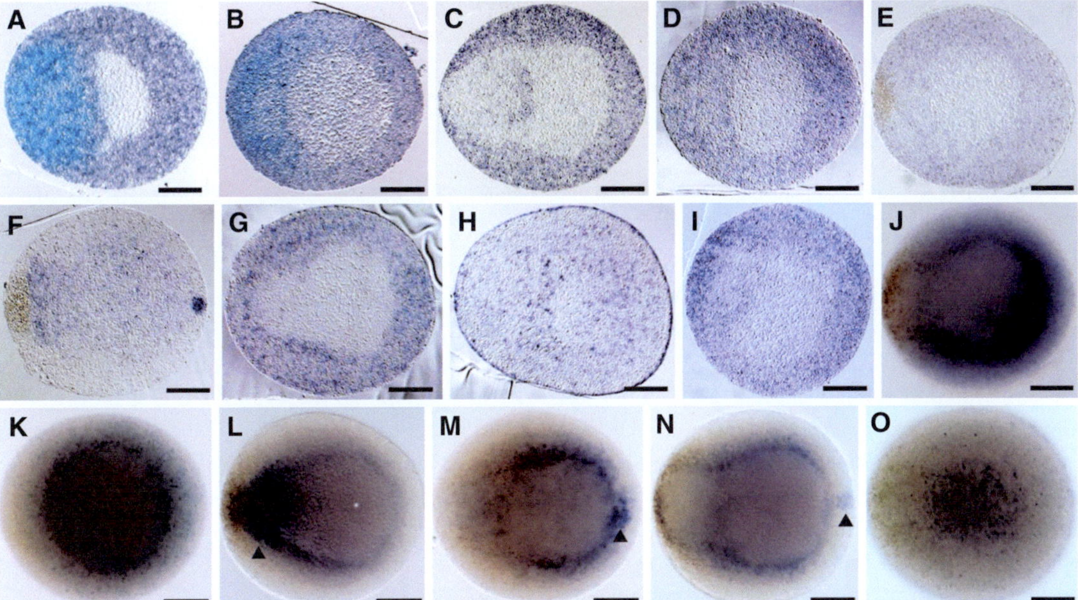

Fig. 4.20 Examples of differential gene expression of Wnt and Notch pathway components during cell layer and posterior spot formation. Posterior to the left in all micrographs. (**A–I**) Wnt pathway components (sections). (**J–O**) *Notch* and *Delta* (whole mounts) (**A**) *FzdA*. Double in situ hybridisation with *WntA* (*light blue*). (**B**) *β-catenin*. Double in situ hybridisation with *WntA* (*light blue*). (**C**) *Gsk*. (**D**) *Dvl*. (**E**) *Gro*. (**F**) *FzdB*. (**G**) *LRP4/6*. (**H**) *TCF*. (**I**) *APC*. (**J**) *Notch*. (**K**) *Delta1*. (**L**) *Delta2*. *Arrowhead points to expression under the forming pigment spot.* (**M**) *Delta3*. *Arrowhead* points to expression in a group of cells located towards the anterior of the embryo. (**N**) *Delta4*. *Arrowhead* points to expression in a group of cells located towards the anterior of the embryo. (**O**) *Delta5*. Scale bars: 100 μm. (**A–J**) are (From Adamska et al. 2010); (**J–O**) are (From Richards and Degnan 2012)

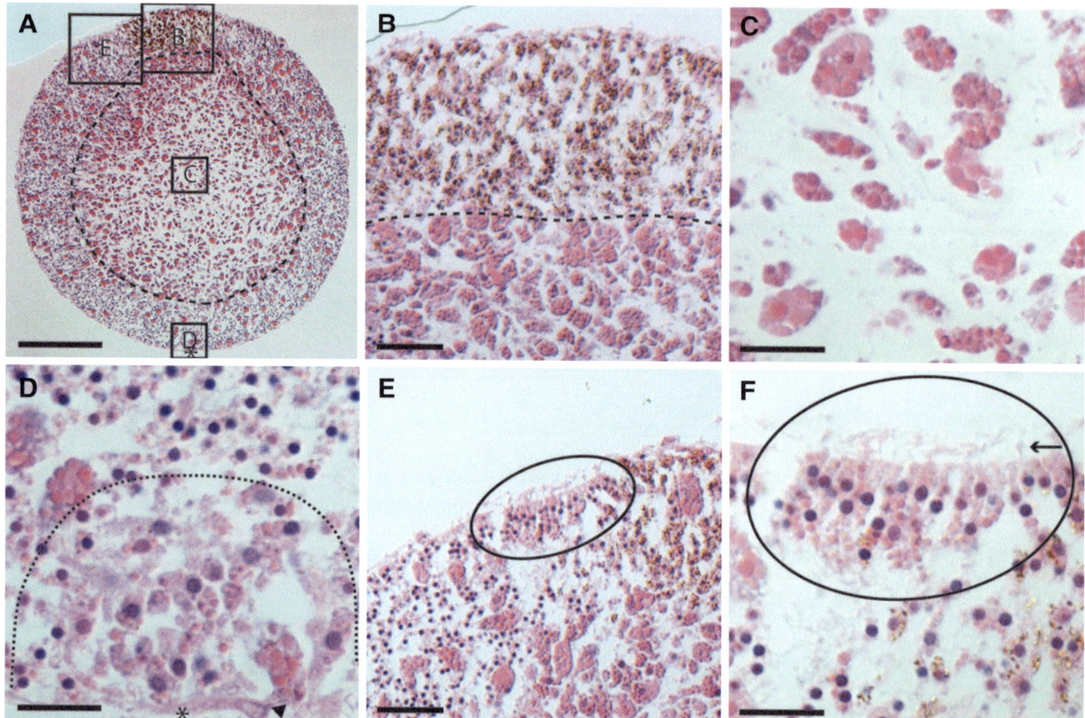

Fig. 4.21 Early spot stage embryos. (**A**) Whole embryo. (**B**) Pigment spot. (**C**) Inner layer. (**D**) Anterior pole cells (*dotted line*). (**E**) Posterior ciliated cells (*oval*). (**F**) Higher magnification of (**E**). All panels show H+E-stained median sections; *dashed line*, inner/outer cell layer boundary; *asterisk*, anterior pole; *arrowhead*, maternal layer; *arrow*, cilia. Scale bars: (**A**), 100 µm; (**B**, **E**), 20 µm; (**C–D**, **F**), 10 µm

LATE EMBRYONIC DEVELOPMENT IN *AMPHIMEDON QUEENSLANDICA*

In *Amphimedon queenslandica*, late embryonic development is considered to start at the spot stage, when the bi-layered embryo has an obvious AP axis and when the larval body plan is completely formed. In some cases, cells combine with other cells of the same type to form simple tissues, including the ciliated epithelium, the anterior cuboidal cell cluster, the pigment ring, and the ciliated ring. The latter two tissues combine to form a functional photosensory organ (Rivera et al. 2012).

Spot Stage and Commencement of Larval Tissue Formation

The early spot stage is characterised by the coalescence of the pigment cells at the posterior pole (Fig. 4.21A, B). Directly beneath the pigment spot, a dense group of type I macromeres remains, as seen in earlier stages, and the inner layer contains an increasing amount of extracellular material (Fig. 4.21C). The cell population at the anterior pole has formed a compact group (Fig. 4.21D). At the posterior, a group of columnar cells with an apical cilium and basal inclusions are aligned adjacent to the pigment spot (Fig. 4.21E, F).

In later spot stage embryos (Fig. 4.22A), a population of micromeres (non-pigmented) at the posterior pole becomes apparent; these form a 'cap' in the centre of the pigment spot (Fig. 4.22B). Directly opposite, the group of cells at the anterior pole is more condensed than previously (Fig. 4.22C). The outer epithelium of the embryo forms in a posterior to anterior progression (Fig. 4.22D), and the follicle layer separates from the embryo as the outer epithelium gains integrity (white arrows, Fig. 4.22D). Immediately

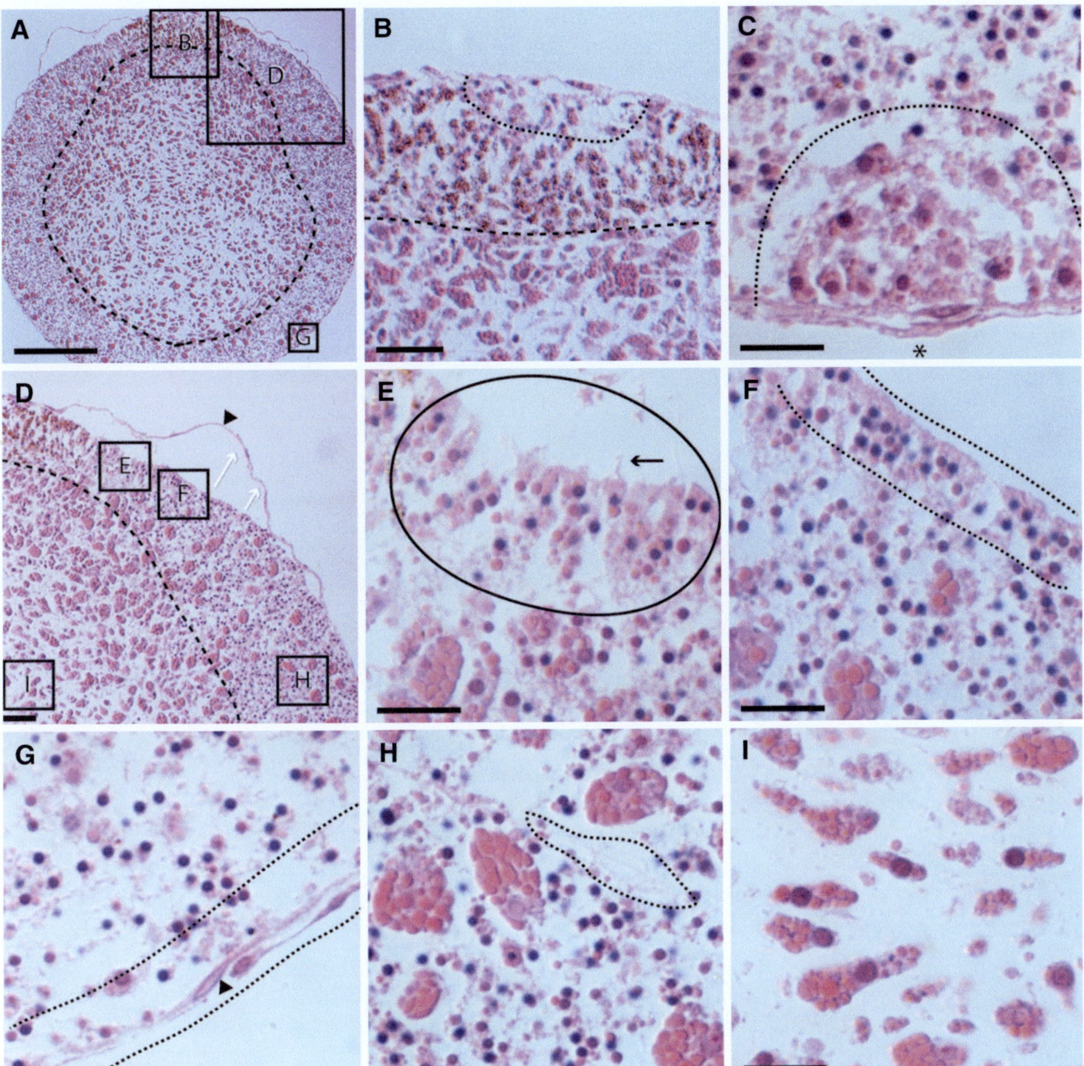

Fig. 4.22 Late spot stage embryos. (**A**) Whole embryo. (**B**) Pigment spot with small cap of non-pigmented cells (*dotted line*). (**C**) Anterior pole cells (*dotted line*). (**D**) Separation of maternal layer (*white arrows*). (**E**) Posterior ciliated cells (*oval*). (**F**) Forming border of posterior epithelium (*dotted lines*). (**G**) Unformed border of anterior outer layer (*dotted line*). (**H**) Outer layer with sclerocyte (*dotted lines*). (**I**) Inner layer cells and extracellular matrix. All panels show H+E-stained median sections; *dashed line*, inner/outer cell layer boundary; *asterisk*, anterior pole; *arrowhead*, maternal layer; *black arrow*; cilia. Scale bars: (**A**), 100 µm; (**B, D**) 20 µm; (**C, E–J**), 10 µm

adjacent to and anterior of the pigment spot is a population of ciliated epithelial cells (Fig. 4.22E). Immediately anterior to these, micromeres of the outer layer are lined up along the margin of the embryo and are thickened apically, forming a distinct boundary edge (Fig. 4.22F). Closer to the anterior pole, the micromeres of the outer layer are not organised into a distinct layer (Fig. 4.22G).

Pigment Ring Formation

The ring stage is identified by the wrinkled appearance of the embryo and the transformation of the pigment spot into a pigment ring at the posterior pole (Fig. 4.23A). Ring formation occurs via an increase in the non-pigmented cells found at the posterior pole and the migration of pigment

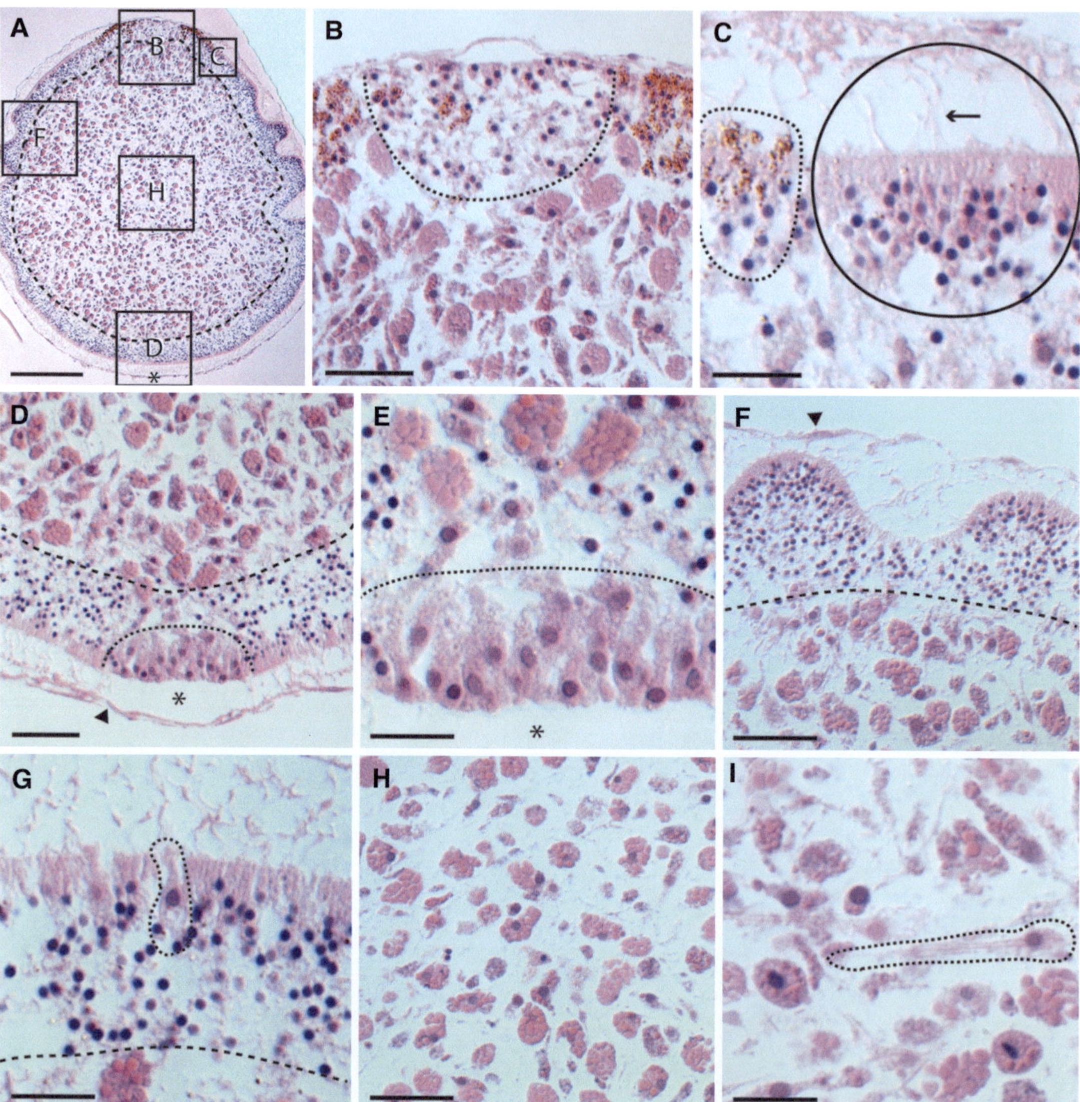

Fig. 4.23 Early ring stage embryos. (**A**) Whole embryo. (**B**) Posterior pole, non-pigmented cells (*dotted line*). (**C**) Posterior ciliated cells (*oval*) and pigment cells (*dotted line*). (**D**) Anterior pole cells (*dotted line*). (**E**) Higher magnification of (**D**). (**F**) Wrinkled outer epithelium. (**G**) Flask cell (*dotted line*). (**H**) Inner layer cells and extracellular matrix. (**I**) Inner layer sclerocyte (*dotted line*). (**E, G,** and **I**) are higher magnification images from the same regions as (**D, F,** and **H**), respectively. All panels show H+E-stained median sections; *Dashed line*, inner/outer cell layer boundary; *arrowhead*, maternal layer; *asterisk*, anterior pole; *arrow*, cilia. Scale bars: (**A**), 100 μm; (**B, D, F, H**), 20 μm; (**C, E, G, I**), 10 μm

cells from a central position at the posterior pole, to being more laterally located at the surface (Fig. 4.23B). In addition, the pigment cells are polarised, with the pigment granules located apically, and the nucleus assuming a basal position in each cell (Fig. 4.23C). Adjacent to the pigment cells, the posterior ciliated cells remain a distinct group, packed in a tight cluster to the exclusion of other cells in the outer layer (Fig. 4.23C). The cells at the anterior pole by now are organised into a single layer, with the nucleus assuming a more apical position in each cell (Fig. 4.23D, E).

The outer layer of the embryo is ciliated (except at the anterior and posterior poles), and, as a consequence, the follicle layer is no longer closely associated with the embryonic surface

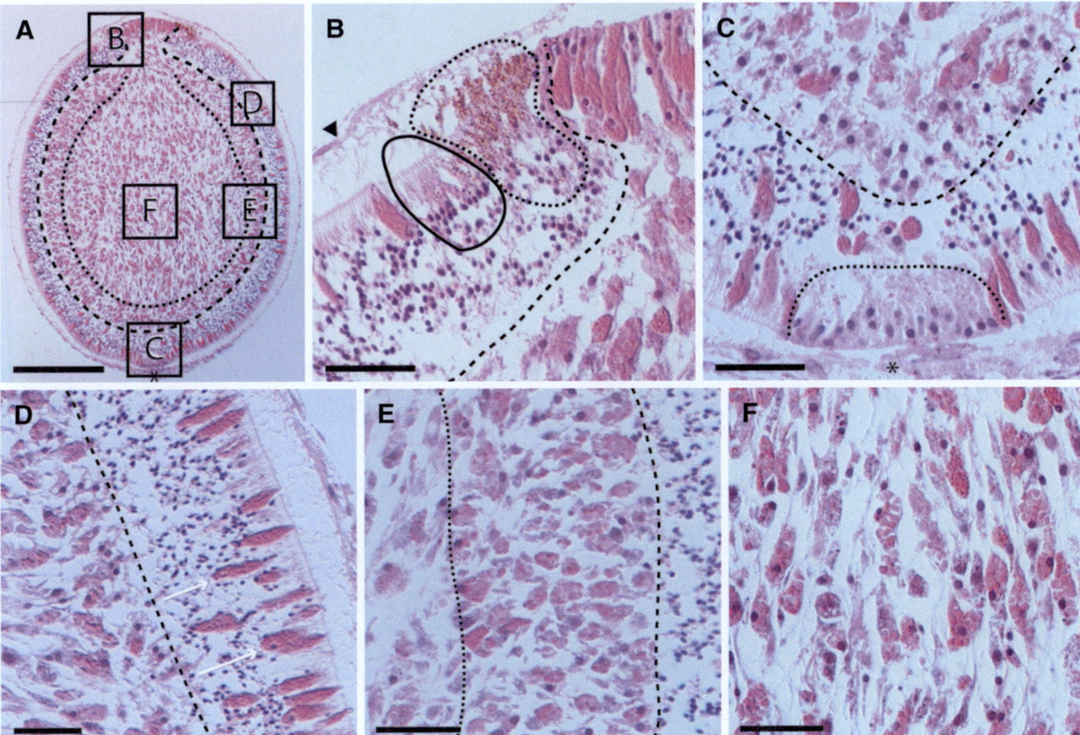

Fig. 4.24 Late ring stage embryos. (**A**) Whole embryo. (**B**) Cells of the pigment ring (*dotted line*) and associated posterior ciliated cells (*solid line*). (**C**) Anterior pole cells (*dotted line*). (**D**) Migration of globular cells (*white arrows*). (**E**) Middle layer, boundary with inner cell mass (*dotted line*). (**F**) Inner cell mass. All panels show H+E-stained median sections; *dashed line*, inner/mid cell layer boundary; *arrowhead*, maternal layer; *asterisk*, anterior pole. Scale bars: (**A**), 100 μm; (**B–F**), 20 μm

(e.g., Fig. 4.23D). The ciliated cells of the outer layer are polarised, with nuclei located basally and the apical region of each cell bearing a cilium (Fig. 4.23F, G). All macromeres have left the outer layer by this stage and are either located in the inner cell mass or at the boundary between the inner and outer layers (Fig. 4.23F). A new cell type – the flask cell (Leys and Degnan 2001) – is now identifiable amongst the ciliated epithelial cells towards the anterior of the embryo; they are ciliated with a centrally located nucleus and numerous small basally located vesicles (Fig. 4.23G). The inner layer of the embryo contains a diversity of unidentified cell types that are embedded in extracellular material (Fig. 4.23H). Numerous sclerocytes are also present (Fig. 4.23I).

The late ring embryo is elongated in comparison to earlier stages and is morphologically very similar to the larval form (Fig. 4.24A). At the posterior, the pigment cells are organised into a symmetrical ring around the pole with the apical region of each pigment cell protruding from the embryo (Fig. 4.24B). The posterior ciliated cells that lie adjacent to the pigment ring are also polarised, with basal nuclei, and the cells now appear to be clustered into small groups (Fig. 4.24B).

At the anterior pole, the cells are organised into a single layer and, in contrast to the surrounding epithelium, are non-ciliated (Fig. 4.24C). In the outer layer, which had previously been empty of macromeres, a population of globular cells (putatively derived from the type II macromere population) is now evident. These cells appear to migrate outwards from the forming subepithelial layer, through the outer layer to the periphery of the embryo (white arrows, Fig. 4.24D). A further group of globular macromeres is found at the posterior pole, within the ring of pigment cells (Fig. 4.24B). Between the ICM and the outer

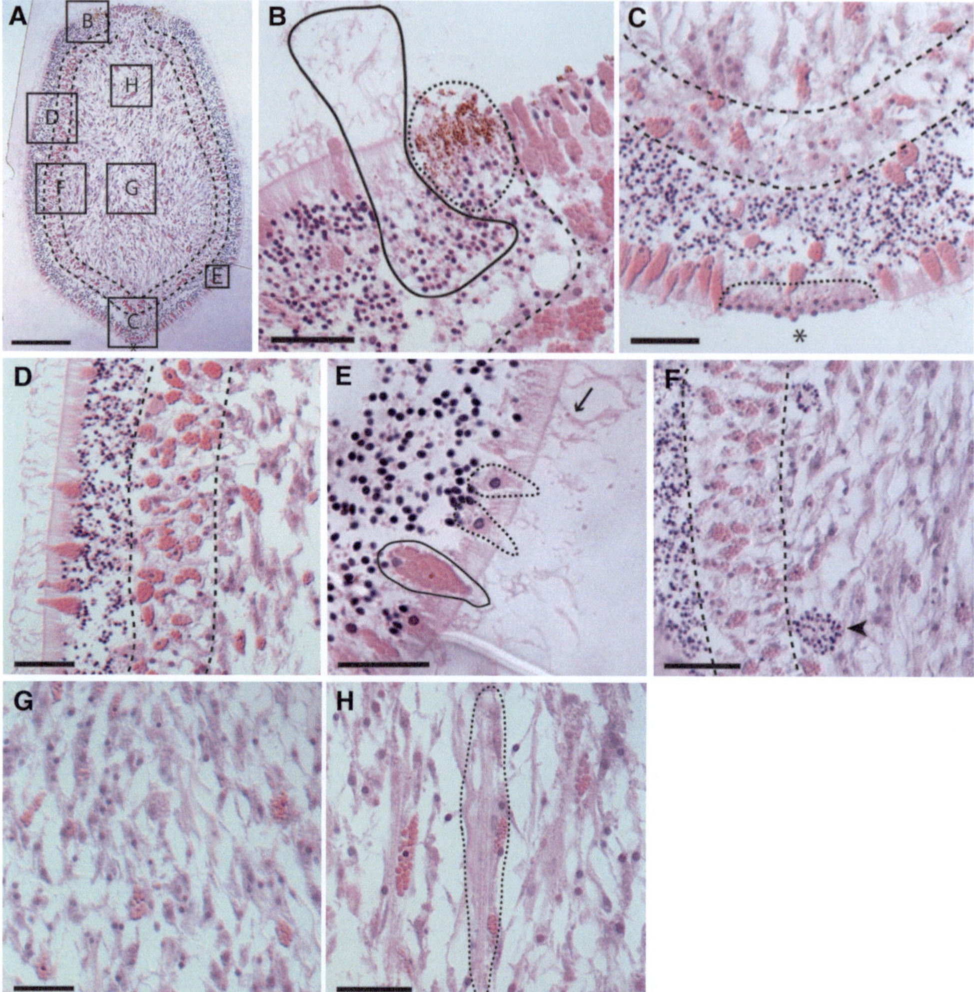

Fig. 4.25 The *Amphimedon queenslandica* larva. (**A**) Whole larva. (**B**) Cells of the pigment ring (*dotted line*) and associated posterior ciliated cells (*solid line*). (**C**) Anterior pole cells (*dotted line*). (**d**) Larval cell layers. (**E**) Intraepithelial cells: flask cells (*dotted line*) and globular cells (*solid line*). (**F**) Precocious choanocyte chambers (*arrowhead*). (**G**) Inner cell mass. (**H**) Sclerocyte bundle (*dotted line*). All panels show H+E-stained median sections; *dashed line*, cell layer boundaries; *asterisk*, anterior pole; *black arrow*, cilia. Scale bars: (**A**), 100 µm; (**B–D, F–H**), 20 µm; (**E**), 10 µm

epithelium, a third cell layer – the subepithelial layer – is now evident; it is composed of spherulous cells interspersed with a number of smaller, unidentified cells (Fig. 4.24E). Cells of the ICM are positioned with their long axes aligned to the anterior-posterior axis of the embryo (Fig. 4.24F). The majority of sclerocytes are now located either within the ICM or at the boundary between the ICM and the subepithelial layer (not shown).

At this late stage, some cells combine with other cells of the same type to form simple tissues, including ciliated epithelium, anterior cuboidal cells, pigment ring cells, and ciliated ring cells. Pigment ring cells and ciliated ring cells combine to form a functional photosensory organ (Fig. 4.25). Figure 4.26 summarises the stages of *Amphimedon queenslandica* embryonic development, tracing the genesis of larval cell types.

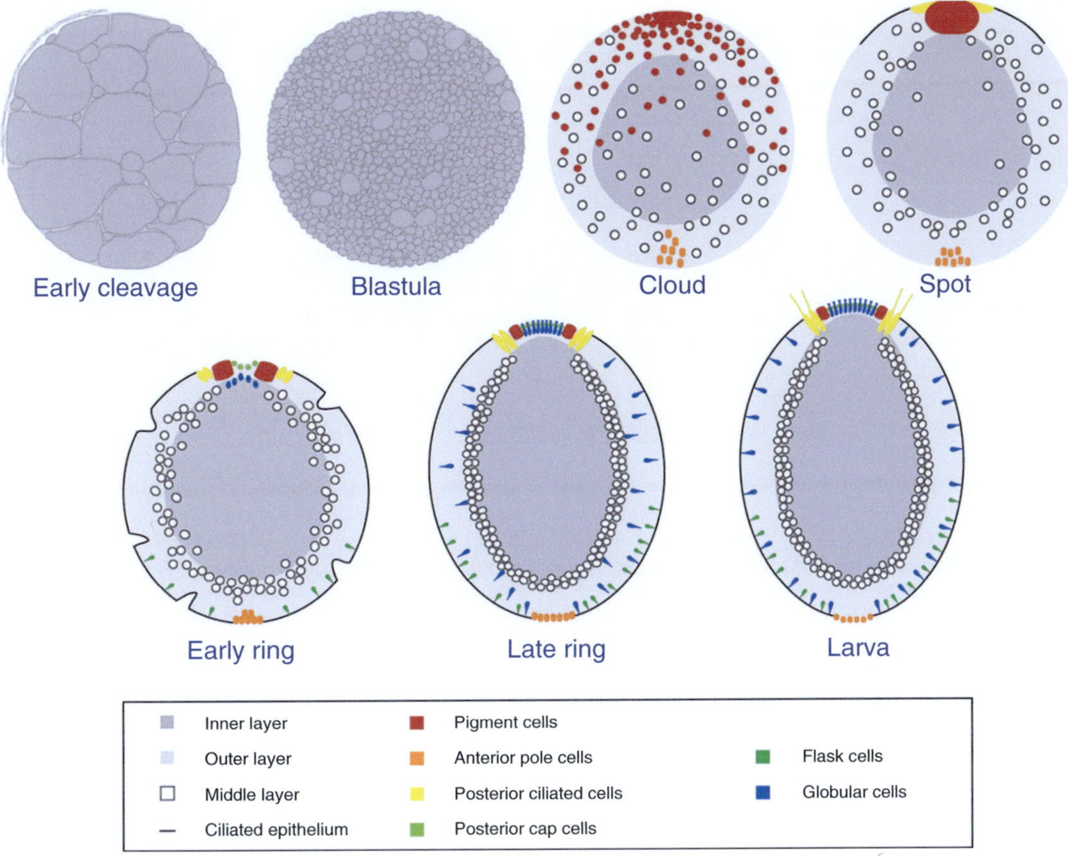

Fig. 4.26 Summary of *Amphimedon queenslandica* development highlighting the ontogeny of selected cell types and regions

Autonomous Formation of Pigment Spots and Rings

The formation of a pigment ring at the posterior end of the larva is essential to the photosensory capabilities of the larva (Leys and Degnan 2001; Rivera et al. 2012). The ring must be in a near-perfect circular pattern for the larva to swim away from light. The developmental mechanisms underlying the formation of this and other sponge tissues remain largely unknown.

Grafting of pigment cells from cloud, spot, and early ring stage embryos to another embryo results in the ectopic formation of pigment spots and rings in the new location (Fig. 4.27). These structures develop in accordance with the location from which they originated, and not the position to which they were transplanted, indicating that the fates of these cells (in larvae) have been determined earlier in development. The ability of these heterotopic grafts to form rings out of the normal spatial context suggests that the formation of pigment rings from spots relies on an intrinsic signalling system and that pigment cells of different ages – from cloud to early ring, at least – have self-organising ability to form an ectopic ring.

Localised and Cell Type-Specific Gene Expression in Late Stage Embryos and Larvae

Many of the genes studied to date in *Amphimedon queenslandica* are differentially expressed in specific cell types or cell layers in spot and ring stage embryos and swimming larvae. The reader is directed to specific publications for detailed descriptions of specific genes (Table 4.3). From late spot/early ring stage to the newly hatched

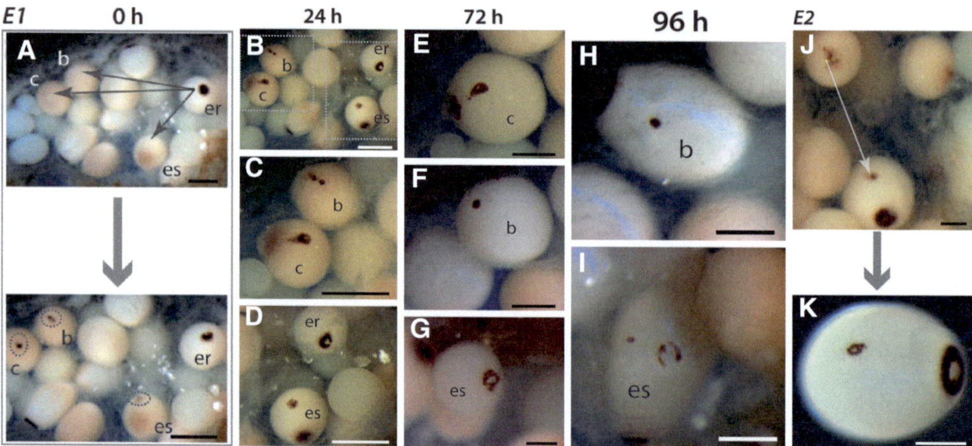

Fig. 4.27 Ectopic formation of pigment spots and rings in heterotopic grafts of pigment cells. Two experiments are shown, *E1* and *E2*. (A–I) In *E1*, pigment and possibly other cells from within the pigment ring of an early ring (*er*) stage embryo are transplanted onto the side of a brown stage embryo (*b*), cloud stage embryo (*c*), and early spot stage embryo (*es*). Manipulated embryos continue to develop in the brood chamber. (**A**) Source and grafted embryos (*arrows*). (**B–D**) 24 h post-graft. (**B**) General view of the brood chamber. (**C, D**) Higher magnification views of the boxed areas in **B**. Some, but not all, of the transplanted pigment cells in brown and cloud embryos are migrating posteriorly. Pigment cells transplanted into the early spot embryo do not migrate posteriorly. (**E–G**) 72 h post-graft. (**E**) Cells that do not migrate posteriorly in brown embryo graft form a small, tight spot. (**F**) Cells that do not migrate posteriorly in cloud embryo graft form a poorly defined spot/ring, which does not change by the end of the experiment. (**G**) Cells transplanted into the early spot embryo form a small ring. (**H, I**) 96 h post-graft. (**H**) The pigment spot formed in the brown embryo graft remains small, tight spot. (**I**) The ectopic ring formed on the side of the early spot embryo continues to expand to the point of becoming fragmented. (**J, K**) In *E2*, pigment and associated cells from a cloud stage embryo are transplanted onto the side of an early ring stage embryo (**J**) and allowed to develop and hatch as a swimming larva (**K**). Scale bars: (**A–D**), 500 µm; (**E–K**), 250 µm

Table 4.3 Genes with localised developmental expression patterns in *Amphimedon queenslandica*

Gene and gene family	References
Transcription factors	
Homeobox: *Bsh*; *NK2/3/4*; *NK5/6/7B*; *Tlx*; *Prox2*; *Pax2/5/8*; *POUI*; *Lhx1/5*; *Lhx3/4*	Larroux et al. (2006), Larroux (2007), Fahey et al. (2008), Srivastava et al. (2010b)
GATA: *GATA*	Nakanishi et al. (2014)
Nuclear receptor: *NR1*; *NR2*	Larroux et al. (2006), Bridgham et al. (2010)
Sox: *SoxB, C, F*	Larroux (2007)
Developmental signalling pathways	
Hedgehog: *Hedgling*	Adamska et al. (2007b)
Notch: *Notch*; *Delta1–5*	Richards et al. (2008), Richards (2010), Richards and Degnan (2012)
TGF-β: *TGF-β*	Adamska et al. (2007a)
Wnt: *WntA-C*; *FzdA, B*; *SFRPA, C, D*; *Lrp5/6*; *GSK*; *APC*; *Axin*; *Dsh*; *β-CatA*; *TCF*; *Gro*	Adamska et al. (2007a, 2010)
Toll pathway	
NF-κB; *TIR1*; *TIR2*; *MyD88*; *Tollip*; *Pellino*	Gauthier and Degnan (2008), Gauthier (2010), Gauthier et al. (2010)
Structural genes	
Cryptochrome: *Cry1*; *Cry2*	Rivera et al. (2012)
Epithelial proteins: *MPP5/7*; *ERM*; *Par-1, Par-6*; *Lgl-1*; *p120Catenin*; *Dlg*; *aPKC*	Fahey (2011)
Myosin: *STmyhc*; *NMmyhc*	Steinmetz et al. (2012)
Other: *ferritin*; *procollagen lysyl hydroxylase*; *galectin*	Larroux et al. (2006)
Postsynaptic proteins: *Dlg*; *Grip*; *Homer*; *Gkap*; *Cript*	Sakarya et al. (2007)

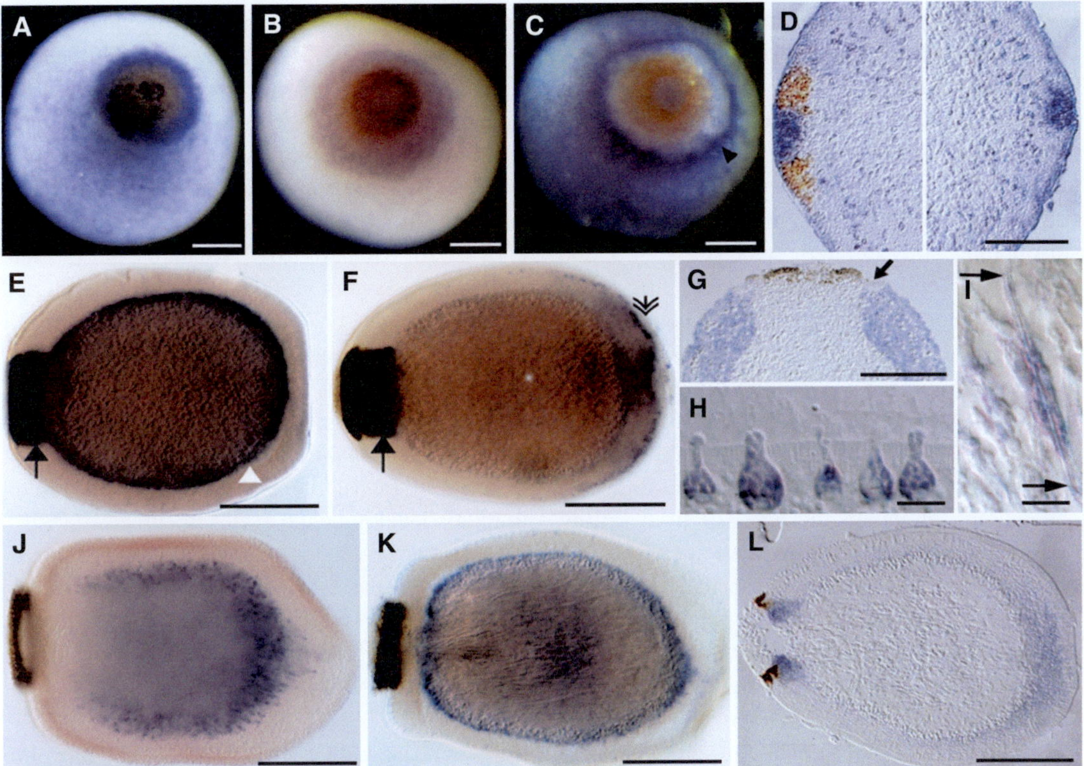

Fig. 4.28 Examples of localised and cell type-specific gene expression in late stage embryos and larvae. (**A–D**) Early ring stage. (**E–G**) Late ring stage. (**H–L**) Swimming larval stage. (**A**) *Cryptochrome 2* (*Cry2*) is expressed around the pigment ring, including long-ciliated cells next to the ring. (**B**) '*Non-muscle*' *type II myosin heavy chain* (*NM myhc*) is more broadly expressed but overlaps with *Cry2* expression. (**C**) '*Striated muscle*' *type II myosin heavy chain* (*ST myhc*) is expressed in the epithelium adjacent to *NM myhc* expression domain. (**D**) *Groucho* (*gro*) expression is enriched in posterior cells inside the ring and cuboidal cells at the most anterior end. (**E, F**) *Delta4* is expressed in a dynamic pattern. First (**E**), overlapping with the pigment ring (*black arrow*) and cells at the boundary between inner and outer cell layers (*white arrowhead*). Then (**F**), expression in boundary cells becomes undetectable while appears in the flask cells, which are enriched in the anterior third of the embryo (*double arrow*). (**G**) *p120catenin* is expressed in the forming epithelium but not in the long-ciliated cells expressing *Cry2* (*arrow*). (**H**) *TLR/ILR1-like receptor* (*IgTIR1*), along with many other genes, is expressed in globular cells that migrate late in development from the middle subepithelial layer to the larval surface. (**I**) *Bsh* homeobox is expressed in sclerocytes (*arrows* mark the cell edge). (**J**) *Sox 2* expression is a subset of cells that line the outer region of the inner cell mass. (**K**) *Delta3* in subepithelial cells. (**L**) *Lrp* in cells underlying the pigment ring and anterior epithelial cells. Scale bars: (**A–G, J–L**), 100 µm; (**H**), 10 µm; (**I**), 5 µm (**A**) (From Rivera et al. 2012); (**B**) and (**C**) (From Steinmetz et al. 2012); (**D, G and L**) (From Adamska et al. 2010); (**E, F and K**) (From Richards et al. 2012); (**H**) (From Gauthier et al. 2010); (**I**) (From Larroux et al. 2006)

larval stage can be considered a second phase of cell differentiation, which follows from the first phase comprising the formation of pigment cells, sclerocytes, and ciliated epithelial cells during cleavage (see above). These later developmental stages are typified by localised and cell type-restricted expression of transcription factor, signalling pathway, and structural genes (Fig. 4.28). This includes enrichment of innate immunity and neuronal genes in globular cells (Fig. 4.28H Sakarya et al. 2007; Gauthier and Degnan 2008; Richards et al. 2008; Gauthier et al. 2010) and epithelial genes in the larval epithelium (Fig. 4.28G Fahey and Degnan 2010); other outer layer cell types – anterior cuboidal and flask cells – have cell-specific gene expression patterns (Fig. 4.28D, F; e.g., Adamska et al. 2007a, 2010; Richards and Degnan 2012). Restricted expression patterns in the ICM are consistent with there being a number of cryptic cell types in this layer (e.g., *Sox2* is expressed only in a subset of cells on the periphery of the inner cell mass; Fig. 4.28J).

Localised Expression of Conserved Developmental Genes During Pigment Ring Formation

The photosensory capabilities of the posterior pigment ring in the *Amphimedon queenslandica* larva requires the patterning of at least two cell types, the inner pigment cells and the surrounding long-ciliated cells (Figs. 4.6 and 4.26); other cell types that exist in this larval territory include a cell type that may contain both pigment and a long cilium. The expression of *cryptochrome 2* (*Cry2*) in long-ciliated cells is consistent with these being able to detect light (Leys et al. 2002; Rivera et al. 2012). Presumably the pigment cells shade the *Cry2*-expressing cells and thereby attenuate the level of light hitting these cells. This in turn affects the behaviour of the long cilia by an unknown mechanism.

During the migration of the pigment cells to the posterior pole, and especially during spot and ring formation, a raft of signalling ligand and transcription factor genes are differentially expressed in this region (state of knowledge summarised in Fig. 4.29). In addition to *Wnt* and *TGF-β*, which are activated before spot formation (Fig. 4.19), *Hedgling* and two *Delta* ligands are differentially expressed in patterns overlapping with the pigment spot and with adjacent *Cry2*-expressing cells (Fig. 4.29A). Some expression patterns correspond to the boundaries between spot and *Cry2*-expressing (long-ciliated) cells (*Hedgling* and *Delta4*) and *Cry2*-expressing (long-ciliated) and surrounding epithelial cells (*Hedgling* and *TGF-β*), while others do not correspond perfectly to obvious morphological territories (Adamska et al. 2007a, b, 2010; Richards and Degnan 2012). The overlapping expression patterns of signalling ligands in *Amphimedon queenslandica* are reminiscent of many situations in eumetazoan development, suggesting that combinatorial signalling via Wnt, TGF-β, Hedgehog/Hedgling, and Notch pathways is a crown metazoan synapomorphy (reviewed in Adamska et al. 2011).

At this same developmental stage, a diversity of transcription factor genes are activated in the posterior pole (Fig. 4.28B). Many of these genes are expressed in *Cry2*-expressing cells, although some have broader patterns; some overlap directly with a given signalling ligand gene (*Lhx3/4* and *WntA*) or with a combination of signalling ligand genes (*POUI* and *WntA* + *TGF-β* + *Delta4*). Of the conserved transcription factor genes expressed in the vicinity of the *Cry2-expressing* cells, most have eumetazoan orthologs involved in neurogenesis and sensory cell specification (Larroux et al. 2006; Larroux 2007; Richards et al. 2008; Adamska et al. 2010; Richards 2010; Richards and Degnan 2012; Srivastava et al. 2010b). Between spot and ring stages, the posterior expression patterns of many of these genes change, often into broader domains (Fig. 4.28B, C). Although the specific role of these developmental genes is currently unknown in *Amphimedon queenslandica*, their restricted expression in particular cell types is akin to many developmental events in eumetazoans.

SETTLEMENT AND METAMORPHOSIS IN *AMPHIMEDON QUEENSLANDICA*

As is the case with embryogenesis, metamorphosis varies markedly between sponges, although a common set of morphogenetic mechanisms tend to be deployed, e.g., epithelial mesenchyme transition (EMT; see Ereskovsky 2010 for a systematic analysis of sponge metamorphosis).

Competent *Amphimedon queenslandica* larvae undergo rapid metamorphosis when they come in contact with an inductive environmental cue. Typically, larvae require at least 4 h of further development (at 24 °C) after emerging from the mother sponge before they are able to respond to this cue. During this time they are negatively phototactic (Leys and Degnan 2001; Leys et al. 2002), although during the first 2 h they can be observed on occasion swimming upwards towards the surface, which may facilitate dispersal (Degnan and Degnan 2010). A strong inductive cue is associated with the surface of live encrusting and articulated coralline algae (Degnan and Degnan 2010). Upon settling on the algae, larvae undergo a rapid and dramatic reorganisation of the body plan (Fig. 4.30). In *Amphimedon queenslandica*, a functional feeding juvenile is formed in about 3 days after the initiation of metamorphosis. As is the case with most other marine invertebrates, *Amphimedon*

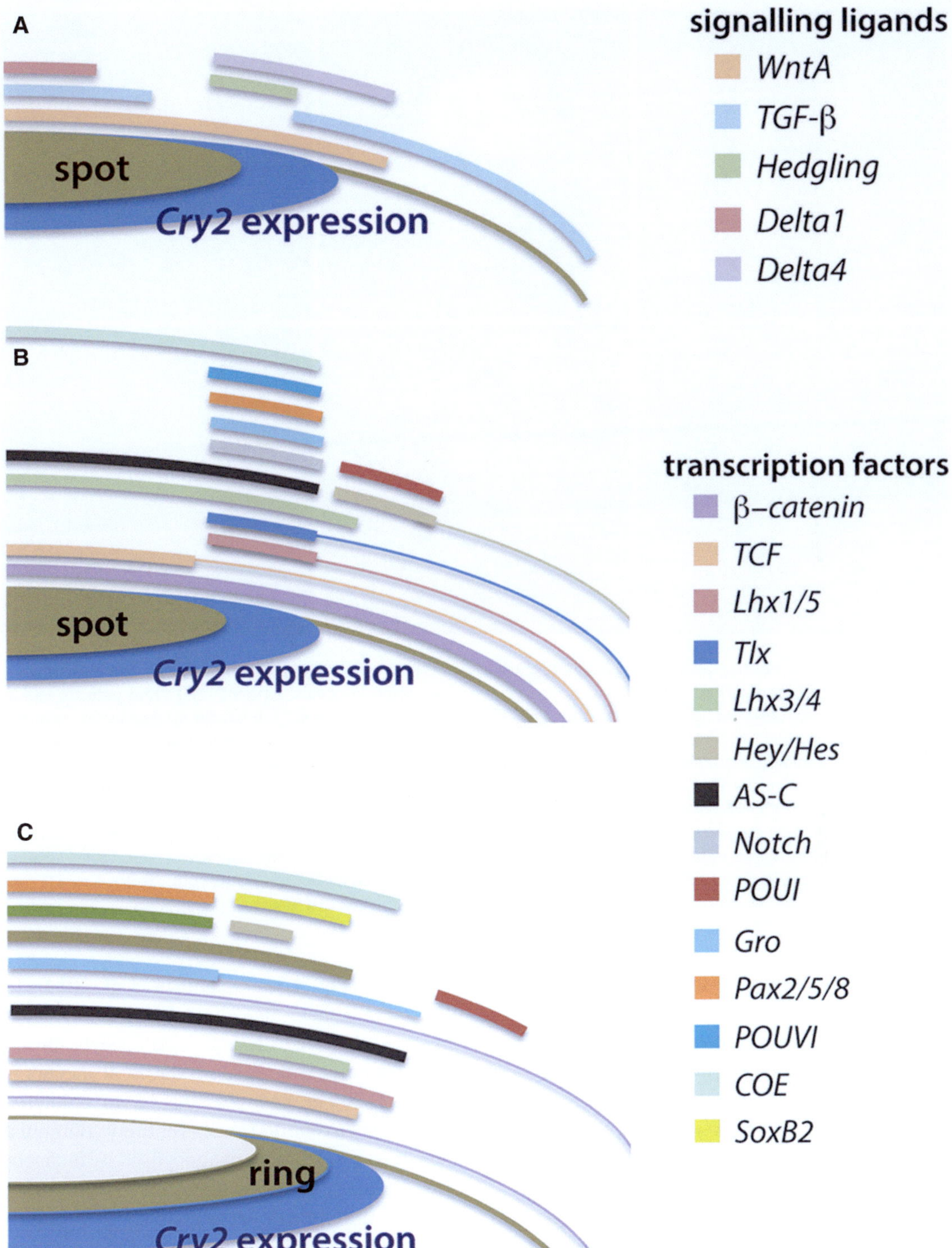

Fig. 4.29 Diagram of localised expression of signalling ligand and transcription factor genes in the vicinity of the pigment spot and ring. One half of the posterior end of spot and ring stage embryos are depicted. (**A**, **B**) Spot stage. (**C**) Ring stage. (**A**) Ligands of Wnt, TGF-β, Notch, and Hedgling pathways are expressed in overlapping pat- terns with the pigment spot and adjacent *Cry2*-expressing cells. (**B**) Multiple conserved transcription factor genes are expressed in this region, many overlapping with *Cry2*-expressing cells. (**C**) These and other transcription factor genes are co-expressed in this region as the pigment ring forms

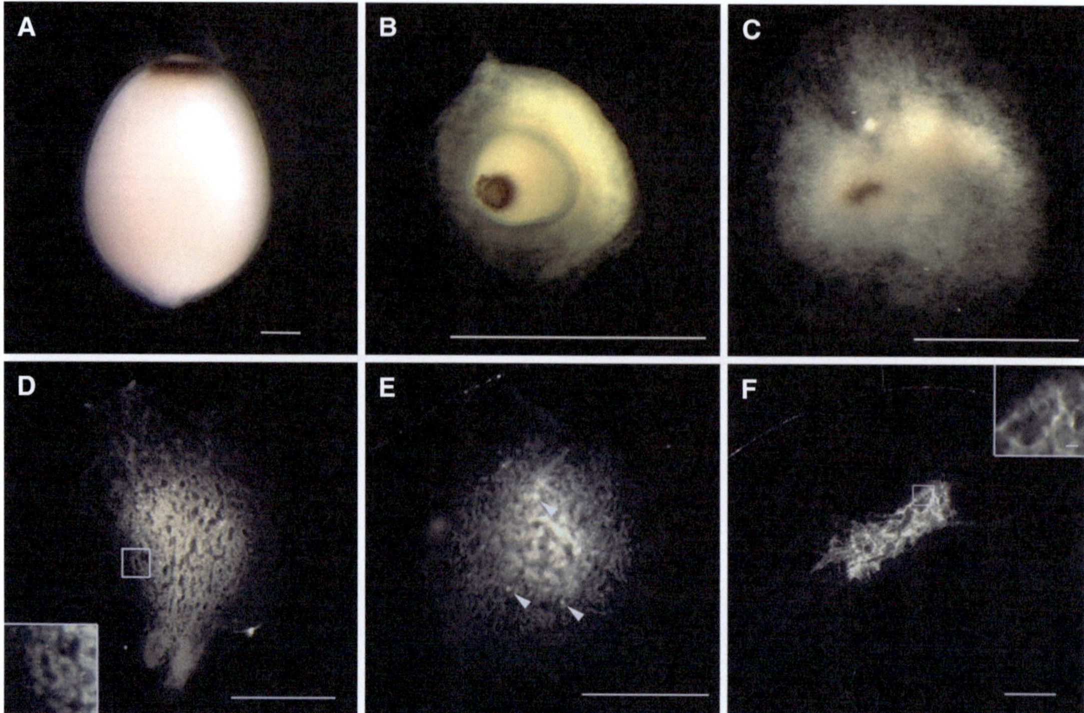

Fig. 4.30 Stages of development during metamorphosis in *Amphimedon queenslandica*. (**A**) Free-swimming parenchymella larva; posterior pigment ring up. (**B–F**) Metamorphosing postlarvae viewed from the top. (**B**) Within 1 h of initiating settlement and metamorphosis. The anterior region of the larva is attached to the substrate, onto which the larva flattens. (**C**) Mat formation, approximately 6 h post-settlement (hps). Cells of the metamorphosing postlarva migrate laterally on the substrate to form a mat-like structure. Note that former posterior pigment ring is diffuse and disappearing. (**D**) Chamber postlarval stage (~48 hps). The aquiferous system composed of canals lined by choanocytes and endopinacocytes first becomes evident. (**E**) Tent-pole-formation postlarval stage (48–72 hps); the exopinacotes covering the outer surface of the metamorphosing postlarva are lifted upwards by formation of tent-pole-like siliceous skeletal elements. Arrowheads show the internal tent-pole-like structures, visible here as clustering of cells. (**F**) Juvenile (rhagon) stage with an osculum (inset), marking the establishment of the functional aquiferous system. Scale bars: (**A**), inset in (**F**), 100 μm; (**B–F**), 1 mm (From Nakanishi et al. 2014)

queenslandica exhibits variation between individual larvae in (i) the timing of the onset of developmental competence to be induced to settle and initiate metamorphosis, (ii) the period of negative photosensitivity, and (iii) the responsiveness to specific environmental cues (e.g., different algae) (Leys and Degnan 2001; Degnan and Degnan 2010).

Within hours of settling, the larva changes into an encrusting mat (Fig. 4.30C). Tracing different populations of labelled larval cells – epithelial, flask, and internal archaeocytes – through metamorphosis reveals that there is no constancy in larval and juvenile cell layers, with all larval cell types apparently capable of transdifferentiating into any juvenile cell type. There is also extensive programmed cell death of epithelial cells at meta-

morphosis (Fig. 4.31; Nakanishi et al. 2014). In other words, there appears to be no relationship between the cell layers established during embryogenesis and those produced at metamorphosis. In other sponges, the larval epithelial layer has been reported to shed entirely (Bergquist and Green 1977), to be phagocytised by archaeocytes (Meewis 1939; Misevic and Burger 1982, 1990), to differentiate into choanocytes through a non-ciliated amoebocyte intermediate (Amano and Hori 1993, 2001), or to directly differentiate into choanocytes without loss of cilia (Ereskovsky et al. 2007; reviewed in Ereskovsky 2010). Interestingly, the endomesoderm gene *GATA* is consistently expressed in the inner layer of both larvae and juveniles, despite the extensive reorganisation of the body plan at meta-

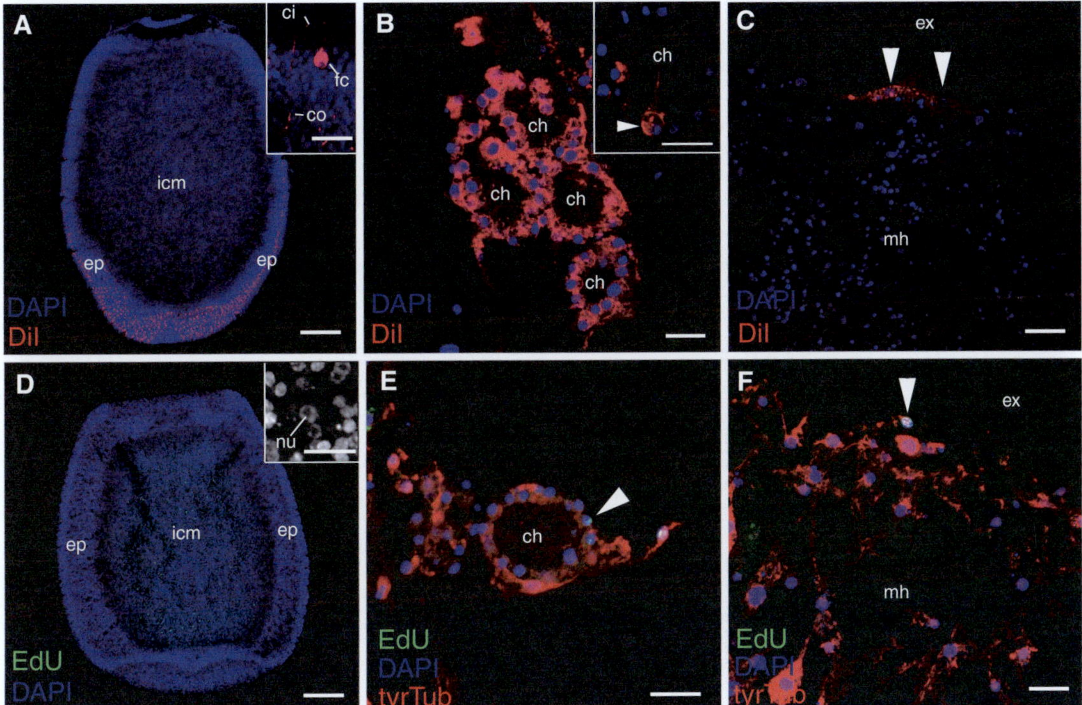

Fig. 4.31 Cell labeling and lineage-tracing through *Amphimedon queenslandica* metamorphosis. (**A**, **D**) Swimming larvae with subsets of cells labelled; anterior is to the *bottom*. (**B, C, E, F**) Descendants of the labelled cells in 3-day-old juveniles. Nuclei are stained with DAPI in all micrographs (*blue*), and in (**E, F**), the juveniles are labelled with an antibody against tyrosinated tubulin (tyr-Tub; *red*). (**A**) Longitudinal confocal sections through the centre of a larva incubated with CM-DiI, showing strong labeling in ciliated epidermal cell types, the columnar epithelial cell (*co*), and the flask cell (*fc*) (*inset*), with little labeling in inner cell mass (*icm*). (**B**) Choanocytes in chambers (*ch*). In some cases, a subset of choanocytes is CM-DiI-labelled in a single choanocyte chamber (*arrow-head* in inset), suggesting that multiple precursor cells can be involved in development of a single chamber. (**C**) Labelled exopinacocytes (*arrowheads*). (**D**) A confocal longitudinal section through the centre of a free-swimming larva pulse-labelled with EdU. Note that the labelled cells localise in the inner cell mass (*icm*) and are likely to be proliferating archaeocytes with characteristic large nucleoli (*nu*). (**E**) An EdU-positive choanocyte in the chamber (*arrowhead*). (**F**) An EdU-positive exopinacocyte (*arrowhead*). Other abbreviations: *ep* outer layer epithelium, *ci* cilium, *ex* external environment, *mh* mesohyl. Scale bars: (**A, D**), 100 μm; (**B, C, E, F**), inset in (**D**), 10 μm; inset in (**A, B**), 5 μm (From Nakanishi et al. 2014)

morphosis. Labeling of juvenile choanocytes reveals a further lack of cell layer and identity permanency, with these cells dedifferentiating into archaeocytes and transdifferentiating into a range of juvenile cell types, including the outer exopinacoderm (Nakanishi et al. 2014).

THE CALCAREOUS SPONGE *SYCON CILIATUM*

Although *Amphimedon queenslandica* serves as an excellent demosponge model in evolutionary and developmental biology research, the vast evolutionary distance between sponge lineages makes it necessary to include additional model species to represent the remaining lineages. Calcareous sponges have been intensively studied in the past centuries, and analysis of their development has significantly influenced evolutionary theory. For example, Ernst Haeckel coined the term gastrulation and formulated the Gastrea theory, after investigation of development and metamorphosis of a syconoid species from the class Calcaronea (Haeckel 1874; revisited by Leys and Eerkes-Medrano 2005). *Sycon ciliatum*, an abundant North East Atlantic calcaronean sponge, is now emerging as a calcisponge model species, with extensive sequence resources and protocols for gene expression utilised in a variety of studies

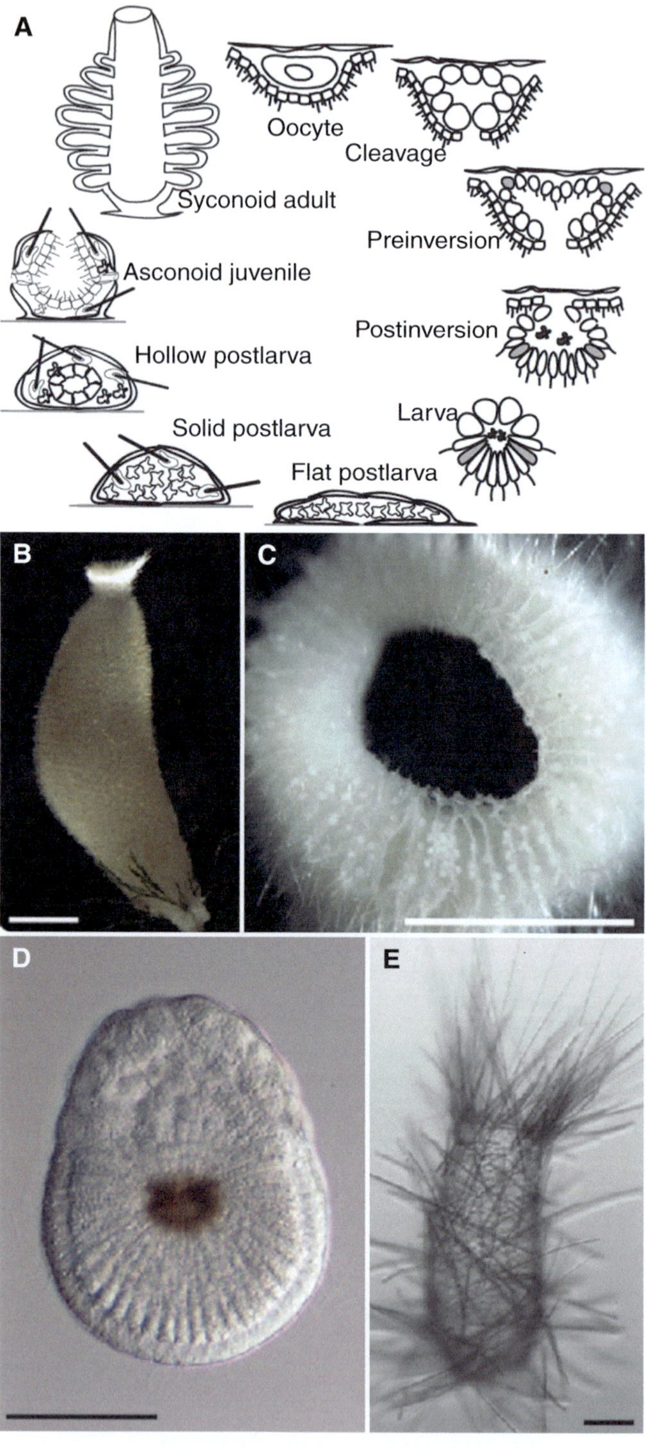

Fig. 4.32 *Sycon ciliatum*, an emerging calcisponge model species. (**A**) Schematic representation of the lifecycle, clockwise from the oocyte to adult. (**B**) Adult specimen. (**C**) Transverse section through a reproductive adult specimen. (**D**) Larva, anterior half composed of micromeres pointing down. (**E**) Juvenile (olynthus). Scale bars: B, C, 5 mm; D, E, 50 μm (Modified from Leininger et al. (2014) and Fortunato et al. (2014))

(Adamska et al. 2011; Fortunato et al. 2012, 2014; Nosenko et al. 2013; Robinson et al. 2013; Sebé-Pedrós et al. 2013; Fortunato 2014; Leininger et al. 2014; Zakrzewski et al. 2014).

The adult specimens of *Sycon ciliatum* are barrel shaped and usually reach 5 cm in length and have a typical syconoid body organisation: choanocyte chambers arranged around the

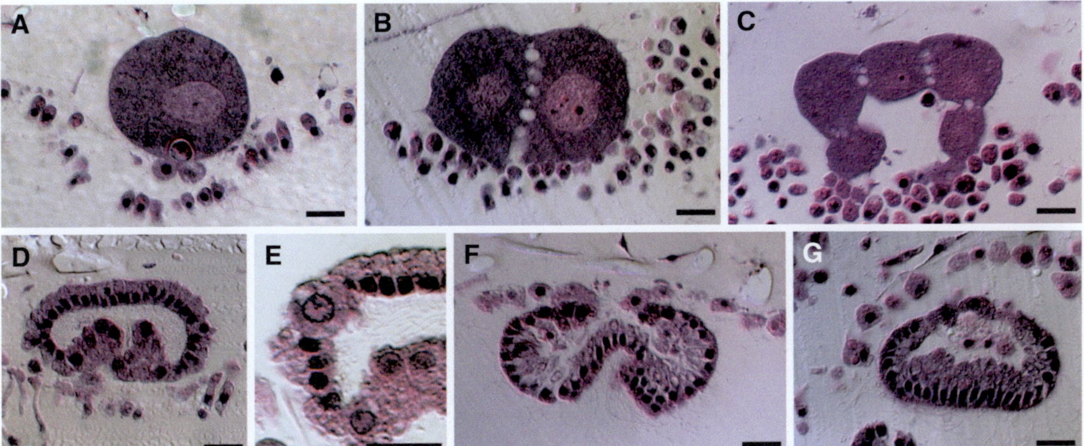

Fig. 4.33 Histological sections through embryonic stages of *Sycon ciliatum*. (**A**) Fertilisation complex. (**B, C**) Cleavage. (**D, E**) Preinversion. (**E**) Higher magnification demonstrating one of four cruciform cells located between micromeres. (**F**) Embryo soon after inversion, the opening between macromeres still communicated with the gap between accessory cells. (**G**) Postinversion stage, several maternal cells present inside of the embryonic cavity. Scale bars: 10 μm (Modified from Leininger et al. (2014))

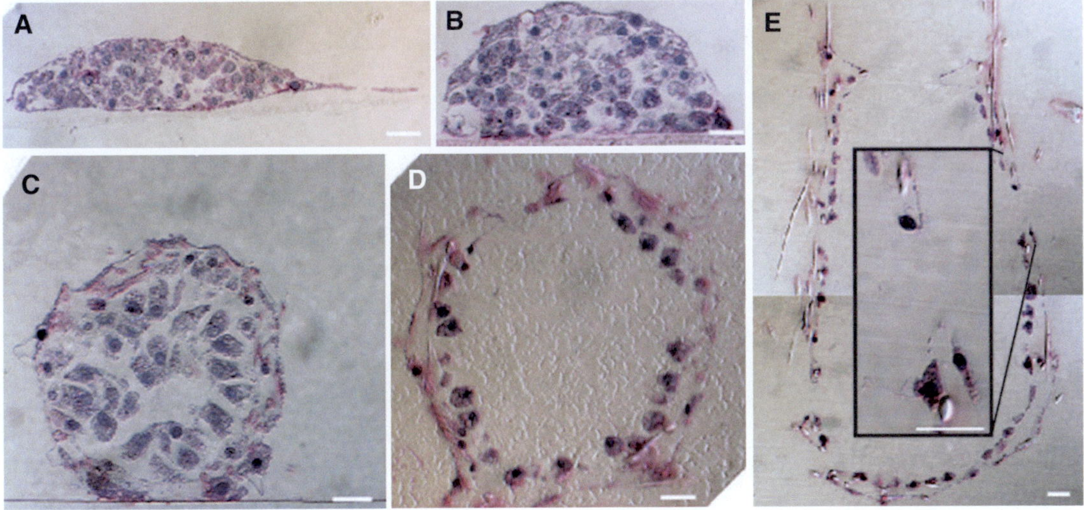

Fig. 4.34 Histological sections through metamorphosis stages of *Sycon ciliatum*. (**A**) Postlarva soon after settlement. (**B–D**) Formation of the choanocyte chamber. (**E**) Juvenile, with magnification showing three principal cell types: choanocyte, pinacocyte, and porocyte. Scale bars: 10 μm

central atrium (Fig. 4.32A–C). As for all calcisponges, it is viviparous with embryogenesis occurring in the narrow mesohyl layer between the outer pinacoderm and the inner choanoderm (Figs. 4.32C and 4.33). The larva ('amphiblastula') is approximately 100 μm long and is transparent, except for pigment deposited within the basal (inner) tips of the micromeres (Fig. 4.32D). It is composed of two major cell types: the numerous micromeres comprising the anterior part of the larva have flagella, in contrast to the larger and less numerous macromeres at the posterior pole (Fig. 4.32A, D). In contrast to the rhagon, the putative phylotypic juvenile stage of demosponges, the calcisponge juvenile has only a single choanocyte chamber (Figs. 4.32A, E and 4.34) and is referred to as the olynthus (Ereskovsky 2010). *S. ciliatum* is one of the few sponge species that maintains radial symmetry throughout its life.

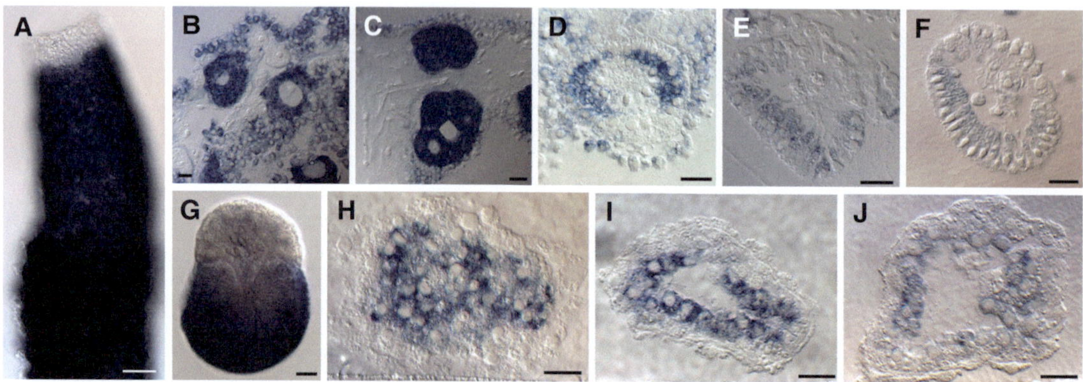

Fig. 4.35 Expression of *Tubulin-β* in *Sycon ciliatum* marks ciliated cells and early embryos. (**A**) Young adult with strongly labelled choanocytes. (**B–F**) Sections showing oocytes, cleavage, preinversion, and early and late postinversion stage embryos. (**G**) Larva with strong expression in micromeres and negative macromeres. (**H–J**) Sections showing expression in the inner cell mass and differentiating choanocytes during metamorphosis. Scale bars: (**A**), 100 µm; (**B–J**), 10 µm

Recent and ongoing studies demonstrate significant differences in the content of developmental regulatory genes (i.e., the developmental toolkit) between *Amphimedon queenslandica* and *Sycon ciliatum* (Fortunato et al. 2012, 2014; Sebé-Pedrós et al. 2013; Fortunato 2014; Leininger et al. 2014). In a majority of cases, the genome of *S. ciliatum* contains more protein family members than *A. queenslandica*: 21 versus 3 Wnt ligands, 22 versus 8 TGF-β ligands (Leininger et al. 2014), and 7 versus 4 Sox transcription factors (Fortunato et al. 2012). In addition, *S. ciliatum* possesses several developmental genes that are absent from *A. queenslandica*, for example, *Eyes absent* (Fortunato et al. 2014). In other gene families, the two species share different paralogs with bilaterians: for example, T-box family genes *Brachyury* and *Eomes* are present in *S. ciliatum*, while *Tbox4/5* and *TboxPor* are present in *A. queenslandica* (Sebé-Pedrós et al. 2013). This complex picture appears more consistent with multiple independent gene family expansions and gene losses in sponge lineages than with being simply explained by sponge paraphyly (Fortunato 2014).

SYCON CILIATUM DEVELOPMENT

Development of calcaronean sponges has several unique features, beginning with fertilisation that is assisted by specialised cells of the mother sponge, called carrier cells (reviewed in Ereskovsky 2010).

The oocytes are positioned between the pinacocyte and choanocyte layers; choanocytes directly overlying the oocytes lose their collars and become accessory cells. As a sperm cell penetrates into one of the accessory cells, this cell becomes a carrier cell. The sperm cell inside of the carrier cell is referred to as a spermiocyst and is transferred into the oocyte to complete fertilisation (Fig. 4.33A). Intriguingly, while 100 % of *Sycon ciliatum* specimens collected in May in the Norwegian fjords contain oocytes and a majority of those collected over a few days in late May contained fertilisation complexes, spermatogenesis was not observed despite frequent sampling over several years (Leininger et al. 2014 and unpublished observations). The development that follows fertilisation is also semi-synchronous across the *S. ciliatum* population, with individual sponges 'lagging behind' no more than a few days. This leads to the release of larvae at the end of June and beginning of July (Leininger et al. 2014).

Embryogenesis of calcaronean sponges is well described on light and electron microscopy levels (Amano and Hori 1992, 1993; Franzen 1988; Eerkes-Medrano and Leys 2006; reviewed in Ereskovsky 2010). Early cleavage is highly stereotypic, and up to the eight-cell stage, the embryo has a rhomboid shape with all blastomeres positioned in the plane defined by the choanocyte and pinacocyte layers (Figs. 4.32A, 4.33B, 4.35C, and 4.36A, B, G, I, N). Cytoplasmic bridges are initially maintained between

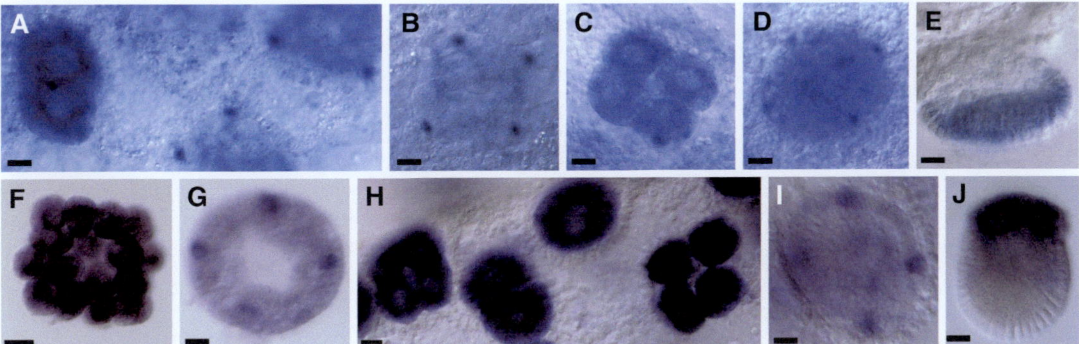

Fig. 4.36 Expression of cruciform cell markers in *Sycon ciliatum*. (**A**, **B**) *DvlB* transcripts become localised to cross cells during cleavage. (**C–E**) *DvlA* transcripts are stronger in the cross cells than in the remaining blastomeres and are then detectable in micromeres of postinversion stage embryos. (**F**, **G**) *SoxL1* transcripts and (**H**, **I**) *SoxB* transcripts are uniformly distributed during cleavage and become specific to the cross cells in preinversion stage embryos. (**J**) In the larvae, *SoxB* is detectable in the macromeres only. Scale bars: 10 µm (Modified from Leininger et al. (2014) and Fortunato et al. (2012))

blastomeres (Figs. 4.33B, C, 4.35C, and 4.36A, B). Subsequent divisions result in formation of a cup-shaped 'stomoblastula' embryo, its opening communicating with an opening formed between the accessory cells. When cell differentiation is completed, the embryo is composed of three cell types: large, granular, non-ciliated macromeres adjacent to the choanocytes; smaller and more numerous micromeres, which have cilia pointing into the embryonic cavity; and four cruciform cells, which convey a unique tetraradial symmetry to the embryos (Figs. 4.32A, 4.33D, E, and 4.36). The embryo then undergoes inversion, which will both translocate it into the radial chamber and position the cilia on the outer surface of the larva. During this stage, a small number of maternally derived cells crawl into the larval cavity (Figs. 4.32A, 4.33G, 4.35E, F, and 4.37A, I, D, E, M, O, P; Franzen 1988; Ereskovsky 2010; Leininger et al. 2014).

Larvae released in laboratory conditions swim close to the water surface during the first 12–24 h post spawning and subsequently begin to search for an appropriate substrate for settlement. During metamorphosis, the larva settles on the anterior pole; within minutes the ciliated cells of the anterior half undergo epithelial-to-mesenchymal transition and form the inner cell mass (Figs. 4.32A and 4.34A). In contrast, the macromeres maintain their epithelial organisation, completely enclose the micromeres, and become the pinacocytes of the forming juvenile.

The cross cells and the maternal cells degenerate soon after settlement (Amano and Hori 1993). Sclerocytes differentiate quickly within the inner cell mass and spicule production starts approximately 12 h after settlement. A single choanocyte chamber forms, and the postlarva expands by increasing the volume of the chamber and thinning its walls, so they are finally composed of two epithelial layers – the outer pinacoderm and the inner choanoderm, with narrow mesohyl sparsely populated with sclerocytes and other not well-characterised cell types in between (Figs. 4.32A, 4.34B–D, and 4.35H–J). Finally, the osculum opens, and the juvenile sponge acquires ascon-level organisation with porocytes providing connections (ostia) between choanoderm and pinacoderm (Figs. 4.32A, E and 4.34E). As the asconoid body plan gives rise to the syconoid body plan during subsequent growth, choanocytes of the original choanocyte chamber become replaced with endopinacocytes in the region where radial chambers form. In terms of morphology and directionality of the water flow, the radial chambers are reminiscent of the original juvenile and can be treated as serially homologous to the olynthus (Manuel 2001; Leininger et al. 2014).

Extensive gene expression analyses, based on a combination of quantitative transcriptome analysis and in situ hybridisation studies, have provided important clues regarding the homology of cell types and body plans between sponges and

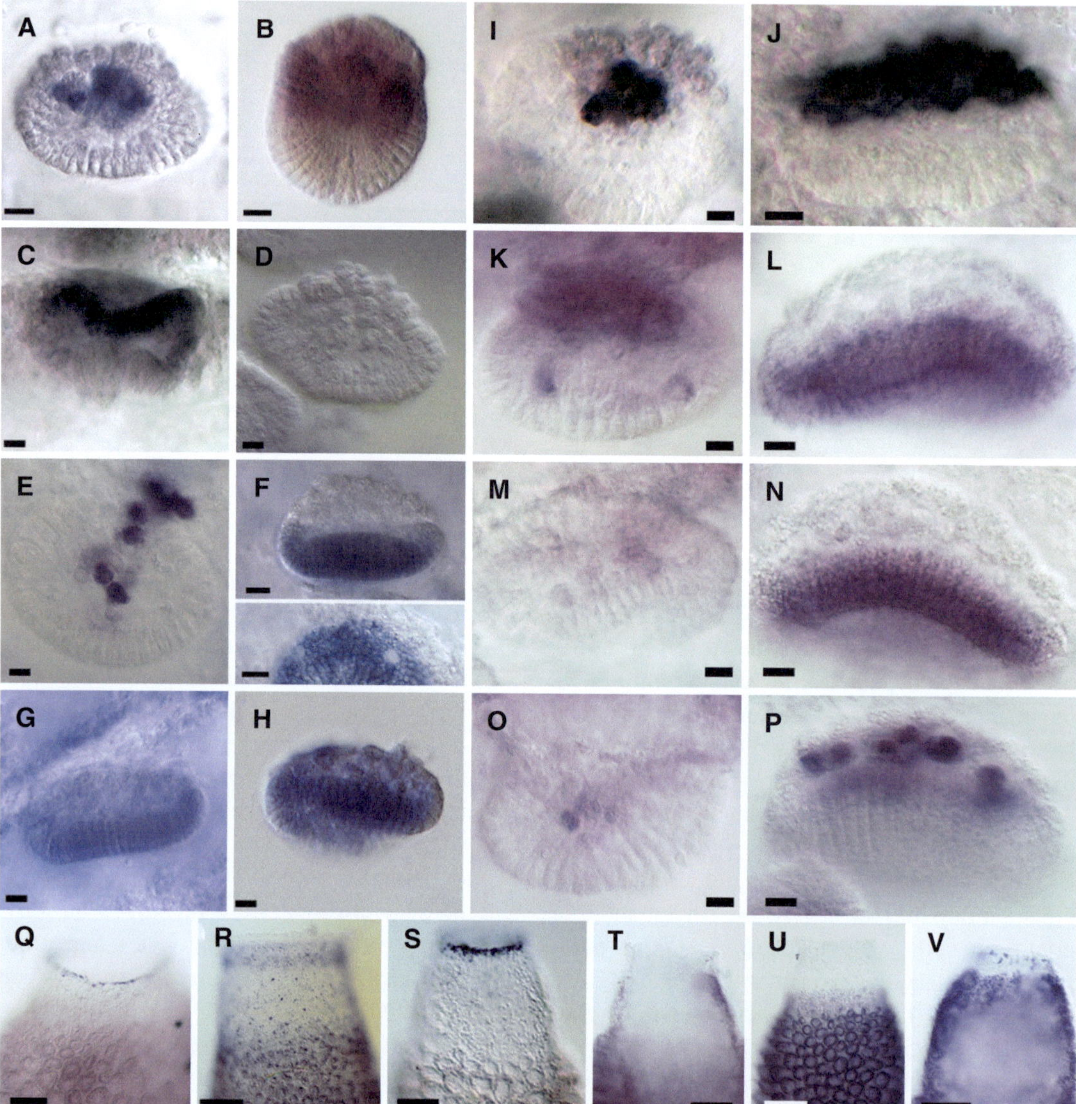

Fig. 4.37 Expression of Wnt and TGF-β pathways components in *Sycon ciliatum*.(**A**) Expression of *WntF* in the maternal cells within postinversion stage embryo. (**B**) Expression of *WntA* in macromeres and posterior micromeres of the larva. (**C, D**) Transient expression of β-*cateninA* in micromeres of preinversion stage embryos. (**E**) Expression of β-*cateninB* in maternal cells migrating into embryonic cavity. (**F, G**) Expression of *FzdA* and *FzdD* in micromeres. (**H**) Uniform embryonic expression of *TcfB*. (**I**) Strong expression of *TGF-βF* in maternal cells and its weaker expression in the macromeres. (**J**) Strong macromere expression of *TGF-βD*. (**K, L**) Expression of *Smad1/5* in macromeres plus cross cells and micromeres of early and late postinversion stage embryos, respectively. (**M, N**) Maternal cells and micromere expression of *Smad4* in early and late postinversion stage embryos, respectively. (**O, P**) Maternal cell expression of *R-Smad*. (**G–V**) *WntG, FzdD, TGF-βU, Smad4, Smad15,* and *SmadR* expression in the upper parts of young adult sponges. Scale bars: (**A–P**), 10 μm; (**Q–V**), 100 μm (Modified from Leininger et al. (2014))

eumetazoans. Several genes involved in specification of neuronal and sensory cells in cnidarians and bilaterians are expressed during differentiation of the cruciform cells, which are suggested to be the sensory cells of the calcaronean larvae (Tuzet 1973). These genes include

SoxB, *PaxB*, *SixC*, *Elav*, *Msi* and *Nanos*, *Hmx*, and other NK-related homeobox transcription factors, as well as several components of the Wnt signalling pathway (Fig. 4.36; Fortunato et al. 2012, 2014; Fortunato 2014; Leininger et al. 2014). Genes involved in specification of the cnidarian and bilaterian endomesoderm are expressed in the embryonic micromeres (which give rise to the choanoderm) and choanoderm of the adult sponges. These include downstream components of the Wnt and TGF-β signalling pathways as well as *Brachyury* and *GATA* transcription factors (Fig. 4.37; Leininger et al. 2014). Finally, numerous Wnt and TGF-β ligands are expressed in the posterior region of the larvae and around the osculum of the adults, highly reminiscent of expression patterns observed in cnidarians and supporting homology of the larval and adult body axes as postulated by Haeckel (1870) (Fig. 4.37; Leininger et al. 2014).

OPEN QUESTIONS

- Is Porifera monophyletic and the sister phylum to all other extant metazoans? Particularly intriguing is the inability to convincingly determine whether sponges or ctenophores are the earliest branching phyletic lineage. One of the standout features when comparing the *Amphimedon queenslandica* genome with ctenophore genomes is the remarkable similarity in developmental and other (e.g., neuronal) gene repertoires. Indeed, the gene content similarity between ctenophores and *Amphimedon queenslandica* might be greater than that between *A. queenslandica* and *Sycon ciliatum*.
- How does sponge embryogenesis and metamorphosis relate to hallmarks of eumetazoan and bilaterian development, including gastrulation and germ layers?
- Are the cell layers observed in sponges homologous to bilaterian germ layers, and if so, is the generative mechanism for the establishment of these layers conserved across the animal kingdom? Even amongst the co-authors of this chapter, there is no agreement.

Regardless, it is clear there exists an ancient developmental gene toolkit that is still in use in all animals. This includes conserved signalling pathways whose differential expression contributes to define body plan axes and embryonic territories (e.g., Wnt, Notch) and transcription factors whose expression correlates with the establishment of a cell layer or type (e.g., GATA). The level at which these developmental similarities are homologous to eumetazoan processes remains an open question.

Acknowledgements Because of space limitations, we are unable to cite many important contributions to the field of sponge developmental biology – we acknowledge these here. We also acknowledge the fine contributions of past and present members of the laboratories of B. Degnan, S. Degnan, and M. Adamska towards our understanding of *Amphimedon* and *Sycon* biology. Research presented in this chapter was made possible by the generous support of the Australian Research Council to BMD, SMD, and MA and the Sars International Centre for Marine Molecular Biology to MA.

References

Adamska M, Degnan SM, Green KM, Adamski M, Craigie A, Larroux C, Degnan BM (2007a) Wnt and TGF-β expression in the sponge *Amphimedon queenslandica* and the origin of metazoan embryonic patterning. PLoS ONE 2:e1031

Adamska M, Matus DQ, Adamski M, Green K, Rokhsar DS, Martindale MQ, Degnan BM (2007b) The evolutionary origin of hedgehog proteins. Curr Biol 17:R836–R837

Adamska M, Larroux C, Adamski M, Green K, Lovas E, Koop D, Richards GS, Zwafink C, Degnan BM (2010) Structure and expression of conserved Wnt pathway components in the demosponge *Amphimedon queenslandica*. Evol Dev 12:492–518

Adamska M, Zwafink C, Green K, Degnan BM (2011) What sponges can tell us about the evolution of developmental processes. Zoology 114:1–10

Amano S, Hori I (1992) Metamorphosis of calcareous sponges. 1. Ultrastructure of free-swimming larvae. Invert Reprod Dev 21:81–90

Amano S, Hori I (1993) Metamorphosis of calcareous sponges. 2. Cell rearrangement and differentiation in metamorphosis. Invert Reprod Dev 24:13–26

Amano S, Hori I (2001) Metamorphosis of coeloblastula performed by multipotential larval flagellated cells in the calcareous sponge *Leucosolenia laxa*. Biol Bull 200:20–32

Anavy L, Levin M, Khair S, Nakanishi N, Fernandez-Valverde SL, Degnan BM, Yanai I (2014) BLIND ordering of large-scale transcriptomic developmental timecourses. Development 141:1161–1166

Bergquist PR, Green CR (1977) Ultrastructural-study of settlement and metamorphosis in sponge larvae. Cahiers Biol Mar 18:289–302

Boury-Esnault N, Efremova S, Bézac C, Vacelet J (1999) Reproduction of a hexactinellid sponge: first description of gastrulation by cellular delamination in the Porifera. Invert Reprod Dev 35:187–201

Bridgham JT, Eick GN, Larroux C, Deshpande K, Harms MJ, Gauthier ME, Ortlund EA, Degnan BM, Thornton JW (2010) Protein evolution by molecular tinkering: diversification of the nuclear receptor superfamily from a ligand-dependent ancestor. PLoS Biol 8:e1000497

Degnan SM, Degnan BM (2006) The origin of the pelago-benthic metazoan life cycle: what's sex got to do with it? Integr Comp Biol 46:683–690

Degnan SM, Degnan BM (2010) The initiation of metamorphosis as an ancient polyphenic trait and its role in metazoan life cycle evolution. Phil Trans R Soc B 365:641–651

Degnan BM, Leys SP, Larroux C (2005) Sponge development and antiquity of animal pattern formation. Integr Comp Biol 45:335–341

De Vos L, Rutzler K, Boury-Esnault N, Donadey C, Vacelet J (1991) Atlas of sponge morphology. Smithsonian Institution Press, Washington D.C. 128 pp

Eerkes-Medrano DI, Leys SP (2006) Ultrastructure and embryonic development of a syconoid calcareous sponge. Invert Biol 125:177–194

Ereskovsky AV, Boury-Esnault N (2002) Cleavage pattern in Oscarella species (Porifera, Demospongiae, Homoscleromorpha): transmission of maternal cells and symbiotic bacteria. J Nat Hist 36:1761–1775

Ereskovsky A (2010) The comparative embryology of sponges. Springer, Netherlands

Ereskovsky AV, Tokina DB, Bezac C, Boury-Esnault N (2007) Metamorphosis of cinctoblastula larvae (Homoscleromorpha, Porifera). J Morphol 268:518–528

Erwin DH, Laflamme M, Tweedt SM, Sperling EA, Pisani D, Peterson KJ (2011) The Cambrian conundrum: early divergence and later ecological success in the early history of animals. Science 334:1091–1097

Fahey B (2011) Origin of animal epithelia: insights from the genome of the demosponge *Amphimedon queenslandica*. PhD Thesis The University of Queensland

Fahey B, Degnan BM (2010) Origin of animal epithelia: insights from the sponge genome. Evol Dev 12:601–617

Fahey B, Larroux C, Woodcroft BJ, Degnan BM (2008) Does the high gene density in the sponge NK homeobox gene cluster reflect limited regulatory capacity? Biol Bull 214:205–217

Fell PE (1969) The involvement of nurse cells in oogenesis and embryonic development in the marine sponge, Haliclona ecbasis. J Morph 127:133–150.

Fortunato SAV (2014) Gene loss, lineage specific expansions and dynamic expression of developmental transcription factors in calcareous sponges. PhD Thesis University of Bergen

Fortunato S, Adamski M, Bergum B, Guder C, Jordal S, Leininger S, Zwafink C, Rapp HT, Adamska M (2012) Genome-wide analysis of the sox family in the calcareous sponge *Sycon ciliatum*: multiple genes with unique expression patterns. EvoDevo 23:142012

Fortunato S, Leininger S, Adamska M (2014) Evolution of the Pax-Six-Eya-Dach network: the calcisponge case study. EvoDevo 5:23

Fortunato S, Adamski M, Mendivil O, Leininger S, Liu J, Ferrier DEK, Adamska M. Calcisponges have a ParaHox gene and dynamic expression of dispersed NK homeobox genes. Nature 514:620–623

Franzen W (1988) Oogenesis and larval development of *Scypha ciliata* (Porifera, Calcarea). Zoomorphology 107:349–357

Funayama N (2012) The stem cell system in demo-sponges: suggested involvement of two types of cells: archeocytes (active stem cells) and choanocytes (food-entrapping flagellated cells). Dev Genes Evol 223:23–38

Gauthier MEA (2010) Developing a sense of self: exploring the evolution of immune and allorecognition mechanisms in metazoans using the demosponge *Amphimedon queenslandica*. PhD Thesis The University of Queensland

Gauthier M, Degnan BM (2008) The transcription factor NF-κB in the sponge *Amphimedon queenslandica*: insights into the evolutionary origin of the Rel homology domain. Dev Genes Evol 218:23–32

Gauthier MEA, Du Pasquier L, Degnan BM (2010) The genome of the sponge *Amphimedon queenslandica* provides new perspectives into the origin of Toll-like and Interleukin1 receptor pathways. Evol Dev 12:519–533

Gazave E, Lapebie P, Ereskovsky AV, Vacelet J, Renard E, Cardenas P, Borchiellini C (2012) No longer demospongiae: homoscleromorpha formal nomination as a fourth class of Porifera. Hydrobiologia 687:3–10

Gonobobleva EL, Ereskovsky AV (2004) Metamorphosis of the larva of Halisarca dujardini (Demospongiae, Halisarcida). Bull Inst R Sci Nat.Belg 74:101–114

Haeckel E (1870). Ueber den Organismus der Schwame und ihre Verwndtschaft mit den Corallen. Jena Zeitsch Naturwiss 5:207–235

Haeckel E (1874) Die Gastrae Theorie, die phylogenetische Classification des Thierreichs und die Homologie der Keimblatter. Jena Zeitsch Naturwiss 8:1–55

Hayward DC, Samuel G, Pontynen PC, Catmull J, Saint R, Miller DJ, Ball EE (2002) Localized expression of a dpp/BMP2/4 ortholog in a coral embryo. Proc Natl Acad Sci USA 99:8106–8111

Hentschel U, Piel J, Degnan SM, Taylor MW (2012) Genomic insights into the marine sponge microbiome. Nature Rev Microbiol 10:641–654

Hill MS, Hill AL, Lopez J, Peterson KJ, Pomponi S, Diaz MC, Thacker RW, Adamska M, Boury-Esnault N, Cárdenas P, Chaves-Fonnegra A, Danka E, De Laine BO, Formica D, Hajdu E, Lobo-Hajdu G, Klontz S, Morrow CC, Patel J, Picton B, Pisani D, Pohlmann D, Redmond NE, Reed J, Richey S, Riesgo A, Rubin E, Russell Z, Rützler K, Sperling EA, di Stefano M, Tarver JE, Collins AG (2013) Reconstruction of family-level phylogenetic relationships within Demospongiae (Porifera) using nuclear encoded housekeeping genes. PLoS ONE 8:e50437

Hooper JNA, Van Soest RWM (eds) (2002) Systema Porifera, vol 1, A guide to the classification of sponges. Kluwer Academic/Plenum Publishers, New York, xlvii, 1708

Kaye HR (1990) Reproduction in West Indian commercial sponges: oogenesis, larval development, and behavior. In: new Perspectives in sponge biology. Rützler K, ed., Smithsonian Institution Press, Washington DC:161–169

Kaye HR, Reiswig HM (1991) Sexual reproduction in four Caribbean commercial sponges. II. Oogenesis and transfer of bacterial symbionts. lnvert Reprod Dev 19:1–11

Knoblich JA (2010) Asymmetric cell division: recent developments and their implications for tumour biology. Nat Rev Mol Cell Biol 11:849–860

Larroux C (2007) Genome content and developmental expression of transcription factor genes in the demosponge Amphimedon queenslandica: insights into the Ancestral Metazoan Developmental Program. PhD Thesis The University of Queensland

Larroux C, Fahey B, Liubicich D, Hinman VF, Gauthier M, Gongora M, Green K, Wörheide G, Leys SP, Degnan BM (2006) Developmental expression of transcription factor genes in a demosponge: insights into the origin of metazoan multicellularity. Evol Dev 8:150–173

Larroux C, Fahey B, Degnan SM, Adamski M, Rokhsar DS, Degnan BM (2007) The NK homeobox gene cluster predates the origin of Hox genes. Curr Biol 17:706–710

Leininger S, Adamski M, Bergum B, Guder C, Liu J, Laplante M, Bråte J, Hoffmann F, Fortunato S, Jordal S, Rapp HT, Adamska M (2014) Developmental gene expression provides clues to relationships between sponge and eumetazoan body plans. Nat Commun 5:3905

Leys SP (2004) Gastrulation in sponges. In Gastrulation, From Cells to Embryo. Edited by Stern CD. New York: cold Spring Harbor Laboratory Press:23–31

Leys SP, Degnan BM (2001) The cytological basis of photoresponsive behavior in a sponge larva. Biol Bull 201:323–338

Leys SP, Degnan BM (2002) Embryogenesis and metamorphosis in a haplosclerid demosponge: gastrulation and transdifferentiation of larval ciliated cells to choanocytes. Invert Biol 121:171–189

Leys SP, Eerkes-Medrano D (2005) Gastrulation in calcareous sponges: in search of Haeckel's gastraea. Integ Comp Biol 45:342–351

Leys SP, Ereskovsky AV (2006) Embryogenesis and larval differentiation in sponges. Can. J. Zool. 84:262–287

Leys SP, Hill A (2012) The physiology and molecular biology of sponge tissues. Adv Mar Biol 62:1–56

Leys SP, Cronin TW, Degnan BM, Marshall JN (2002) Spectral sensitivity in a sponge larva. J Comp Physiol A 188:199–202

Maldonado M, Bergquist PR (2002) Phylum Porifera. In: atlas of marine invertebrate larvae. Young CM, ed, Academic press, London:21–50

Manuel M (2001) Origine et évolution des mécanismes moléculaires contrôlant la morphogenèse chez les Métazoaires: un nouveau modèle spongiaire, Sycon raphanus (Calcispongia, Calcaronea). PhD Thesis Université de Paris XI

Maritz K, Calcino A, Fahey B, Degnan BM, Degnan SM (2010) Remarkable consistency of larval supply in the spermcast-mating demosponge Amphimedon queenslandica (Hooper and van Soest). Open Mar Biol J 4:57–64

Matus DQ, Pang K, Marlow H, Dunn CW, Thomsen GH, Martindale MQ (2006) Molecular evidence for deep evolutionary roots of bilaterality in animal development. Proc Natl Acad Sci USA 103:11195–11200

McFall-Ngai M, Hadfield MG, Bosch TC, Carey HV, Domazet-Lošo T, Douglas AE, Dubilier N, Eberl G, Fukami T, Gilbert SF, Hentschel U, King N, Kjelleberg S, Knoll AH, Kremer N, Mazmanian SK, Metcalf JL, Nealson K, Pierce NE, Rawls JF, Reid A, Ruby EG, Rumpho M, Sanders JG, Tautz D, Wernegreen JJ (2013) Animals in a bacterial world, a new imperative for the life sciences. Proc Natl Acad Sci U S A 110:3229–3236

Medioni C, Mowry K, Besse F (2012) Principles and roles of mRNA localization in animal development. Development 139:3263–3276

Meewis H (1939) Contribution a l'étude de l'embryogenése de Chalinulidae: haliclona limbata. Ann Soc R Zool Belg 70:201–243

Misevic GN SV, Burger MM (1990) Larval metamorphosis of Microciona prolifera: evidence against the reversal of layers. In: Rt K (ed) New perspectives in sponge biology. Smithsonian Institution Press, Washington, DC, pp 182–187

Misevic GN, Burger MM (1982) The molecular basis of species specific cell-cell recognition in marine sponges, and a study on organogenesis during metamorphosis. Prog Clin Biol Res B 85:193–209

Moroz LL, Kocot KM, Citarella MR, Dosung S, Norekian TP, Povolotskaya IS, Grigorenko AP, Dailey C, Berezikov E, Buckley KM, Ptitsyn A, Reshetov D, Mukherjee K, Moroz TP, Bobkova Y, Yu F, Kapitonov VV, Jurka J, Bobkov YV, Swore JJ, Girardo DO, Fodor A, Gusev F, Sanford R, Bruders R, Kittler E, Mills CE, Rast JP, Derelle R, Solovyev VV, Kondrashov FA, Swalla BJ, Sweedler JV, Rogaev EI,

Halanych KM, Kohn AB (2014) The ctenophore genome and the evolutionary origins of neural systems. Nature 510:109–114

Nakanishi N, Sogabe S, Degnan BM (2014) Evolutionary origin of gastrulation: insights from sponge development. BMC Biol 12:26

Nosenko T, Schreiber F, Adamska M, Adamski M, Eitel M, Hammel J, Maldonado M, Müller WE, Nickel M, Schierwater B, Vacelet J, Wiens M, Wörheide G (2013) Deep metazoan phylogeny: when different genes tell different stories. Mol Phylogenet Evol 67:223–233

Philippe H, Derelle R, Lopez P, Pick K, Borchiellini C, Boury-Esnault N, Vacelet J, Renard E, Houliston E, Quéinnec E, Da Silva C, Wincker P, Le Guyader H, Leys S, Jackson DJ, Schreiber F, Erpenbeck D, Morgenstern B, Wörheide G, Manuel M (2009) Phylogenomics revives traditional views on deep animal relationships. Curr Biol 19:706–712

Redmond NE, Morrow CC, Thacker RW, Diaz MC, Boury-Esnault N, Cárdenas P, Hajdu E, Lôbo-Hajdu G, Picton BE, Pomponi SA, Kayal E, Collins AG (2013) Phylogeny and systematics of Demospongiae in light of new small-subunit ribosomal DNA (18S) sequences. Integ Comp Biol 53:388–415

Richards GS (2010) The origins of cell communication in the animal kingdom: notch signalling during embryogenesis and metamorphosis of the demosponge *Amphimedon queenslandica*. PhD Thesis The University of Queensland

Richards GS, Degnan BM (2012) The expression of Delta ligands in the sponge *Amphimedon queenslandica* suggests an ancient role for Notch signaling in metazoan development. EvoDevo 3:e15

Richards GS, Simionato E, Perrron M, Adamska M, Vervoort M, Degnan BM (2008) Sponge genes provide new insight into the evolutionary origin of the neurogenic circuit. Curr Biol 18:1156–1161

Rivera AS, Ozturk N, Fahey B, Plachetzki DC, Degnan BM, Sancar A, Oakley TH (2012) Blue-light-receptive cryptochrome is expressed in a sponge eye lacking neurons and opsin. J Exp Biol 215:1278–1286

Robinson JM, Sperling EA, Bergum B, Adamski M, Nichols SA, Adamska M, Peterson KJ (2013) The identification of microRNAs in calcisponges: independent evolution of microRNAs in basal metazoans. J Exp Zool B Mol Dev Evol 320:84–93

Ryan JF, Pang K, Schnitzler CE, Nguyen AD, Moreland RT, Simmons DK, Koch BJ, Francis WR, Havlak P, NISC Comparative Sequencing Program, Smith SA, Putnam NH, Haddock SH, Dunn CW, Wolfsberg TG, Mullikin JC, Martindale MQ, Baxevanis AD (2013) The genome of the ctenophore *Mnemiopsis leidyi* and its implications for cell type evolution. Science 342:1242592

Sakarya O, Armstrong KA, Adamska M, Adamski M, Wang IF, Tidor B, Degnan BM, Oakley TH, Kosik KS (2007) A post-synaptic scaffold at the origin of the animal kingdom. PLoS ONE 2:e506

Saller U, Weissenfels N (1985) The development of Spongilla lacustris from the oocyte to the free larva (Porifera, Spongillidae). Zoomorph 105:252–277

Schierwater B, Eitel M, Jakob W (2009) Concatenated analysis sheds light on early metazoan evolution and fuels a modern "urmetazoon" hypothesis. PLoS Biol 7:e20

Sebé-Pedrós A, Ariza-Cosano A, Weirauch MT, Leininger S, Yang A, Torruella G, Adamski M, Adamska M, Hughes TR, Gómez-Skarmeta JL, Ruiz-Trillo I (2013) Early evolution of the T-box transcription factor family. Proc Natl Acad Sci U S A 110: 16050–16055

Simpson TL (1984) The cell biology of sponges. Springer, New York

Sperling EA, Peterson KJ, Pisani D (2009) Phylogenetic-signal dissection of nuclear housekeeping genes supports the paraphyly of sponges and the monophyly of Eumetazoa. Mol Biol Evol 26:2261–2274

Srivastava M, Simakov O, Chapman J, Fahey B, Gauthier ME, Mitros T, Richards GS, Conaco C, Dacre M, Hellsten U, Larroux C, Putnam NH, Stanke M, Adamska M, Darling A, Degnan SM, Oakley TH, Plachetzki DC, Zhai Y, Adamski M, Calcino A, Cummins SF, Goodstein DM, Harris C, Jackson DJ, Leys SP, Shu S, Woodcroft BJ, Vervoort M, Kosik KS, Manning G, Degnan BM, Rokhsar DS (2010a) The *Amphimedon queenslandica* genome and the evolution of animal complexity. Nature 466:720–726

Srivastava M, Larroux C, Lu DR, Mohanty K, Chapman J, Degnan BM, Rokhsar DS (2010b) Early evolution of the LIM homeobox gene family. BMC Biol 8:e4

Steinmetz PRH, Kraus JEM, Larroux C, Hammel JU, Amon-Hassenzahl A, Houliston E, Wörheide G, Nickel M, Degnan BM, Technau U (2012) Independent evolution of striated muscles in cnidarians and bilaterians. Nature 487:231–234

Thacker RW, Freeman CJ (2012) Sponge-microbe symbioses: recent advances and new directions. Adv Mar Biol 62:57–111

Thacker RW, Hill AL, Hill MS, Redmond NE, Collins AG, Morrow CC, Spicer L, Carmack CA, Zappe ME, Pohlmann D, Hall C, Diaz MC, Bangalore PV (2013) Nearly complete 28S rRNA gene sequences confirm new hypotheses of sponge evolution. Integ Comp Biol 53:373–387

Tuzet O (1973) Éponges calcaires. In: Grassé P-P (ed) Traité de Zoologie Anatomie, Systématique, Biologie Spongiaires, vol 3. Masson et Cie, Paris, pp 27–132

Worheide G, Dohrmann M, Erpenbeck D, Larroux C, Maldonado M, Voigt O, Borchiellini C, Lavrov DV (2012) Deep phylogeny and evolution of sponges (phylum Porifera). Adv Mar Biol 61:1–78

Zakrzewski A-C, Weigert A, Helm C, Adamski M, Adamska M, Bleidorn C, Raible F, Hausen H (2014) Early divergence, broad distribution and high diversity of animal chitin synthases. Genome Biol Evol 6:316–325

Placozoa

Bernd Schierwater and Michael Eitel

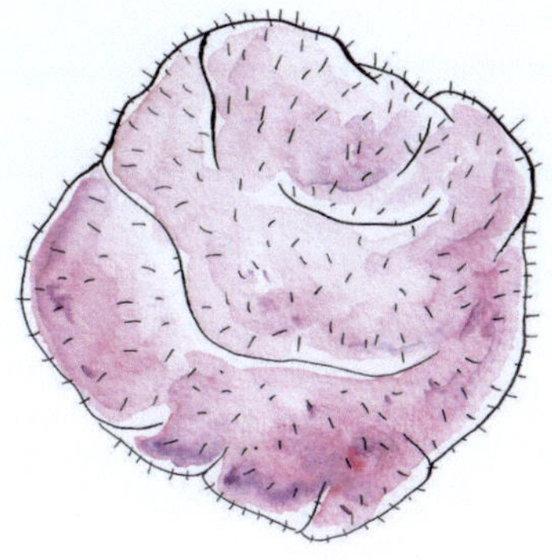

Chapter vignette artwork by Brigitte Baldrian.
© Brigitte Baldrian and Andreas Wanninger.

B. Schierwater (✉) • M. Eitel
ITZ – Institut für Tierökologie und Zellbiologie,
Stiftung Tierärztliche Hochschule,
Bünteweg 17d, Hannover 30559, Germany
e-mail: bernd.schierwater@ecolevol.de

A. Wanninger (ed.), *Evolutionary Developmental Biology of Invertebrates 1:*
Introduction, Non-Bilateria, Acoelomorpha, Xenoturbellida, Chaetognatha
DOI 10.1007/978-3-7091-1862-7_5, © Springer-Verlag Wien 2015

INTRODUCTION

The most primitive metazoan animal phylum Placozoa presently harbors a single named species, the enigmatic *Trichoplax adhaerens*. In 1883, the German zoologist Franz Eilhard Schulze discovered this microscopic marine animal on the glass walls of a seawater aquarium at the University of Graz, Austria (Schulze 1883). The animal, usually measuring less than 5 mm in diameter and less than 20 μm in thickness, looked like an irregular hairy plate sticking to the glass surface (Fig. 5.1) and was thus named *Trichoplax adhaerens* (Greek for "sticky hairy plate") (see Schierwater 2005 for historical overview). Recent genetic analysis of placozoan specimens from different ocean waters around the world, including the Mediterranean Sea, revealed the presence of several cryptic species (Eitel et al. 2013), i.e., species, which are morphologically *cum grano salis* undistinguishable. The real placozoan biodiversity is estimated to include several dozen genetically, developmentally, and ecologically distinguishable species.

In contrast to the "typical" multicellular animal, *Trichoplax* does not show anything like an oral-aboral axis, nor does the animal possess any organs, nerve or muscle cells, basal lamina, or extracellular matrix (Schierwater 2013). Because of the lack of any axis, placozoans also lack any type of symmetry. The defining characteristic that separates placozoans (and any other metazoan) from protozoans is the number of somatic cell types. In contrast to protozoans, which consist of either a single cell or several cells of the same somatic cell type, Placozoa possess at least five defined somatic cell types: lower epithelia cells, upper epithelia cells, gland cells, fiber cells, and small potentially "omnipotent" cells (Jakob et al. 2004; Guidi et al. 2011). The epithelia cells are arranged in a sandwich-like manner, with the lower epithelia and gland cells at the bottom, the upper epithelia cells at the top, the fiber cells in between, and the "omnipotent" cells at the margin between the upper and lower epithelium (Fig. 5.2). Cells of the lower epithelium attach the animal to a solid substrate, enable the animal to crawl (with the aid of cilia) and allow

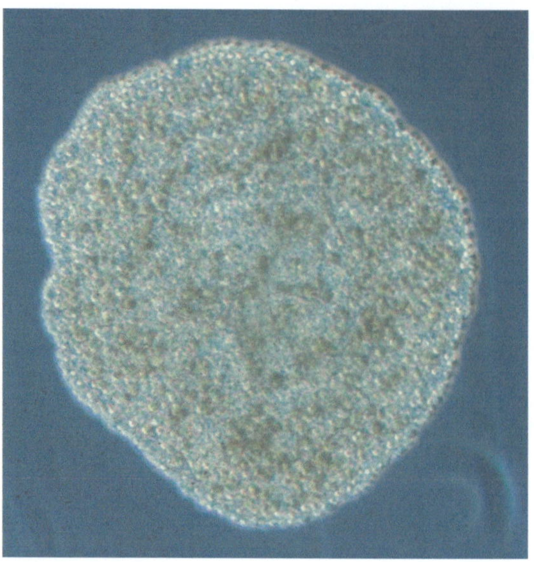

Fig. 5.1 The placozoan *Trichoplax adhaerens*, yet the only described species of the phylum Placozoa. The flat animal, which represents the most simple (not secondarily reduced) body plan of all metazoans, is found in tropical, subtropical, and certain temperate waters around the world ((Photograph by Bernd Schierwater) © Bernd Schierwater All Rights Reserved)

feeding. During feeding, the animal lifts up the center region of its body to form an external digestive cavity between the substrate and lower epithelium (see Schierwater 2013 for details). Interestingly, the upper epithelium is also capable of feeding. Algae and other food particles are trapped in a slime layer coating the upper epithelium and are subsequently taken up (phagocytized) by the inner fiber cells; this unique mode of feeding is called "transepithelial cytophagy" (Wenderoth 1986). Placozoa presumably harbor endosymbiotic bacteria in the endoplasmic reticulum of the fiber cells (Grell and Benwitz 1971; Eitel et al. 2011). A possible role for these endosymbionts in feeding is not yet understood.

In general, very little is known about the biology of Placozoa, and almost all current knowledge derives from laboratory observations. Ecological data are limited to records of finding *Trichoplax* on hard substrate surfaces from tropical and subtropical marine waters around the world (Eitel and Schierwater 2010; Eitel et al. 2013).

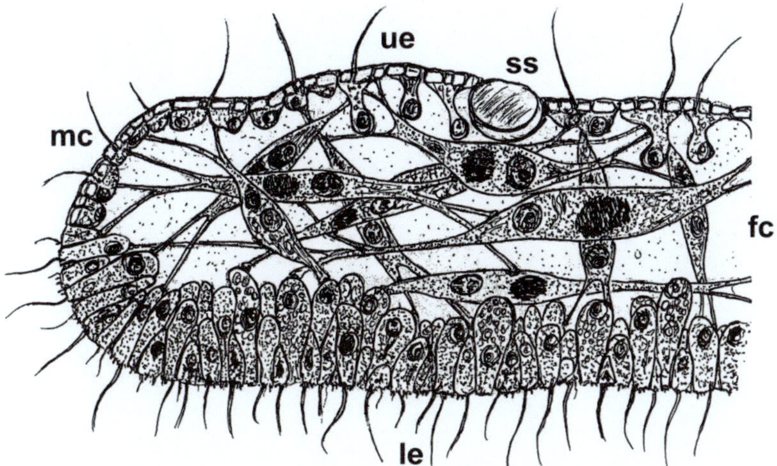

Fig. 5.2 Cross section of *Trichoplax adhaerens* showing the sandwich-like organization of the body plan: the ciliated upper epithelium, the ciliated lower epithelium, and the fiber cells in between. The interior fiber cells form a contractile, three-dimensional meshwork. The lower epithelium serves as the nutritive body region with gland cells incorporated into the epithelium. The upper epithelium shows no specializations, with the exception of the shiny spheres (originally named "Glanzkugeln"). These are lipid droplets, which are usually understood as residues of degenerated epithelial cells. A remarkable and exclusive feature of Placozoa is the lack of both an extracellular matrix and a basal lamina. *mc* marginal cells, *ue* upper epithelium, *le* lower epithelium, *fc* contractile fiber cell, *ss* shiny sphere (Modified from Eitel et al. (2011))

After its original description in 1883, *Trichoplax* attracted particular attention because it possibly mirrored the basic and ancestral state of metazoan organization. Almost a hundred years later, the German zoologist Karl Grell further highlighted this view and created the new, and monotypic, phylum Placozoa (Grell 1971). The phylum name refers to Bütschli's placula hypothesis (Bütschli 1884), which sees a placozoan-like animal as the Urmetazoon (Schierwater et al. 2010). Although a variety of molecular data support the traditional view that Placozoa are closest to the very root of metazoan origin, quantitative molecular systematics overall has created more confusion than resolution yet. We believe that eventually Haeckel's biogenetic rule (c.f. Schierwater et al. 2010) will provide the answer we are looking for.

For a comprehensive comparative analysis of the development of all non-bilaterian and some basal bilaterian phyla, however, the recent bottleneck is the Placozoa. A lot more work on the development of placozoans is needed. Given the importance of this enigmatic phylum Placozoa, any investment in developmental research seems to be justified.

SEXUAL REPRODUCTION

Despite more than half a century of research efforts, the complete life cycle of Placozoa can only be suspected. Most likely, the adult placozoon – after a series of vegetative reproductions – becomes sexually mature either as protandrous or simultaneous hermaphrodite. If the sperm is released into the open water, outcrossing might be possible, but in most cases selfing may occur. Sexual reproduction has been studied in the laboratory using different placozoan species. Experiments have identified specific environmental conditions that are required for the generation and maturation of oocytes. These include high animal density, food scarceness, and temperatures above 23 °C (see Eitel et al. 2011).

Oogenesis

Female gametocytes (oocytes) are presumably produced in the lower epithelium (Grell and Benwitz 1974), while maturation and fertilization occur somewhere in the center of the body (Fig. 5.3). During oocyte maturation, the mother

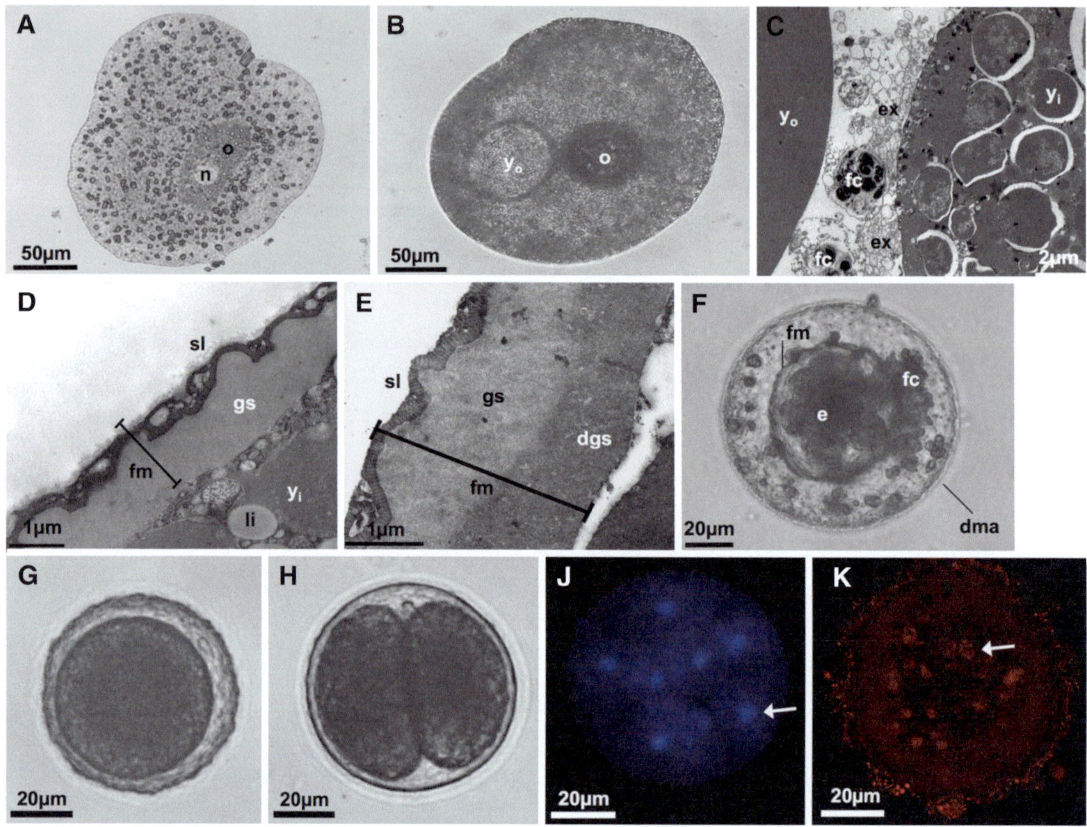

Fig. 5.3 Oogenesis and early embryonic development in the undescribed Placozoa sp. H2. Shown are light microscopy (**A**, **B**, **F–H**), transmission electron microscopy (**C–E**), and fluorescence microscopy (**J**, **K**) images of oocytes and embryos. (**A**) An oocyte with a large nucleus grows in a flat animal without any signs of degeneration. (**B**) Accompanied by the generation of yolk droplets, the animal enters the degeneration phase after 5–6 weeks. (**C**) By incorporating extensions from fiber cells through pores, the oocyte grows. (**D**) After fertilization, the "fertilization membrane" is built around the zygote. This protective egg shell drives from the fusion of cortex granules in the margin of the oocyte. It has a characteristic two-layered appearance in early stages. (**E**) Later embryo with three-layered egg shell. (**F**) By lifting the upper and condensing the lower epithelium, the mother animal rounds up and builds a "brood chamber" for the embryo. (**G–K**) The first and all later cleavages are total-equal. Fluorescence microscopy (DAPI in **J** and propidium iodide in **K**) is used to count nuclei and chromosomes in embryos (*arrows* in **J** and **K**, respectively). *n* nucleus, *o* oocyte, *yo* yolk outside oocyte, *yi* yolk inside oocyte, *fc* fiber cells, *ex* fiber cell extensions, *fm* fertilization membrane, *sl* striped layer, *gs* ground substance, *dgs* dense ground substance, *li* lipid droplet, *e* embryo, *dma* degenerating mother animal (Figure modified after Eitel et al. (2011))

animal enters the so-called D phase (degeneration phase) in which the upper epithelium lifts and the lower epithelium condenses. A key feature of this phase is the generation of yolk droplets, not only inside but also outside the oocyte. The outside yolk droplets accumulate to form a large cluster, possibly an energy source for the growing embryo while still inside the mother organism. To grow, the oocyte incorporates extensions from nursing fiber cells through pores on its surface (Grell and Benwitz 1974, 1981; Eitel et al. 2011). In the maturation process, specific granules are formed throughout the oocyte. These structures of unknown material look strikingly similar to – and likely resemble – cortical granules known from other marine invertebrates. During maturation, these granules increase in number and are transported toward the oocyte's surface. In addition to yolk droplets and cortical granules, the oocyte stores lipid droplets and glycogen granules. A fully mature oocyte reaches a variable size of 70–120 μm, depending on the size of the mother animal and the number of oocytes that are built.

From electron and fluorescence microscopy, it is known that a vast amount of bacteria are transferred into the oocyte during maturation. These are vertically transmitted from the nursing fiber cells (Grell and Benwitz 1974, 1981; Eitel et al. 2011).

Spermatogenesis

The existence of male gametocytes (sperm) was claimed based on ultrastructural observations (Grell and Benwitz 1974; Eitel et al. 2011), but their functionality has not been confirmed yet. Sperm is probably produced in the center of the animal, but the exact location and the progenitor cells are unknown. The expression of sperm-associated marker genes strongly suggests spermatogenesis and sperm maturation in placozoans. According to transcriptome analyses of three placozoan species, the potential sperm markers cover various stages of spermatogenesis, ranging from early meiosis to sperm maturation (Eitel et al. 2011). Even markers known to encode proteins for functional sperm flagella and sperm-oocyte recognition proteins used in fertilization were identified. Sperm markers were found expressed in adult, healthy growing animals that did not show any sign of degradation. This indicates production and storage of sperm before the animal experiences unfavorable conditions. It is thus likely that only oocytes are produced at that stage and that placozoans are possibly protandrous hermaphrodites.

Early Embryonic Development

Mature oocytes are fertilized internally. Subsequently, the "fertilization membrane" is built by fusion of accumulated granulae on the oocyte's surface. The "fertilization membrane" serves as a protective egg shell and resembles the cortex of other invertebrates (Grell and Benwitz 1974; Eitel et al. 2011). Early embryos grow inside the mother animal until the latter completely degenerates and releases the embryo. Despite great investigator efforts, the embryonic development was never completed under laboratory conditions. Here, embryos do not develop past the 128- to 256-cell stage. The reasons are unknown, but it must be speculated that the water chemistry (including microflora) under laboratory conditions does not meet the specific requirements for the developing embryos. What we do know from the early development is that cleavage is total and equal from the zygote to the 128-cell stage.

VEGETATIVE REPRODUCTION

Besides the sexual reproduction outlined above, two modes of vegetative reproduction are known: (1) fission (normal type of vegetative reproduction in Placozoa) and (2) swarmer formation (occasional type of vegetative reproduction). In fission, animals grow to a certain size and divide into two approximately equally sized daughter individuals, which then regrow to "normal" size. This mode of vegetative reproduction can go on *ad infinitum*, and it is conceivable that there might be placozoan species and populations out there that only reproduce vegetatively. The second mode of vegetative reproduction has only been observed in the laboratory when environmental conditions become unfavorable. Under such conditions, placozoans may develop small spherical swarmers, which are planktonic (free-floating) and thus are taken by water currents to new habitats. Several different swarmer types have been characterized among which the "hollow spheres" have been shown to open at one point to create flattened animals (Thiemann and Ruthmann 1988, 1989). These spheres are made up of an outer layer of upper epithelial cells, a fiber cell layer, and an inner layer of lower epithelial cells. After opening, the spheres will fully rebuild the normal adult habitus within a day.

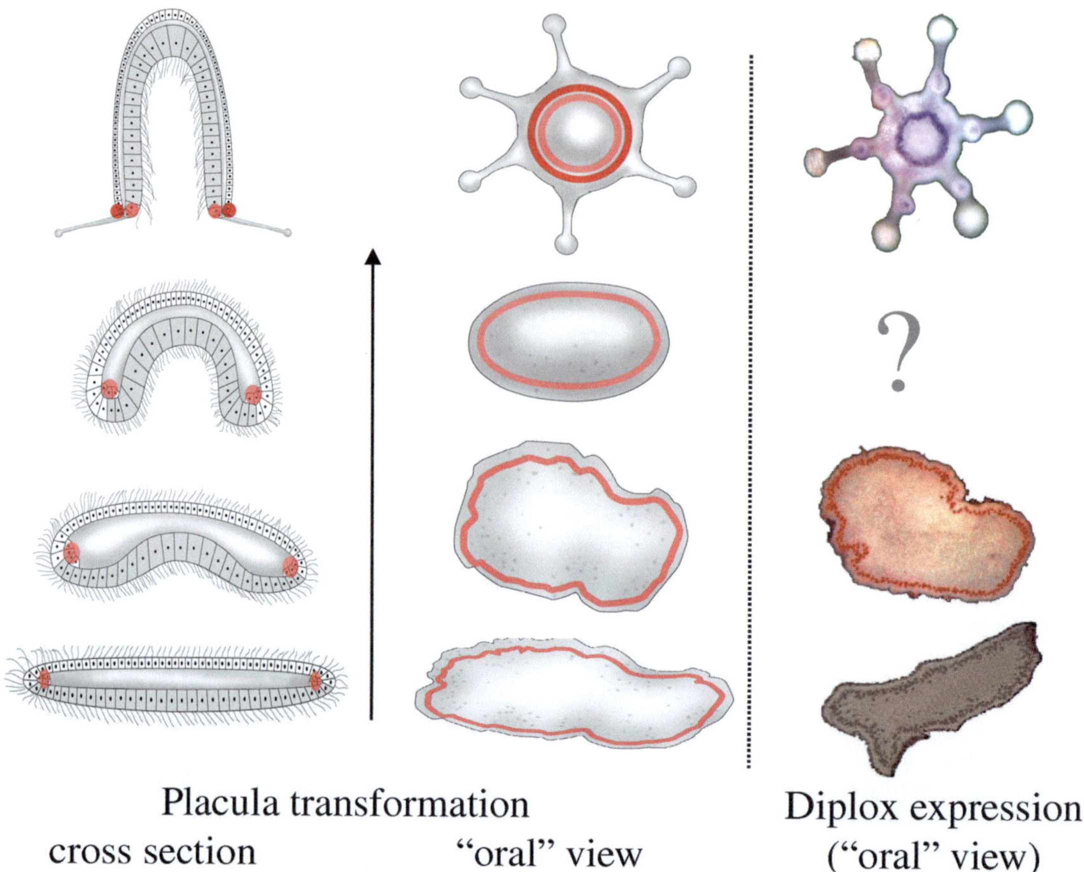

Placula transformation
cross section   "oral" view

Diplox expression
("oral" view)

Fig. 5.4 The "new Placula hypothesis of metazoan origin" (Schierwater et al. 2009b). A nonsymmetric and axis-lacking body plan (placula) transforms into a typical symmetric metazoan body plan with a defined oral-aboral (or anterior-posterior) body axis (indicated by the *arrow*). This original idea from Bütschli (1884) has recently been complemented by expression patterns of the putative Proto-Hox/ParaHox gene, *Trox-2* (in *red*). A single regulatory gene, like *Trox-2*, can control the separation between lower and upper epithelium (*three lower rows*), i.e., create polarity as a precursor of symmetry. Once a main body axis, like oral-aboral (e.g., in cnidarians) has developed, duplication of the Proto-Hox/ParaHox gene could aid the invention and organization of new head structures originating from the ectoderm-endoderm boundary of the oral pole (*upper row*). Indeed, two putative descendants of the *Trox-2* gene, *Cnox-1*, and *Cnox-3* show these hypothesized expression patterns in the hydrozoan cnidarian *Eleutheria dichotoma* (*upper row*, for simplicity, only the ring for *Cnox-1* expression is shown) (See Schierwater et al. (2009b) for details © Bernd Schierwater. (All Rights Reserved)

DEVELOPMENTAL GENES

The study of developmental genes in placozoans is as exciting as explosive, since the interpretation of any kind of gene expression data depends on the evolutionary perspective one looks from. Those scientists believing in so-called quality data (cf. Osigus et al. 2013), which put placozoans in an ancestral position to other diploblastic animals, see a clear picture for, e.g., ancestral Hox-like genes in placozoans. Here, Hox-like genes are responsible for coordinating an ancestral symmetry pattern which is called "polarity". The putative Proto-Hox/ParaHox gene, *Trox-2*, determines the setup of polarity in *Trichoplax* (Jakob et al. 2004). Since polarity is the first step for creating symmetry, the "new placula hypothesis" (Fig. 5.4; Schierwater et al. 2009b) derives as naturally as a baby's smile. The possibly ancestral Hox-like gene fulfills a logical and predicted ancestral function, and animals, which give up symmetry and strict polarity (like sponges), do no longer need Hox-like genes. Scientists believing in one of the several contradictive evolutionary scenarios arising from so-called quantity data (Osigus et al. 2013) may have a completely different view. They may see placozoans somewhere in a derived position in the tree of life and may interpret the *Trox-2* expression patterns and Hox-like gene presence in any evolutionary and less parsimonious way they want. Nothing to blame, this is scientific freedom.

When talking about developmental genes, one automatically comes to signaling pathways. In placozoans, representatives of all important components are present for the *BMB/TGF* beta, *Wnt*, and *Notch* signaling pathways. In the Hedgehog pathway, however, only *Fused* was found in the genome (Srivastava et al. 2008; Schierwater et al. 2009a). We do not know anything about the expression of these pathway genes neither in the embryo nor as maternal factors in the oocyte. Given that basically all major developmental gene families are present in placozoans, it seems disappointing how few expression and functional data we have here. Clearly, much more work is needed to obtain comparative developmental genetic data from placozoans. These data will not solve but contribute to the debate of ancestral versus derived developmental patterns to be found in placozoans.

OPEN QUESTIONS

- Type of hermaphroditism
- Selfing versus outcrossing
- Completion of the life cycle
- Function and expression of developmental genes

References

Bütschli O (1884) Bemerkungen zur Gastraea-Theorie. Morphologisches Jahrblatt 9:415–427

Eitel M, Schierwater B (2010) The phylogeography of the placozoa suggests a taxon rich phylum in tropical and subtropical waters. Mol Ecol 19:2315–2327

Eitel M, Guidi L, Hadrys H, Balsamo M, Schierwater B (2011) New insights into placozoan sexual reproduction and development. PLoS One 6:e19639

Eitel M, Osigus HJ, DeSalle R, Schierwater B (2013) Global diversity of the placozoa. PLoS One 8:e57131

Grell KG (1971) *Trichoplax adhaerens* F.E. Schulze und die Entstehung der Metazoan. Naturwissenschaftliche Rundschau 24:160–161

Grell KG, Benwitz G (1971) Die Ultrastruktur von *Trichoplax adhaerens* F.E. Schulze. Cytobiologie 4:216–240

Grell KG, Benwitz G (1974) Elektronenmikroskopische Beobachtungen über das Wachstum der Eizelle und die Bildung der "Befruchtungsmembran" von *Trichoplax adhaerens* F.E. Schulze (placozoa). Z für Morphol Tiere 79:295–310

Grell KG, Benwitz G (1981) Ergänzende Untersuchungen zur Ultrastruktur von *Trichoplax adhaerens* F.E. Schulze (placozoa). Zoomorphology 98:47–67

Grell KG, Ruthmann A (1991) Placozoa. In: Harrison FW, Westfall JA (eds) Microscopic anatomy of invertebrates, placozoa, porifera, cnidaria, and ctenophora. Wiley-Liss, New York, pp 13–28

Guidi L, Eitel M, Cesarini E, Schierwater B, Balsamo M (2011) Ultrastructural analyses support different morphological lineages in the placozoa, Grell 1971. J Morphol 272:371–378

Jakob W, Sagasser S, Dellaporta S, Holland P, Kuhn K, Schierwater B (2004) The *Trox-2* Hox/ParaHox gene of *Trichoplax* (placozoa) marks an epithelial boundary. Dev Genes Evol 214:170–175

Osigus HJ, Eitel M, Schierwater B (2013) Chasing the urmetazoon: striking a blow for quality data? Mol Phylogenet Evol 66:551–557

Schierwater B (2005) My favorite animal, *Trichoplax adhaerens*. Bioessays 27:1294–1302

Schierwater B (2013) Placozoa, Plattentiere. In: Westheide W, Rieger R (eds) Spezielle Zoologie, Teil 1: einzeller und Wirbellose Tiere, 3rd edn. Springer-Spektrum, Berlin, pp 103–107

Schierwater B, de Jong D, DeSalle R (2009a) Placozoa and the evolution of Metazoa and intrasomatic cell differentiation. Int J Biochem Cell Biol 41: 370–379

Schierwater B, Eitel M, Jakob W, Osigus HJ, Hadrys H, Dellaporta SL, Kolokotronis SO, DeSalle R (2009b) Concatenated analysis sheds light on early metazoan evolution and fuels a modern "Urmetazoon" hypothesis. Plos Biol 7:36–44

Schierwater B, Eitel M, Osigus HJ, von der Chevallerie K, Bergmann T, Hadrys H, Cramm M, Heck, L, MR L, DeSalle R (2010) *Trichoplax* and placozoa: one of the crucial keys to understanding metazoan evolution. In: DeSalle R, Schierwater B (eds) Key transitions in animal evolution. CRC Press, Enfield, USA. pp 289–326

Schulze FE (1883) *Trichoplax adhaerens*, nov. gen., nov. spec. Zool Anz 6:92–97

Srivastava M, Begovic E, Chapman J, Putnam NH, Hellsten U, Kawashima T, Kuo A, Mitros T, Salamov A, Carpenter ML, Signorovitch AY, Moreno MA, Kamm K, Grimwood J, Schmutz J, Shapiro H,

Grigoriev IV, Buss LW, Schierwater B, Dellaporta SL, Rokhsar DS (2008) The *Trichoplax* genome and the nature of placozoans. Nature 454:955–960

Thiemann M, Ruthmann A (1988) *Trichoplax adhaerens* Schulze, F. E. (placozoa) – the formation of swarmers. Z Naturforsch C 43:955–957

Thiemann M, Ruthmann A (1989) Microfilaments and microtubules in isolated fiber cells of *Trichoplax adhaerens* (placozoa). Zoomorphology 109:89–96

Wenderoth H (1986) Transepithelial cytophagy by *Trichoplax adhaerens* F.E. Schulze (placozoa) feeding on yeast. Z Naturforsch C 41:343–347

Cnidaria

Ulrich Technau, Grigory Genikhovich, and Johanna E.M. Kraus

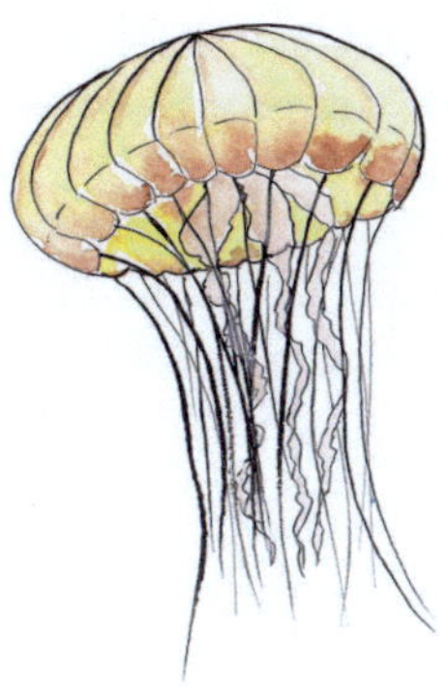

Chapter vignette artwork by Brigitte Baldrian.
© Brigitte Baldrian and Andreas Wanninger.

Although recognized as a subtaxon of Cnidaria by most recent phylogenetic analyses, the Myxozoa are covered separately in the following chapter of this volume.

U. Technau (✉) • G. Genikhovich • J.E.M. Kraus
Department of Molecular Evolution and Development, Faculty of Life Sciences, University of Vienna, Althanstrasse 14, Wien 1090, Austria
e-mail: ulrich.technau@univie.ac.at

A. Wanninger (ed.), *Evolutionary Developmental Biology of Invertebrates 1: Introduction, Non-Bilateria, Acoelomorpha, Xenoturbellida, Chaetognatha*
DOI 10.1007/978-3-7091-1862-7_6, © Springer-Verlag Wien 2015

INTRODUCTION

Cnidaria is a large animal phylum comprising around 10,000 species, most of which are marine, with few species that have adapted to freshwater environments. Molecular phylogenies place the Cnidaria as a sister group to the Bilateria. It is less clear whether they share this position with the Ctenophora (reviving the Coelenterata) and the Porifera (Philippe et al. 2009; Pick et al. 2010; Ryan et al. 2013). The position of the ctenophores appears most contentious; however, it seems rather unlikely that the ctenophores and cnidarians are closely related and therefore the concept of the Coelenterata is not well supported (Ryan et al. 2013). Regardless, the sister group relationship between Cnidaria and Bilateria is very robust and puts this phylum in a strategic position for the understanding of the evolution of key bilaterian features, such as the third germ layer (the mesoderm), the central nervous system, and bilaterality.

Cnidarians are divided into two major groups, the Anthozoa and the Medusozoa (Fig. 6.1). Within the Anthozoa, two subclasses are distinguished, the Hexacorallia and the Octocorallia. Anthozoans occur only as either solitary or colonial polyps. By contrast, in addition to forming solitary or colonial polyps, medusozoans typically form gamete-bearing medusa. Medusozoans are currently divided into four classes, the Hydrozoa, Scyphozoa, Cubozoa, and Staurozoa. Staurozoa have previously been grouped together with Cubozoa, yet, recent phylogenies place them as a sister group to all other Medusozoa (Fig. 6.1). Recent molecular phylogenetic analyses suggest that the parasitic Myxozoa may form a cnidarian subclade,

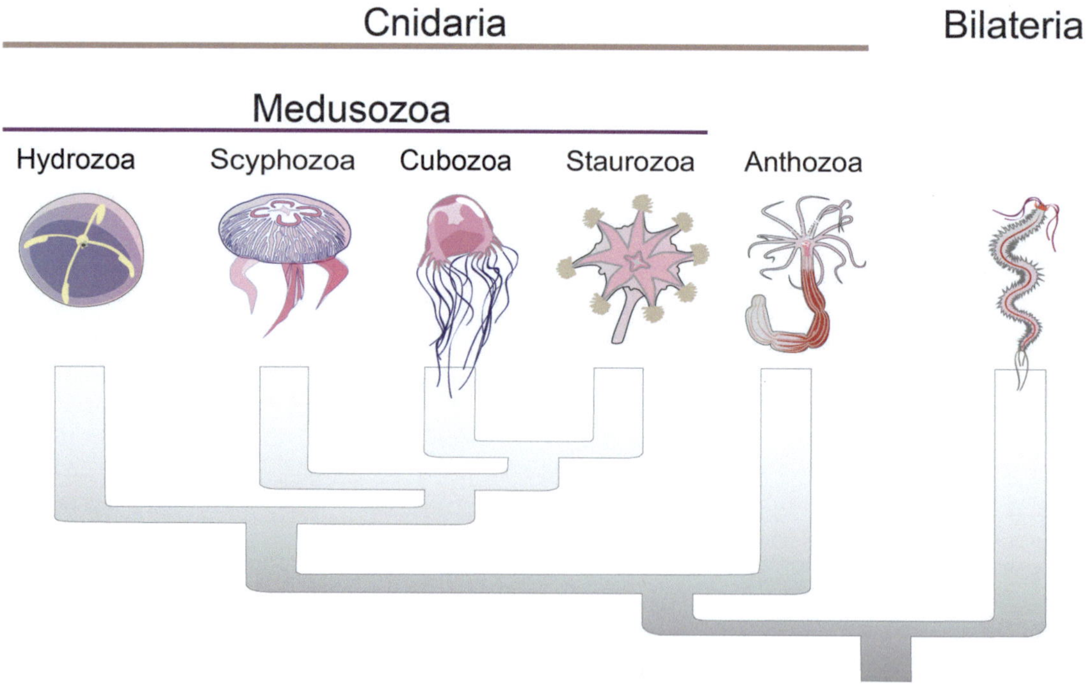

Fig. 6.1 Phylogenetic relationship of cnidarian classes. Cnidarians are a sister group of the bilaterian animals exemplified by a polychaete worm. Within Cnidaria, the medusozoans have complex life cycles where usually both medusa and polyp form, whereas their sister group, the Anthozoa, have polyps only. Phylogeny according to Collins (2002) and Marques and Collins (2004) with Staurozoa as a sister group to the Cubozoa. Note that Collins et al. (2006) position the Staurozoa at the base of the medusozoans (© Ulrich Technau, Grigory Genikhovich, Johanna Kraus, 2015. All Rights Reserved)

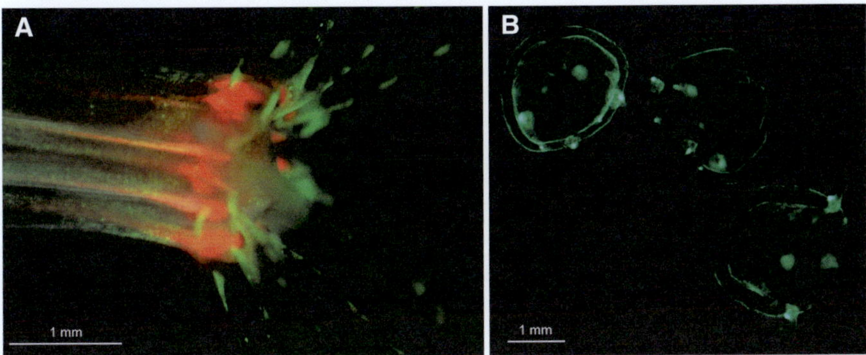

Fig. 6.2 (A) Red and green fluorescent proteins of the sea anemone *Nematostella vectensis* (B) Green fluorescent protein of the young medusa of the hydrozoan *Clytia hemisphaerica*

possibly as sister group to the Medusozoa. Since they have a highly aberrant life different to that of all other cnidarians, they are treated separately in this treatise (Chapter 7).

Although a recent report on mitochondrial sequences proposed a paraphyletic relationship of Hexacorallia and Octocorallia (Kayal et al. 2013), most other phylogenetic analyses agree on the monophyletic status of these two classes of Anthozoa. At the same time, all phylogenies support the monophyly of Cnidaria. If the paraphyletic relationship of Hexacorallia and Octocorallia is true, it would suggest that all medusozoan features that distinguish anthozoans and medusozoans are either derived or have been lost independently in Octocorallia and Hexacorallia.

Cnidarians are characterized by stinging cells, called cnidocytes or nematocytes, which are used to capture prey and to defend the organism. Another characteristic of cnidarians are endogenous fluorescent proteins (Fig. 6.2). Their discovery and subsequent biotechnological development as one of the most important and widely used tools in cell and developmental biology has led to the award of the Nobel Prize in 2008.

Cnidarians are diploblastic, i.e., they are composed of only two cell layers, ectoderm and endoderm (Fig. 6.3). Hence, they lack the third germ layer, the mesoderm, present in all Bilateria. Cnidarian ectoderm and endoderm are epithelial monolayers during development and throughout the life of the animal.

The epithelial cells are apically connected to each other by adherens and septate junctions. Basally, hemidesmosomes attach the two epithelia to a common extracellular matrix, called mesogloea. The mesogloea is secreted by the epithelial cells and is composed of laminin, fibronectin, and collagen IV, very similar to the basal membrane of Bilateria (Sarras 2012).

Gland cells, neurons, and nematocytes intermingle among the epithelial cells. Nematocytes (also called cnidocytes) are the phylum-characteristic cell type only found in Cnidaria. They harbor a large capsule, the nematocyst, an extrusive organelle which can be triggered to discharge and eject a thread and/or a stylet. Cnidocytes are probably built exclusively in the ectodermal layer or its derivatives (e.g., the inverted pharynx). Cnidocytes are considered a specialized sensory cell type (Oliver et al. 2008) which serves mainly for predation and defense but also for locomotion (when attaching to a substrate) (Tardent 1978). Different types of nematocysts, spirocysts, and ptychocysts exist, which differ in the shape of the capsule, the thread, and the stylet. Depending on this structure, they may penetrate a prey and release toxins attached to the thread (e.g., stenoteles in *Hydra*) or they may entangle the prey's bristles (e.g., desmonemes and isorhizas). The thread can be trichous or atrichous (with or without spines). The different structures and shapes of the capsules serve as taxonomic traits to distinguish between cnidarian

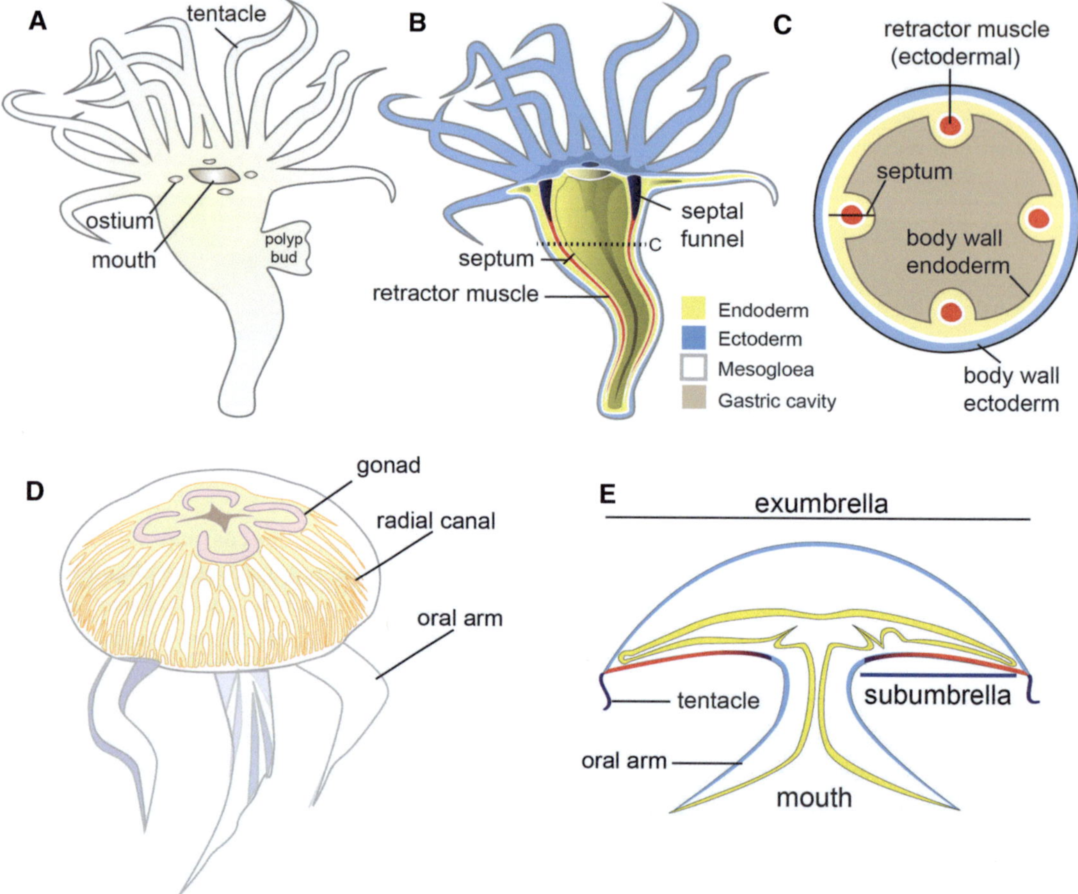

Fig. 6.3 Typical organization of a cnidarian polyp and a medusa exemplified by a scyphozoan (*Aurelia aurita*). (**A**) Outer appearance of a polyp (scyphistoma). (**B**) Inner organization of a polyp shown in a longitudinal section. (**C**) Schematic cross section through the midbody region of a polyp. (**D**) Outer appearance of an *A. aurita* jellyfish. (**E**) Inner organization of a jellyfish. Areas with striated muscles in the subumbrella are highlighted in *red* (© Ulrich Technau, Grigory Genikhovich, Johanna Kraus, 2015. All Rights Reserved)

species (Weill 1934; Zenkert et al. 2011). The best-studied nematocyte is the stenotele of *Hydra*, where the protein composition of the capsule and the discharge mechanism have been studied in detail. The capsule wall primarily consists of minicollagens, very short cnidarian-specific variants of collagens forming intermolecular bridges ensuring an impressive tensile strength (Kurz et al. 1991; Holstein et al. 1994). Inside the capsule, the main component of the matrix is γ-polyglutamate. It is thought to act as a mineral gel, responsible for the enormous internal pressure of about 150 bar inside the capsule (Weber 1990). Stenotele discharge (Fig. 6.4) has been monitored by high-speed cameras (Holstein and Tardent 1984; Nüchter et al. 2006) and found to be the fastest movement in the animal world, albeit over only a short distance. During the first phase of stylet ejection, it accelerates with over 53 Mio m/s (Nüchter et al. 2006).

The nervous system of Cnidaria is generally considered to have a diffuse organization, since no brain-like structures exist. However, both in polyps and in medusae, neurons can be concentrated in some regions, sometimes forming nerve rings, for instance, at the margin of the bell of medusae. Neurons are found both in the ectoderm and in the endoderm. Morphologically, they form either different types of ganglion cells, located basiepithelially, or sensory cells, which are intercalated between the epithelial cells, perceiving external mechanical or chemical cues

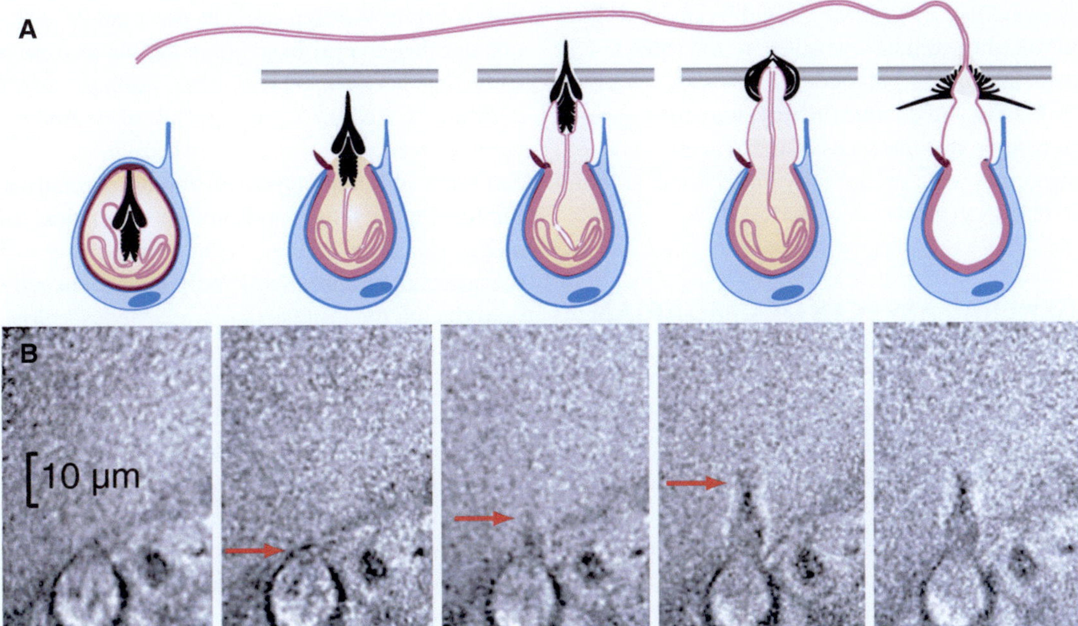

Fig. 6.4 Discharge of a stenotele nematocyst shown schematically (**A**) and recorded with a Hamamatsu C4187 high-speed camera (**B**). (**A**) Nematocytes (*blue*; cell and vesicle membranes in *dark blue*) harbor one cyst (*pink*; minicollagen wall, tubule, and operculum) with stylets (*black*) punching a hole into prey. (**B**) Sequential images, taken 195 µs after triggering at 1.430.000 frames per second (200 ns exposure time, 500 ns frame interval); *arrows* indicate progress of discharge (Reproduced with permission from Nüchter et al. (2006). Copyright Elsevier 2006)

and transmitting them to ganglion neurons or directly to nematocytes in the tentacles (Hobmayer et al. 1990). Neurons and cnidocytes are terminally differentiated, whereas epithelial cells (at least in *Hydra* polyps) are continuously proliferating under optimal feeding conditions. Hence, new neurons and cnidocytes need to differentiate during growth or to replace the discharged cnidocytes. In hydrozoans, both cell types differentiate from a common progenitor, the interstitial stem cell. Since interstitial cells have so far not been identified outside of hydrozoans, it is currently unclear whether neurons and nematocytes differentiate directly and independently from epithelial cells or whether they share a specialized progenitor cell in the representatives of the other cnidarian classes.

Different types of gland cells, present both in ectoderm and endoderm and intercalated between the epithelial cells, secrete mucus and digestive enzymes. In addition to that, they appear to be the source of important regulatory proteins, such as Wnt antagonists (e.g., Dickkopf) (Guder et al. 2006b; Augustin et al. 2006). Little is known about their development except for certain gland cells in *Hydra*, which appear to have a limited cycling capacity, but also arise from interstitial stem cells (Schmidt and David 1986; Bode et al. 1987).

Cnidarians can move with the help of muscles. In many cases, muscles are formed at the base of the epithelial cells (myoepithelial cells and epitheliomuscle cells), but they can also detach from the epithelial junctions and take a basiepithelial position. With very few exceptions, polyps form smooth muscle type, but medusae also form striated, mononuclear muscle cells in the subumbrella.

Most studies on the developmental biology of cnidarians concern adult polyps or medusae. Herein, a general, comparative account of cnidarian embryonic and larval development is first provided, followed by a detailed description about what is known on the development of the better-investigated taxa. It is important to note that extreme deviations from common cnidarian life cycles are also known, as in the abovementioned myxozoans (Chapter 7) or the enigmatic

Polypodium hydriforme, a cnidarian that lives as an intracellular parasite in the oocytes of acipenserid and polyodontid fishes. A review of its bizarre and complex life cycle, which appears to involve two instances of germ layer inversion, a process, which is not fully understood, is found in Raikova (1994).

GENERAL ASPECTS OF CNIDARIAN EARLY DEVELOPMENT

Generally, the first two or three cleavage divisions are radial, but not always complete, while the following cleavage planes are much more irregular and variable from embryo to embryo (Tardent 1978). Notably, gastrulation can occur by any possible mode known from bilaterians, depending on the species: anthozoans and scyphozoans often gastrulate by invagination and hydrozoans by unipolar and multipolar immigration or morula delamination. Epiboly and cellular delamination may also occur in rare cases (Tardent 1978; Byrum and Martindale 2004). Since gastrulation modes are much more diverse among hydrozoans than among anthozoans, one may assume that invagination is the ancestral mode, but that is by no means fully clear. The post-gastrula stage typically develops into a pear-shaped or elongated planula larva that is usually free-swimming by means of ciliary motion. The Staurozoa however, after gastrulation through unipolar ingression, form unusual worm-shaped planulae that creep on the substrate (Kowalewsky 1884; Wietrzykowski 1910, 1912; Hanaoka 1934; Otto 1976). Interestingly, the number of endodermal cells in staurozoan planulae appears to remain constant after gastrulation (Otto 1976), which is the only known example of cell constancy of developmental stages among cnidarians.

GENERAL ASPECTS OF CNIDARIAN LATE DEVELOPMENT

After some time of exploration and dispersal, the planula larva starts to settle, either spontaneously or triggered by an external cue. The planula then undergoes a gradual or dramatic metamorphosis into a primary polyp and, in the case of colonial species, the primary polyp begins to form a spreading network of horizontal stolons, called hydrorhiza, and in some species also produce vertical branching shoots. Hydrozoan colonies can carry different types of zooids specialized for feeding, defense, and medusa formation. In medusozoans, a regular metagenetic change of the asexually reproducing polyps and sexually reproducing medusae is observed. Medusae arise from polyps in various ways: either by strobilation (Scyphozoa, Cubozoa) or by lateral budding (Hydrozoa). Gametes are either formed by the polyp (anthozoans) or by the medusa (medusozoans) and are usually spawned into the water. However, especially hydrozoans show various degrees of reduction of either the medusa generation called hypogenesis, whereby specific reduction stages of ancestrally free-swimming medusae remain attached to the polyps as egg- and sperm-producing "organs." In this case, fertilization and embryonic development (in some cases until the planula stage) take place in the female colony. Forms of "labor division" between specialized polypoid and medusoid zooids within a colony are common among the Hydrozoa. One of the most specialized animal colonies with various organ-like individuals is found in the Siphonophora (e.g., *Physalia physalis*, the "Portuguese Man o' War"). Starting early in larval development, the animals will give rise to a floating colony of hundreds of highly specialized zooids.

GENERAL QUESTIONS OF DEVELOPMENT AND BODY PLAN EVOLUTION

For centuries cnidarians have served as models to study fundamental aspects of developmental biology. It is probably fair to say that the birth of experimental developmental biology started with experiments on *Hydra*. In 1744, Abraham Trembley, a Swiss naturalist and physician from Geneva, published his "Mémoires, pour servir à l'histoire d'un genre de polypes d'eau douce, à bras en forme de cornes." In this volume he reported the first bisection experiments, showing that *Hydra* is able to regenerate the missing part within a few days. Ever since, *Hydra* and other cnidarian species were used to study pattern formation, regeneration, and stem cells. When powerful genetic models such as *Drosophila melanogaster* and *Caenorhabditis elegans* became dominant, cnidarians were less competitive as model organisms, since genetics or functional gene analysis was not readily possible until recently. However, with the rise of EvoDevo, cnidarians became very interesting due to their phylogenetic position as a sister group to the Bilateria and were often referred to as "basal metazoans," a somewhat misleading term, as this refers to animals that lived in the deep past and not to extant animals. Instead, they are representatives of basally or early branching lineages. This distinction is important, as the traits we observe in cnidarians (or in other basally branching lineages) do not necessarily reflect the ancestral state, because these lineages had the same time for evolutionary divergence as any other extant animal lineage. However, in many cases comparative approaches both on the morphological and molecular level allow us to reconstruct the likely ancestral and derived states.

Often, there are crucial differences observed between the main protostome model species (*Drosophila* and *Caenorhabditis*) and main deuterostome model species (e.g., mouse and zebrafish). In such cases, cnidarians have a strategic position as an outgroup to the Bilateria to infer ancestral states.

MEDUSOZOAN MODEL SYSTEMS

In the following major findings and recent advances made by using various cnidarian model species are highlighted.

Hydra

Hydra is a freshwater hydrozoan genus which has lost the medusa and planula stages (Fig. 6.5). Several species are currently predominantly used in experimental research: *Hydra magnipapillata*, the Japanese species, of which the genome has been sequenced (Chapman et al. 2010); the European *Hydra vulgaris*, which is genetically very similar to *H. magnipapillata*; and a

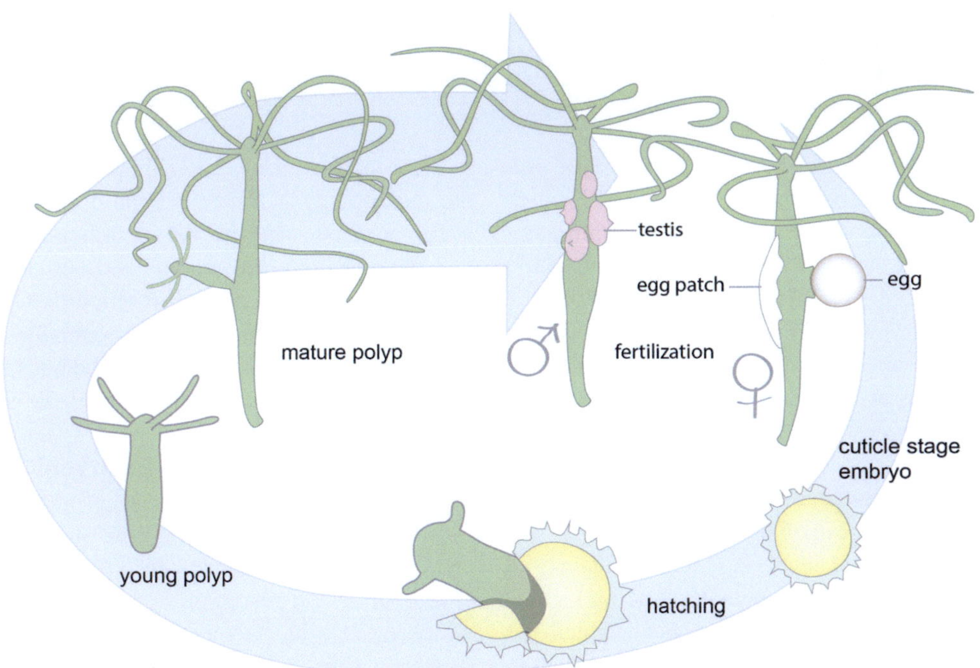

Fig. 6.5 The life cycle of *Hydra* (© Ulrich Technau, Grigory Genikhovich, Johanna Kraus, 2015. All Rights Reserved)

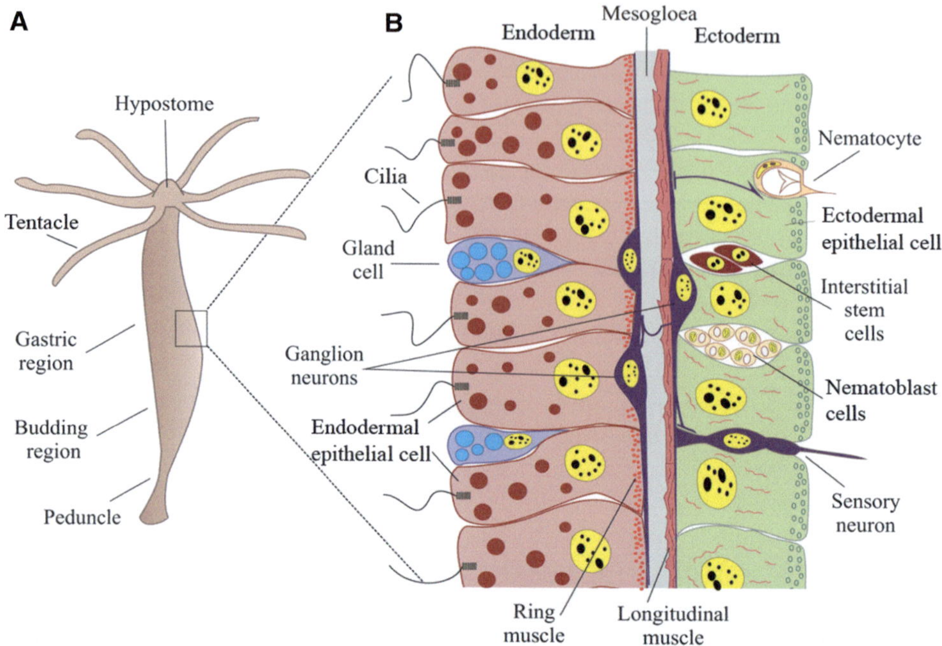

Fig. 6.6 Anatomy of a hydrozoan polyp. (**A**) A *Hydra* polyp is essentially a two-layered tube with a ring of tentacles around the mouth opening. Asexual budding occurs on the lower half of the body column. Interstitial stem cells and nematoblasts are distributed evenly in the body column, below the tentacle ring and above the border of the peduncle, which is the stalk between the budding region and pedal disc. (**B**) The bilayered cellular organization of a *Hydra* polyp. Ectoderm and endoderm are separated by an acellular matrix called the mesogloea (*gray*). All epithelial cells in *Hydra* are myoepithelial, with myofibers on the basal side (*red*). In ectodermal epithelial cells (*green*), the fibers are oriented longitudinally, and in endodermal epithelial cells (*pink*), they are oriented circumferentially (ring muscle). Most interstitial cells and nematoblast clusters are located between ectodermal epithelial cells. Neurons are found in both the endoderm and ectoderm. Sensory neurons are located between epithelial cells and connect to ganglion neurons (*purple*), which are at the base of the epithelium on top of the myofibers and sometimes cross the mesogloea. Different types of gland cells, most of which are found in the endoderm, are intermingled between the epithelial cells (Reproduced with permission from Technau and Steele (2011))

laboratory cross of two American wild-type polyps of *Hydra carnea*, termed "strain AEP," which is used because of the possibility to obtain gametes (Martin et al. 1997; Hemmrich et al. 2007) and generate transgenics (Wittlieb et al. 2006).

Hydra has a simple anatomy: it is basically a tube with a single opening serving both as a mouth and an anus at one end and a basal disc used to attach to the substrate at the other end. Like other cnidarian polyps, it consists of two layers of epithelial cells, separated by an extracellular matrix, the mesogloea (Fig. 6.6). Other cell types are found intercalated in the epithelium: interstitial cells (i-cells), neurons, nematocytes, and gland cells. An important technique that enabled the quantitative analysis of cellular differentiation was maceration (David 1973), which maintains the morphology of the cells upon dissociation in acetic acid, allowing for the identification of all cell types. Except for neurons and nematocytes, all cells in the body column proliferate. Overall, this proliferative activity of the epithelial cells leads to a tissue displacement into the buds, head, or foot (for review, see Watanabe et al. 2009).

Hydra Gametogenesis and Embryonic Development

Hydra oogenesis and embryonic development are very peculiar. Gametes are the product of a sperm- or oocyte-restricted stem cell population, which resides in clusters or strings within the ectoderm (Fig. 6.7). Gametogenesis can either

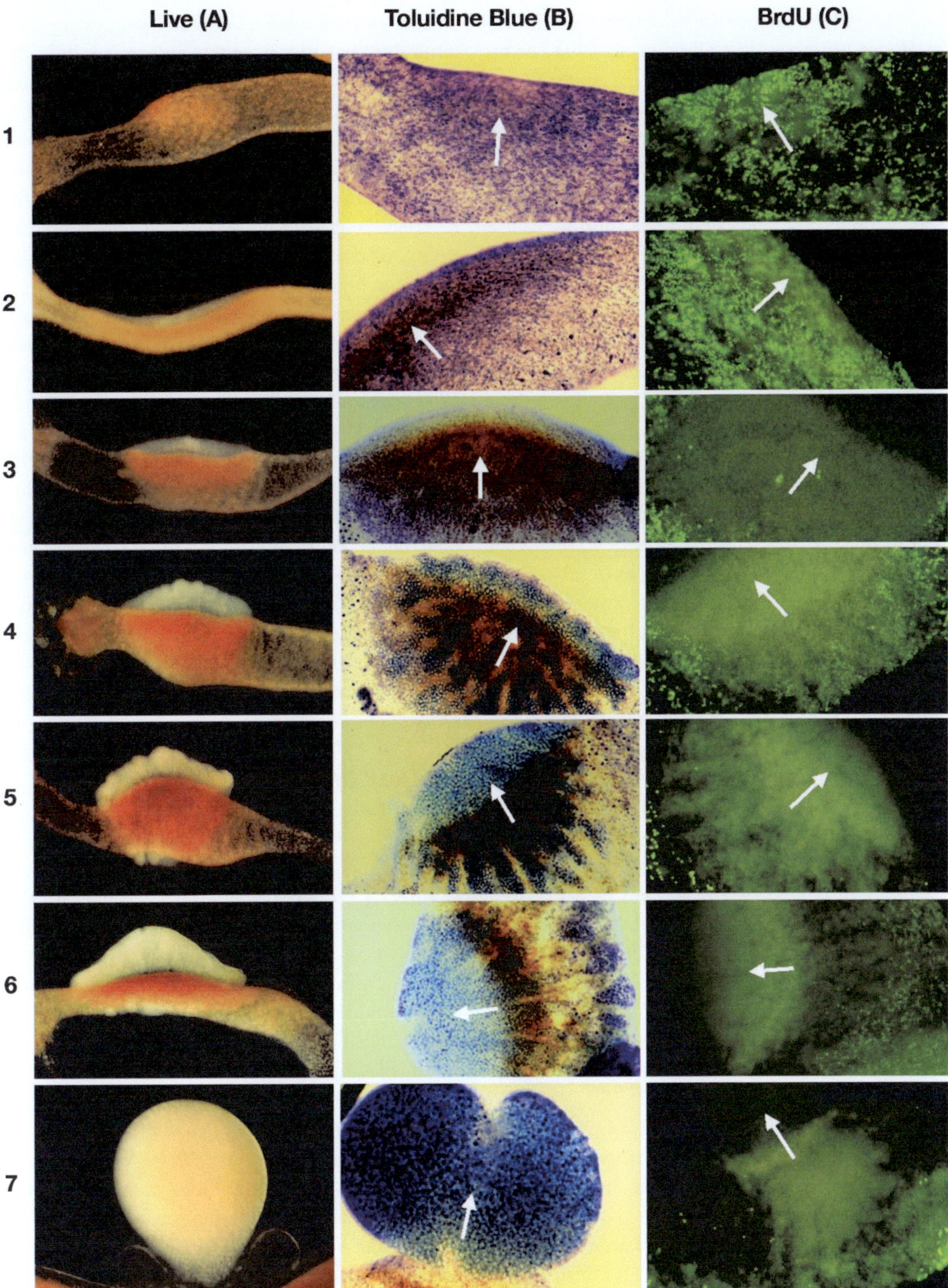

Fig. 6.7 Oogenesis stages in *Hydra carnea* (strain AEP). Animals were monitored and staged live (**A**), stained with toluidine blue to visualize the stem cells and germ cells (**B**) and BrdU (**C**) to visualize cells in S-phase of the cell cycle. The *arrows* point to the position of the determined and growing oocyte. The diameter of the mature egg exposed to the environment is approximately 400–500 µm (Reproduced with permission from Miller et al. (2000))

be induced by a temperature decrease in the case of *Hydra oligactis* (Littlefield et al. 1991) or by a period of starvation in *Hydra carnea* strain AEP (Martin et al. 1997). Gamete-restricted stem cells then start to accumulate massively by migration and proliferation to form future egg patches or testis (Fig. 6.7). During spermatogenesis, gamete-restricted stem cells migrate to the forming testis, formed by ectodermal epithelial cells in the upper third of the body column. Spermatogonia proliferate heavily at the proximal position of the testis and differentiate into spermatids and sperm at more distal position of the testis (Littlefield et al. 1985, 1991; Littlefield and Bode 1986).

During oogenesis, one cell is selected out of a field of competent germ cells by an unknown mechanism, which becomes apparent by the stop of mitosis and entry into meiosis. Rapidly, all germ cells surrounding the oocyte also stop cycling in a spreading concentric wave and start differentiating into "nurse cells," characterized by lipid synthesis (Miller et al. 2000).

As part of the differentiation program, the nurse cells then start an apoptotic program and become engulfed by the growing oocyte. By the time of fertilization, the egg cell contains up to 10,000 endocytes, also called "nurse cells." Remarkably, the phagocytosed nurse cells persist throughout the whole embryonic development despite having engaged in a state of apoptosis, which becomes arrested until hatching, between 1 month and 1 year after fertilization (Technau et al. 2003). In the hatchling all endocytes are rapidly digested within a few hours, probably serving as energy source for the first phase of growth and cellular differentiation (Technau et al. 2003).

After the mature egg breaks through the ectoderm and becomes exposed to the environment, it can be fertilized at the distal, animal pole (Figs. 6.8 and 6.9). Fertilization is followed by two cleavage divisions along the animal-vegetal axis starting at the animal pole. The third cleavage plane is then perpendicular to the first two cleavages, dividing the embryo into an animal and a vegetal half.

Like in most other cnidarians, the cleavage patterns then become irregular and lead within the first 12 h to the formation of a coeloblastula. Gastrulation occurs by multipolar immigration of cells spreading in a wave from the animal pole (Martin et al. 1997). The resulting parenchymula-like gastrula consists of an ectodermal epithelial layer and a mass of non-epithelial cells filling the cavity. The ectoderm develops filopodia, which secrete a hard cuticle, and the embryo enters a long dormant phase of about 1 month to 1 year. Prior to hatching, the endodermal cells organize into an epithelium, i-cells form in the endoderm and migrate to populate the ectoderm, and tentacles form (Genikhovich et al. 2006). The presence of the dormant stage with a cuticle and difficulties in obtaining enough embryonic material has always hindered the use of *Hydra* as an embryonic model. However, strain AEP of *Hydra carnea* is extensively used to produce transgenic polyps, which are then used to study cellular differentiation in the polyp (Wittlieb et al. 2006). Indeed, adult polyps with their unique regeneration capacity and constantly occurring pattern formation provide numerous exciting research questions.

The Head Organizer and Axis Formation

One of the classical experiments for our current understanding of axis formation was carried out in 1909 by Ethel Browne, then a graduate student in the laboratory of Thomas Hunt Morgan. She transplanted a piece of the hypostome (oral dome) from bleached, aposymbiotic *Hydra viridissima* laterally onto green, symbiotic *Hydra viridissima* host polyps. She found that a very small piece of hypostome could induce the outgrowth of a secondary body axis, fulfilling the criterion of an organizer (Browne 1909).

In the 1980s and 1990s, *Hydra* experienced a revival in developmental biology. Above all, the landmark lateral transplantation experiments of Harry MacWilliams led to a deep understanding of the tissue properties and revealed in detail and on a statistical level the head activation and head inhibition gradients (Bode and Bode 1980; MacWilliams 1982, 1983a, b). His experiments

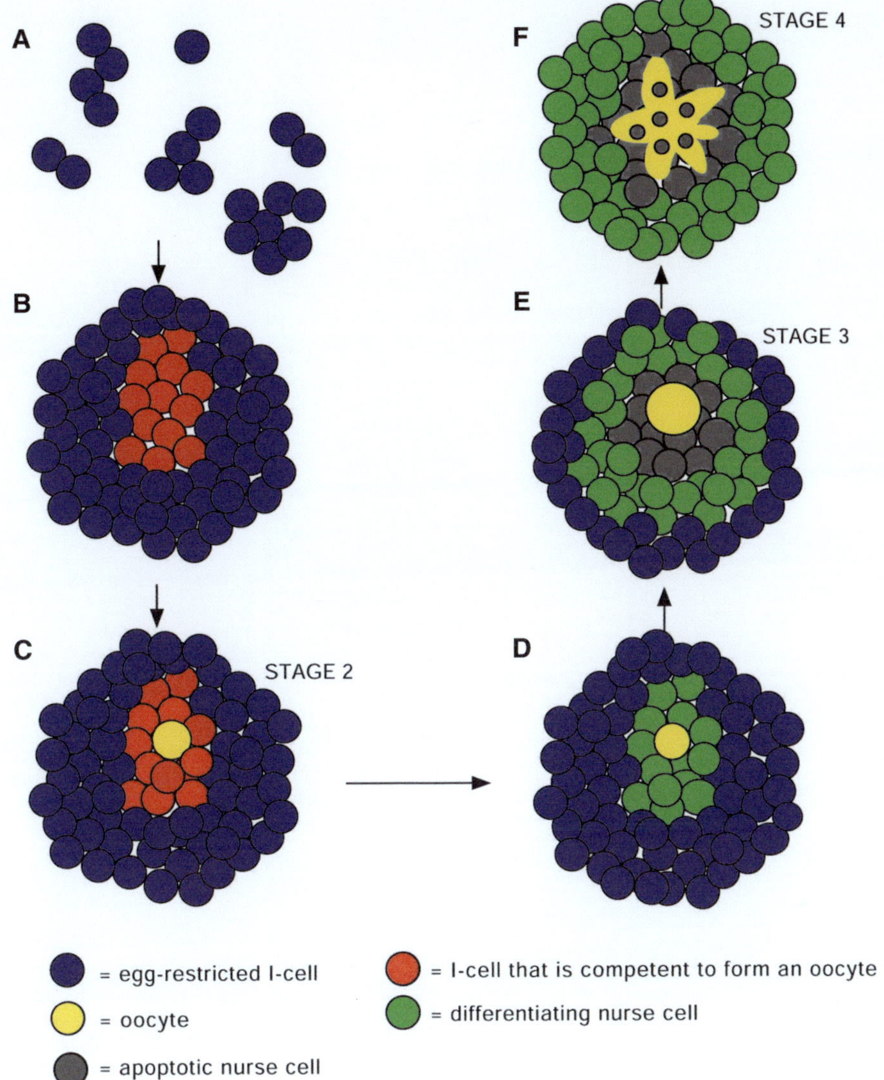

Fig. 6.8 Model of oogenesis. (**A**) Interstitial stem cells and gamete-restricted stem cells accumulate by proliferation and migration. (**B**) A group of oocyte-competent cells in the center is defined (*red*). (**C**) In the center of the competent cells, a single oocyte is singled out (*yellow*). (**D**) The oocyte and competent cells stop cycling and the competent cells start differentiating into nurse cells (*green*). (**E**) The oocyte starts growing and the nurse cells enter the apoptotic program (*gray*). (**F**) The oocyte starts phagocytosing the nurse cells, which remain in a stalled apoptotic state within the oocyte (Reproduced with permission from Miller et al. (2000))

were strongly influenced and inspired by theoretical models that explained how a polarity within a tissue can be generated and regenerated on the basis of a reaction-diffusion mechanism (Gierer and Meinhardt 1972). The Gierer-Meinhardt model is similar to the reaction-diffusion model of Alan Turing (1952), with important differences and modifications to account for biological systems (Meinhardt and Gierer 2000). In simple words, the Gierer-Meinhardt model assumes a short-range activator and a long-range inhibitor, which are coupled by an autocatalytic loop of the

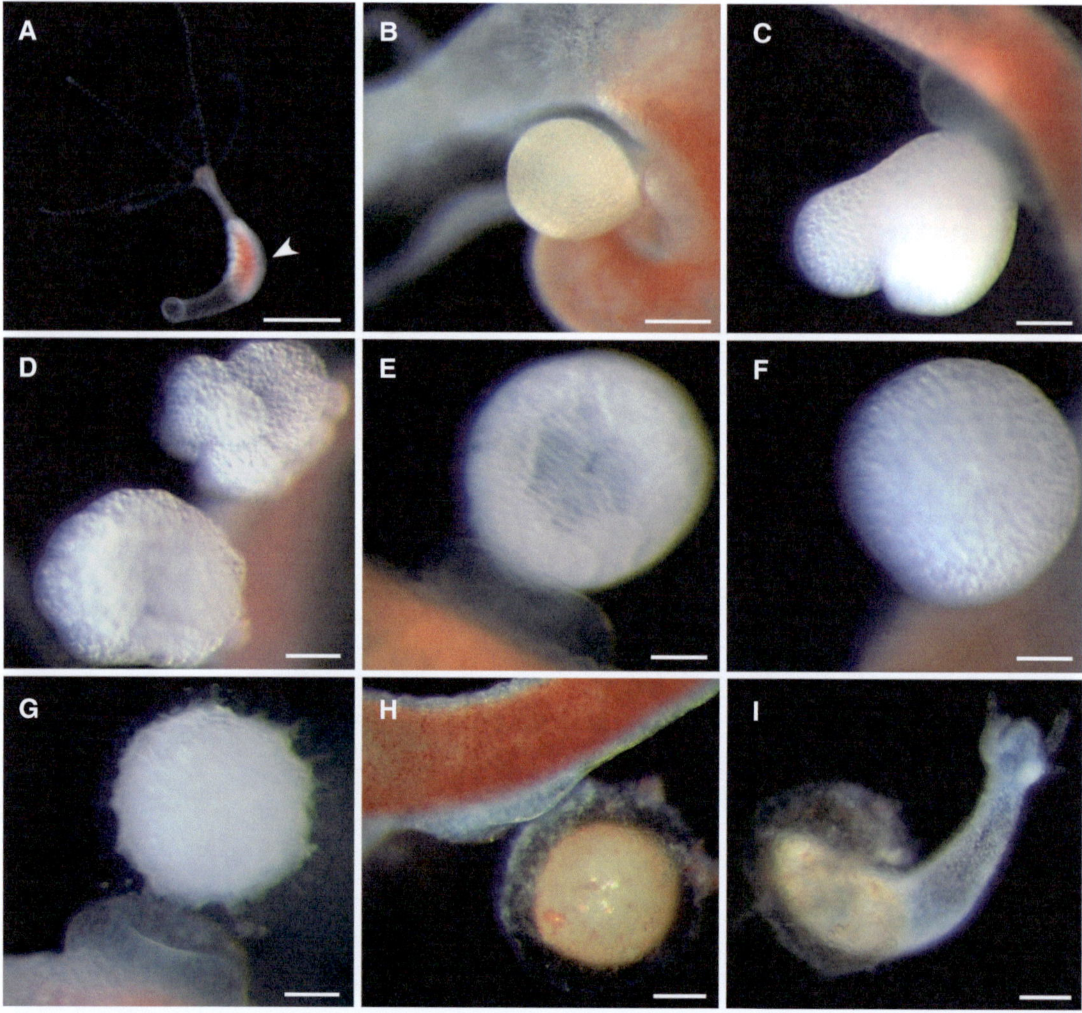

Fig. 6.9 *Hydra* oogenesis and embryogenesis. (**A**) Female *Hydra* polyp. The *arrowhead* indicates the egg fleck – the accumulation of interstitial cells converting into oogonia. (**B**) *Hydra* egg. (**C**) First cleavage furrow. (**D**) Cleaving embryos at different stages. (**E**) Coeloblastula. (**F**) Gastrula. (**G**) Spike stage. (**H**) Cuticle stage embryo, still attached to the mother polyp. (**I**) Hatching polyp. Scale bars: (**A**) 3 mm, (**B**) 200 μm, (**C–I**) 100 μm (© Ulrich Technau, Grigory Genikhovich, Johanna Kraus, 2015. All Rights Reserved)

activator and a cross-activation of the inhibitor by the activator. The inhibitor in turn attenuates the autocatalytic loop of the activator. Due to the coupling and the different diffusion ranges, both activator and inhibitor have their maximum in the same point, yet leading to local activation and lateral inhibition. Unlike many other models, these reaction-diffusion models have the capacity to generate patterns (stripes and points) in a morphogenetic field and are able to regenerate from remnant gradients or even from noise levels of biological fluctuations. Various more specific adaptations of the model were published during

the last two decades and numerous experimental results were in agreement with these models (Bode et al. 1988; Meinhardt 1993, 2012; Technau and Holstein 1995b; Meinhardt and Gierer 2000; Shimizu 2012). In summary, these experiments showed that there is a gradient of head activation and a gradient of head inhibition along the body column, both having their maximum in the hypostome of the animal.

Having understood the general principles of patterning in *Hydra*, researchers started searching for the molecular basis of the observed phenomena. Most of the early studies were guided by the

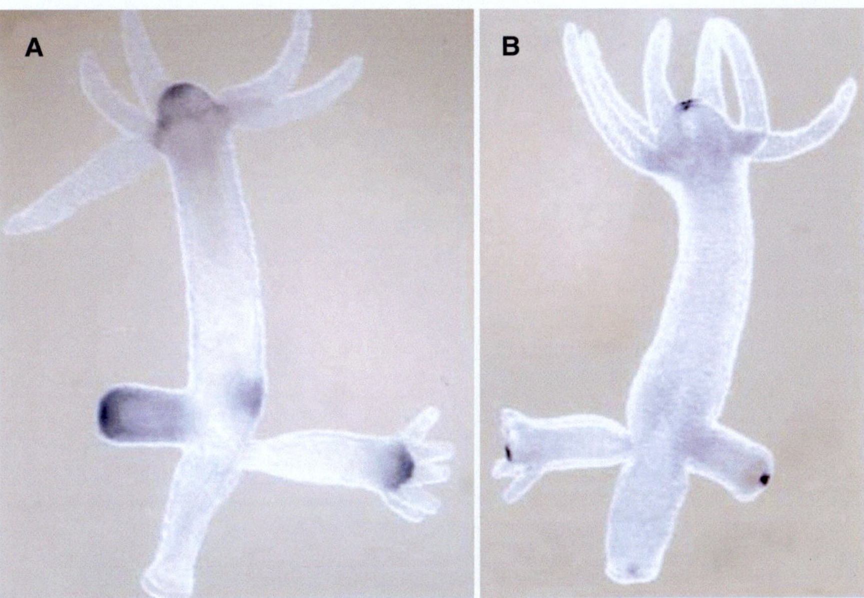

Fig. 6.10 Expression of *HyTcf* in the hypostome (**A**) and *HyWnt3* (**B**) in the oral organizer of *Hydra*. The average size of the *Hydra* polyp is 1 cm (Reprinted with permission from Macmillan Publishers Ltd: Hobmayer et al. (2000))

assumption that only small molecules can have the diffusion properties necessary to act as morphogens along the body axis. This led to the biochemical isolation of an 11 amino acid peptide, called the head activator (Schaller and Gierer 1973), which indeed had reproducible, although subtle, effects on head regeneration and on proliferation (Schaller et al. 1990; Hobmayer et al. 1997). However, a large-scale biochemical approach to isolate as many peptides as possible could not recover the head activator (Takahashi and Fujisawa 2009; Takahashi 2013), nor could it later be found in the genome sequence (Chapman et al. 2010). It is therefore unclear whether the head activator peptide has still been missed – despite considerable depth of sequencing and screening efforts – or whether it does not exist and the biological effects simply mimic the effect of a related endogenous peptide, since many of them have clear effects on morphogenesis and cellular differentiation (Takahashi and Fujisawa 2009; Takahashi 2013). Other molecules that have an effect on the head-forming capacity of body column tissue are diacylglycerol (DAG) and lithium ions (Muller 1990; Hassel et al. 1993; Hassel and Bieller 1996).

It took 90 years after the experiments of Ethel Browne to realize that the genetic basis for this tissue property bears striking similarities to that of the Mangold/Spemann organizer of frogs: in both cases, Wnt/β-catenin signaling is involved in establishing the organizer. A major breakthrough was the identification of all major components of the Wnt/β-catenin pathway and the demonstration that the ligands as well as several of the cytoplasmatic components are expressed in the hypostome (Fig. 6.10; Hobmayer et al. 2000). The identification of further Wnt ligands revealed that all Wnts are expressed in staggered domains, some highly restricted, and others somewhat broader in the hypostome and in the subhypostomal tissue (Lengfeld et al. 2009).

Another gene that is specifically expressed in the hypostome is the homolog of the T-box transcription factor *Brachyury*, *Hybra1* (Fig. 6.11; Technau and Bode 1999). Like *Wnt3* or *β-catenin*, *Hybra1* is expressed at very early time points, whenever a head is formed during regeneration, budding, and embryonic development (Technau and Bode 1999).

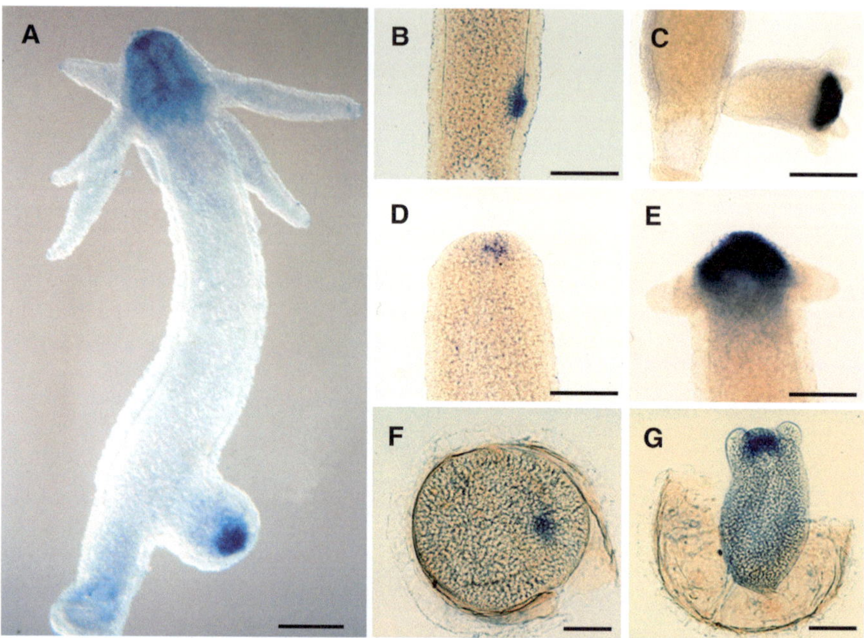

Fig. 6.11 Expression of *Hydra brachyury 1* (*Hybra1*). (**A**) Whole polyp shows expression in the hypostome and the tip of a bud. (**B, C**) Expression during budding. Note that expression starts before any sign of evagination. (**D, E**) Expression during head regeneration starts as early as 1–3 h after decapitation, long before the appearance of tentacles at 36 h. (**F, G**) Embryonic expression starts shortly before hatching, indicating the future hypostome of the primary polyp. Scale bars: (**A–E**) 1 mm, (**F–G**) 100 μm (Modified from Technau (2001), BioEssays with permission)

Notably, a *Hybra1* paralog, *Hybra2*, is also expressed in the hypostome, yet predominantly in the ectoderm, unlike the endodermally expressed *Hybra1*, with a later onset of expression (Bielen et al. 2007).

More recent transplantation experiments confirmed the experiments by Ethel Browne and showed that chemical ectopic activation of the Wnt/β-catenin pathway by alsterpaullone treatment induces the outgrowth of head structures in the body column and alters the head-inducing capacity of the body column tissue (Broun and Bode 2002; Broun et al. 2005). Notably, *Hybra1* becomes expressed throughout the body column, while *Wnt3* expression is found in numerous spots along the body column (Broun et al. 2005), supporting the idea that Wnt signaling and *Brachyury* are engaged in a feedback loop. In vertebrates, there is a feedback loop between *Wnt3* and *Brachyury*, suggesting that this might reflect an ancestral genetic circuit (Yamaguchi et al. 1999; Holstein 2003, 2008,

2012; Guder et al. 2006a; Meinhardt 2012). Recently, taking advantage of the establishment of transgenic *Hydra* (Wittlieb et al. 2006), over-expression of β-catenin in the body column could also induce the outgrowth of ectopic head structures (Gee et al. 2010). Furthermore, when tested in transplantation assays, ectopic β-catenin expression increases the head activation potential (Gee et al. 2010). These data strongly suggest that Wnt/*β-catenin* signaling plays a crucial role in providing an oral identity to the tissue and in conferring the organizing capacity. However, also Wnt/planar cell polarity (PCP) signaling appears to play an important role in morphogenesis, e.g., during budding, and is connected to Wnt/β-catenin signaling (Philipp et al. 2009).

Strikingly, *Hydra* can regenerate not only from dissected parts of the body column but also from dissociated and reaggregated single cell suspensions (Gierer et al. 1972). This process (Fig. 6.12) involves sorting of endodermal and ectodermal cells during the first 12 h, followed by formation

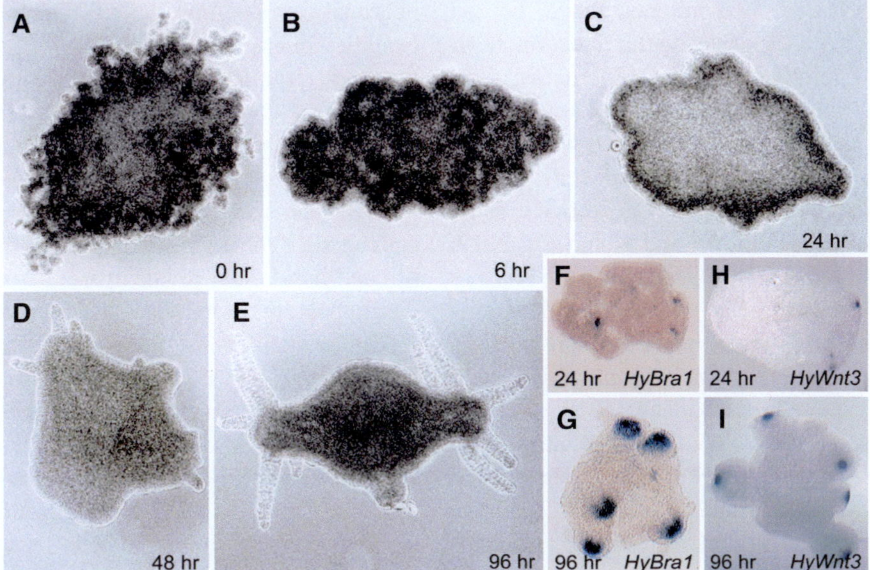

Fig. 6.12 Pattern formation in reaggregated single cell suspension in *Hydra*. (**A–E**) Consecutive stages of the aggregate development in the course of 4 days. Hours indicate time after centrifugation of the cell suspension. (**F–I**) Expression of the head-specific regulatory genes *HyBra1* and *HyWnt3* in the newly forming heads in the aggregates. Average aggregate size is 500 μm ((**F–I**) Reproduced with permission from Technau et al. (2000). Copyright (2000) National Academy of Sciences, USA)

of ectodermal and endodermal epithelia. Since cells remain randomly distributed during this process, this is a de novo pattern formation process (Sato et al. 1990; Technau and Holstein 1992). Molecular and cellular analysis demonstrated that clusters of 5–15 cells but no single cells originating from apical tissue are sufficient to define a new organizer with high efficiency (Technau et al. 2000). These newly defined organizing centers express *HyBra1* and *Wnt3* in spots of few cells (Fig. 6.12f–i), while ubiquitously expressed *tcf* and *β-catenin* become restricted to broader domains surrounding the *HyBra1/Wnt3* spots (Hobmayer et al. 2000; Technau et al. 2000).

Interstitial Stem Cell in Hydra: Maintenance and Differentiation

Interstitial stem cells are small, spindle-like cells that occur as single cells or twins predominantly between the ectodermal epithelial cells. A low number of i-cells are also found in the endoderm (David and Plotnick 1980). Typical for stem cells, they show a certain probability of maintenance by self-renewal and differentiation into

specific derivatives. Cloning experiments in epithelial feeder aggregates without any i-cells have shown that i-cells are multipotent and can give rise to neurons, nematocytes, gland cells, and gametes (David and Murphy 1977; David and MacWilliams 1978; David 2012). These elegant experiments have been recently confirmed using transgenic animals, where only i-cells as well as all their offspring are labeled by the expression of a reporter gene (Khalturin et al. 2007; Hemmrich et al. 2012). Again, no evidence was found that i-cells can give rise to epithelial cells, unlike in *Hydractinia* (see below). Interestingly, stem cell-depleted *Hydra* (e.g., "epithelial *Hydra*"), which have also lost all neurons and nematocytes, can still bud and regenerate, suggesting that i-cells or their derivatives are not required for pattern formation (Campbell 1976; Marcum and Campbell 1978). However, this view has been questioned by new experiments proposing a role for i-cells as the source of the transient Wnt signals necessary for regeneration (Chera et al. 2009).

Since endodermal and ectodermal epithelial cells remain stable lineages, there are three cell lineages in *Hydra*. As mentioned above, the vast majority of i-cells is located in the ectoderm; however, during embryogenesis, i-cells derive from endoderm (Martin et al. 1997; Genikhovich et al. 2006). Interestingly, i-cells express typical marker genes of germ cells as well as stem cells, such as *eed, nanos, vasa,* and *piwi* (Genikhovich et al. 2006; Nishimiya-Fujisawa and Kobayashi 2012; Juliano et al. 2014). This is similar to the pluripotent neoblast stem cells in planarian flatworms (Vol. 2, Chapters 3 and 4) and it strongly suggests a close relationship between multipotent stem cells in *Hydra* and germ cells in *Drosophila* and vertebrates. Of note, no bona fide homologs of the pluripotency marker genes *sox2, klf4, nanog,* and *oct4* that were identified in vertebrates by the Nobel Prize-winning work of Yamanaka and colleagues (Takahashi and Yamanaka 2006) have been found in the *Hydra* genome. However, there are two paralogs of the oncogene *myc*, which indeed appears to be implicated in stem cell maintenance (Hartl et al. 2010; Ambrosone et al. 2012). Recently, another conserved transcription factor *FoxO* has been shown to play an important role in maintenance of stemness (Boehm et al. 2012). Since *FoxO* has been implicated in longevity of bilaterians, its role in *Hydra* stem cells might be associated with the putative immortality of these animals. Using transgenic lines specific for each of the three cell lineages in *Hydra* (endodermal epithelial cells, ectodermal epithelial cells, i-cells), transcriptome profiles have been generated that provide insights into the molecular fingerprints of these stem cell lineages (Hemmrich et al. 2012).

Nematocytes and neurons, the two major derivatives of i-cells, are terminally differentiated and arrested in G0/1. The differentiation of neurons and nematocytes has been studied in some detail on the cellular as well as molecular level. *Hydra* has ganglion (defined by having multiple dendrites) and sensory neurons, which form a diffuse nervous system with much higher densities at both ends, in the head and tentacles, as well as in the peduncle (Figs. 6.13 and 6.14).

Notably, the i-cells are only found in the body column, but largely excluded from the head and the lower peduncle and foot region (Fig. 6.15).

This complementary pattern of precursor stem cells and derivatives is explained by the fact that neuronal progenitor cells, which probably become committed stochastically somewhere in the body column, migrate axially towards the head or the foot and differentiate after a final mitosis at those positions wherever there is a gap in the nervous system (Technau and Holstein 1996; Hager and David 1997). Since epithelial cells (except for cells in the tentacle, hypostome tip, and basal disc) divide every 3–4 days when the animals are fed regularly (David and Campbell 1972; Holstein et al. 1991), the neuron density would diminish by half every 3–4 days. To keep a homeostasis of cell ratios, new neurons need to be supplied by the interstitial stem cells. Neurons are considered to be relatively long-lived and they are displaced passively in axial directions together with the epithelial cells by tissue turnover. This led to the idea that neurons of the body column might change their phenotype when they are displaced into the head or foot region (Koizumi and Bode 1991; Bode 1992). However, at least 95 % of the ganglion neurons of the peduncle nervous system arise by new differentiation from stem cells and not from previously existing body column neurons through phenotypic plasticity (Fig. 6.16; Technau and Holstein 1996). The requirement of new differentiation from stem cells during regeneration appears to be a general feature of the nervous system, since similar findings were made in the head nervous system (Miljkovic-Licina et al. 2007).

By contrast, nematocytes differentiate in clusters throughout the body column, congruent with the distribution of stem cells, and then, upon maturation and disintegration of the cell cluster, rapidly migrate into the tentacles or elsewhere. After commitment, nematoblasts undergo several rounds of divisions, but remain connected through cytoplasmic bridges and form large nests of up to 64 cells. After a final mitosis, the differentiation of the nematocyst begins, which involves the expression of capsule-specific proteins like mini-collagens, spinalin, nematocyst outer wall antigen (NOWA), and nematogalectin (Kurz et al. 1991; Koch et al. 1998; Engel et al. 2002; Hwang et al. 2010). It is still unclear at which stage the four different types of nematocytes are specified.

The molecular control of neurogenesis and nematocyte differentiation is still poorly

Fig. 6.13 Nervous system of *Hydra*. (**A**) Oral view on the RFamide-positive nerve net of the tentacles and the hypostome (*orange*) projecting to the mouth opening, marked by endodermal boundary cells (*green*). (**B**) Sensory neurons detected by monoclonal antibody Nv1 (*red*) innervating individual nematocytes (stenoteles) in the tentacle (*yellow*). (**C**) Overview of the L96-positive nerve net of the peduncle. (**D**) Close-up of the L96-positive nerve net showing the connections of the ganglion cells. Scale bars: (**A, B**) 250 µm, (**C**) 300 µm, (**D**) 50 µm ((**A**) From Technau and Holstein (1995a); Cell and Tissue Research, with permission)

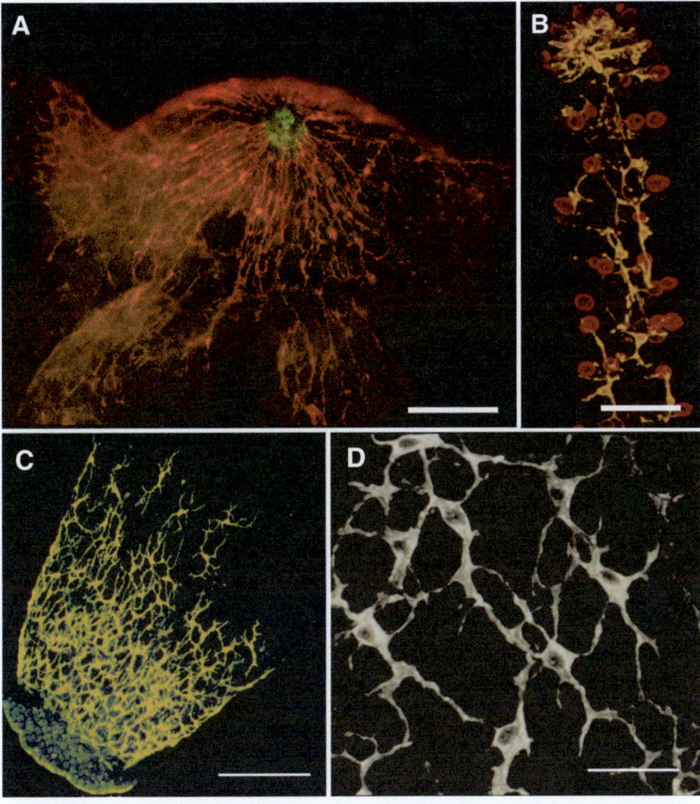

Fig. 6.14 The nervous system has overlapping populations expressing different combinations of marker proteins. Double staining of peduncle nervous system with RFamide antibody (**A**; *red*) and mab L96 (**B**; *green*). Note that L96 neurons are a subset of RFamide neurons in the peduncle. Scale bar 200 µm (Taken from Technau and Holstein (1996) Developmental Biology, with permission)

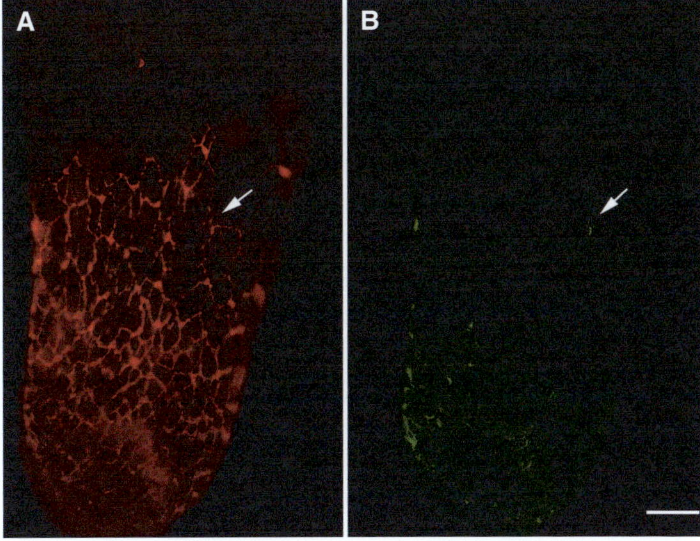

understood. Interestingly, the conserved zinc finger transcription factor gene *Hyzic* is specifically expressed in a subpopulation of i-cells and the proliferating nematoblasts (Lindgens et al. 2004). In vertebrates, *zic* is required to keep committed neuronal precursor cells of the neural crest in proliferative state before final differentiation (Elms et al. 2003). Thus, while functional data are lacking, *zic* could play a similar role in nematocyte differentiation. In

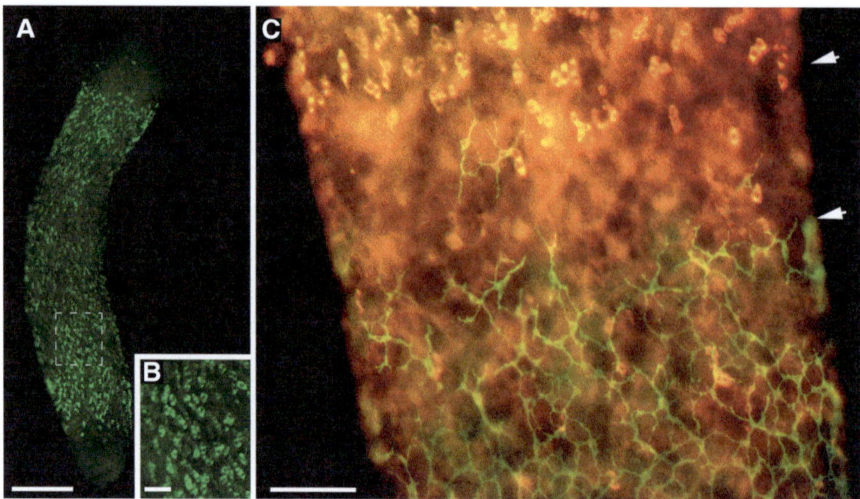

Fig. 6.15 Distribution of interstitial stem cells and neurons is largely complementary. (**A**) Interstitial stem cells in *Hydra* stained with monoclonal antibody C41 are localized to the ectoderm of the body column, but absent from the extremities. (**B**) Double staining of interstitial stem cells (mab C41, *yellow*) and ganglion nerve cells in the peduncle (mab L96; *green*). *Arrowheads* mark the boundaries of the cells. Scale bars: 500 µm (**A**), 30 µm (**B**), 100 µm (**C**) (© Ulrich Technau, Grigory Genikhovich, Johanna Kraus, 2015. All Rights Reserved)

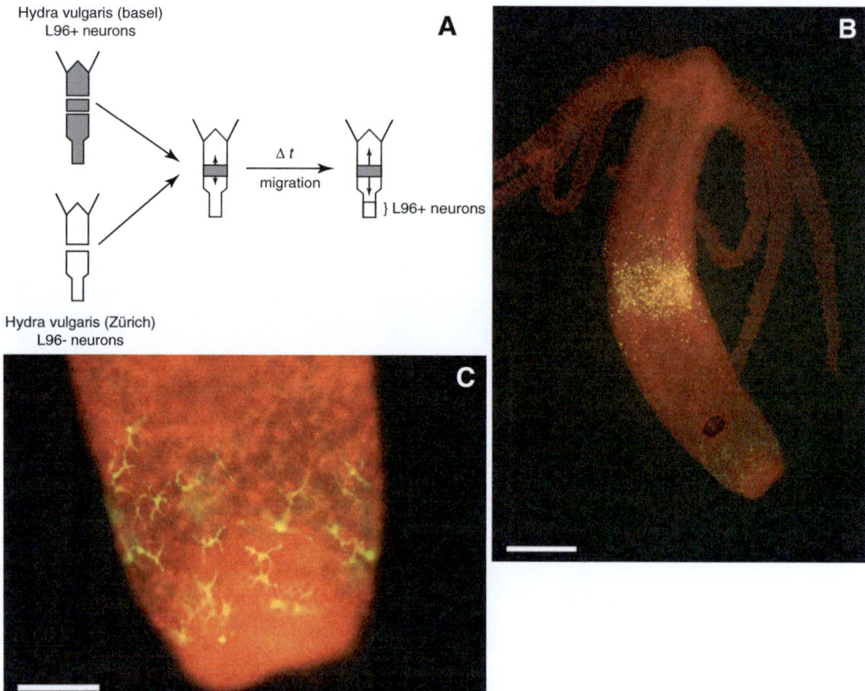

Fig. 6.16 Experiment demonstrating the constant migration and differentiation of neuronal precursor cells in *Hydra*. (**A**) Experimental scheme. A mid-column fragment of a vitally labeled polyp of *H. vulgaris* strain Basel (L96-positive) is transplanted into the equivalent position of a non-labeled polyp of *H. vulgaris* strain Zürich (L96-negative). Assessment of L96+ neurons in the host tissue over time. (**B**) Representative polyp 6 days after transplantation shows L96+ -positive neurons in the host, which have differentiated from migratory precursor cells of the grafted tissue. (**C**) Close-up of the peduncle region showing the differentiated neurons. Scale bars: 1 mm (**B**), 300 µm (**C**) (Taken from Technau and Holstein (1996) Developmental Biology, with permission)

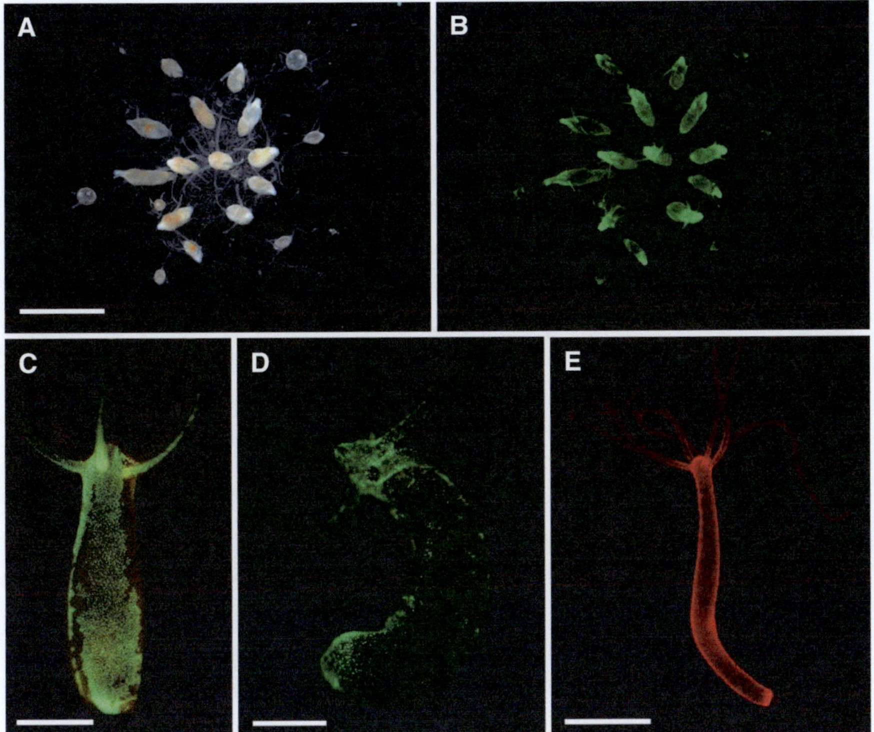

Fig. 6.17 Transgenic *Hydractinia* and *Hydra* polyps. (**A**) *Hydractinia* polyps are expressing GFP (same colony shown in GFP channel in **B**) under the control of a ubiquitous actin promoter. (**C**) *Hydra* expressing GFP under a ubiquitous actin promoter in a contiguous patch of transgenic epithelial cells. (**D**) Interstitial cell-specific transgenic line expressing *actin*: *GFP*. (**E**) *Actin::dsRed2-*expressing ectodermal epithelial cells in a *Hydra* polyp (Images courtesy of Günter Plickert (**A**, **B**), Thomas Bosch (**C**, **D**) and Rob Steele (**E**). Scale bars: 2 mm in (**A**), 500 μm in (**C–E**). Taken from Technau and Steele (2011) Development, with permission)

line with this, the homolog of *achaete-scute* is expressed in postmitotic nematocytes and in a subpopulation of neurons (Grens et al. 1995; Hayakawa et al. 2004; Lindgens et al. 2004). A number of other conserved transcription factors such as *COUP-TF*, *prdl-a* and *prdl-b* (Gauchat et al. 1998, 2004), the *Gsx* homolog *Cnox2* (Miljkovic-Licina et al. 2007), and *cnot* (Galliot et al. 2009) are also expressed during neurogenesis or in early neurons (for review, see Galliot and Quiquand 2011). This suggests a close relationship between nematocytes and neurons and supports the view that nematocytes are a specialized sensory neuronal cell type.

In the future many questions regarding origin, migration, and turnover of various cell populations will be addressed using the newly established transgenic lines in *Hydra* (Fig. 6.17).

Hydractinia echinata and *Hydractinia symbilongicarpus*

Hydractinia echinata and its sister species *H. symbiolongicarpus* are colonial marine hydroids that preferentially grow on gastropod shells inhabited by hermit crabs – hence their common English name "snail fur." The dioecious colonies develop up to four different types of zooids, gastrozooids (also called trophozooids, autozooids, or hydranths), which are used for feeding; gonozooids which produce gametes; sensory spirozooids; and dactylozooids used for defense. Zooid buds form on a branching system of stolons, which is used to transport nutrients throughout the colony via the gastrovascular system and allow exchange of stem cells (see below). Both species have lost a free-living medusa

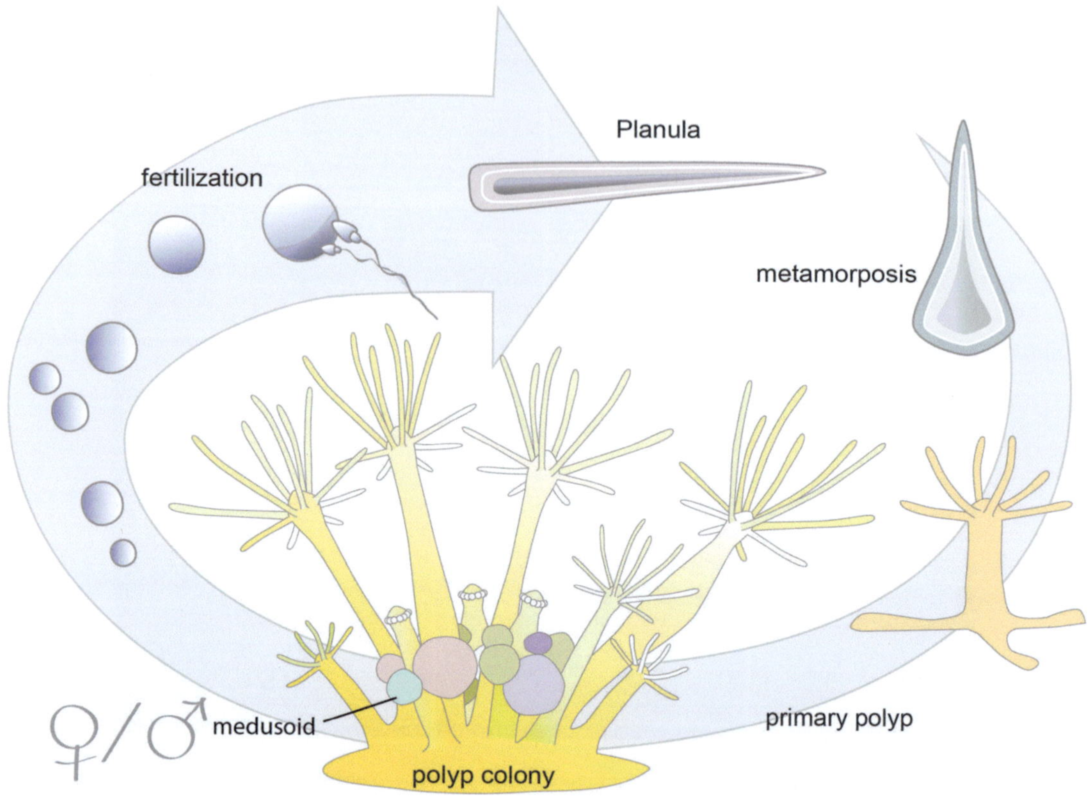

Fig. 6.18 The life cycle of *Hydractinia* (© Ulrich Technau, Grigory Genikhovich, Johanna Kraus, 2015. All Rights Reserved)

generation (Fig. 6.18). Gametes are spawned into the water upon light stimulus, where they are fertilized and develop into a planula larva, which settles to give rise to the first polyp of the new colony (Fig. 6.18; Frank et al. 2001; Plickert et al. 2012).

Hydractinia echinata and its close relative *Hydractinia symbiolongicarpus* have long served as models to investigate metamorphosis (Schmich et al. 1998; Seipp et al. 2007, 2010), colony organization (Müller 1964, 1982), and allorecognition (Mokady and Buss 1996; Poudyal et al. 2007; Nicotra et al. 2009; Rosa et al. 2010; Rosengarten et al. 2011). Recent advances in generating transgenic *Hydractinia* (Kunzel et al. 2010) as well as in gene knockdowns by RNAi (Duffy et al. 2010) have made *Hydractinia* a powerful model.

Hydractinia has attracted a lot of attention as a representative of the earliest branching animal phylum capable of allorecognition. While some *Hydractinia* colonies are histocompatible and can anastomose and, essentially, form genetic chimeras with a continuous gastrovascular system, others cannot. These reject either passively or violently, in the latter case discharging nematocysts at each other. Allorecognition has been shown to be genetically determined by two hypervariable genes, *alr1* and *alr2*, encoding Ig-domain-containing transmembrane proteins (Nicotra et al. 2009; Rosa et al. 2010; Rosengarten et al. 2011).

However, *Hydractinia*'s most striking feature for a developmental biologist is the apparent pluripotency of its interstitial stem cells (Fig. 6.19). Indeed, unlike *Hydra* i-cells, which cannot differentiate epithelial cells under any known conditions, *Hydractinia* i-cells can (Muller et al. 2004). In this study, the interstitial cells of histocompatible mutant colonies were allowed to invade the cell cycle inhibitor mitomycin C-treated i-cell-free wild-type colony or vice versa. This resulted in the conversion of the originally wild-type host into a mutant or, in the reciprocal experiment, in the conversion of the mutant into a wild-type colony. In this process, donor i-cells were shown to differentiate into epithelial cells (Muller et al. 2004). Recently, this result was confirmed in histocompatible non-mutant colonies fusing under normal conditions: i-cells from a transgenic colony immigrated into a non-transgenic colony and differentiated into EGFP-expressing epithelial cells, which gave rise to autozooids (Kunzel et al. 2010). Like in *Hydra*, subsets of *Hydractinia* i-cells express a typical set of germ cell and stem cell markers, such as *vasa*, *piwi*, and *nanos* (Rebscher et al. 2008; Plickert et al. 2012). Interestingly, overexpression of a *Hydractinia* POU domain protein *Pln* in the epithelial cells induced formation of neoplasia comprised of cells resembling i-cells and expressing *piwi*, *myc*, *nanos*, and *vasa*. RNAi knockdown of *Pln*, in contrast, appeared to promote i-cell differentiation into nematocytes (Millane et al. 2011). In the age of comparatively easy and cheap transcriptomics, future research will show what conveys such plasticity in differentiation capacity. Some findings (Teo et al. 2006; Muller et al. 2007) point at a possible role of Wnt signaling in cell fate determination in *Hydractinia*.

Similar to *Hydra*, *Clytia*, and *Nematostella* (see below), a more clear-cut role of Wnt/β-catenin signaling has been demonstrated in the process of oral-aboral axis formation. Wnt signaling components have been shown to be asymmetrically deposited in the *Hydractinia* egg and later associated with the posterior pole (according to the swimming direction) of the planula larva, which eventually gives rise to the oral structures (Plickert et al. 2006).

Ubiquitous activation of Wnt/β-catenin signaling achieved by inhibiting GSK-3β by alsterpaullone resulted, depending on the timing of treatment, in the formation of multipolar or bipolar larvae with two or more posterior ends. Bipolar larvae then were shown to develop into polyps with two hypostomes (Plickert et al. 2006). Later, more experiments confirmed that Wnt signaling

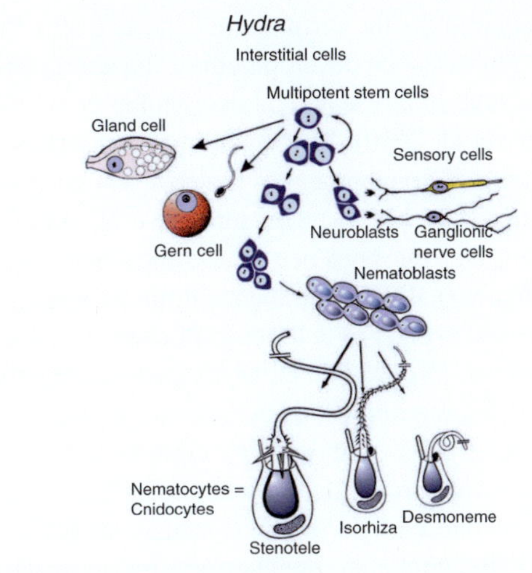

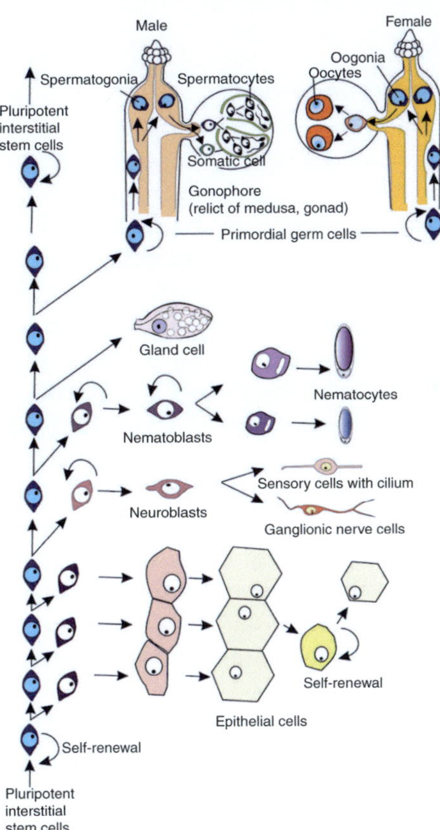

Fig. 6.19 In contrast to the i-cells of *Hydra*, which give rise to gametes, nematocytes, neurons, and gland cells, the interstitial cells of *Hydractinia echinata* also give rise to epithelial cells of the colony (Reproduced with permission from *The International Journal of Developmental Biology* (Int. J. Dev. Biol.) (2012))

is required for the formation of the oral structures and represses the development of the aboral structures both in larvae and adults (Muller et al. 2007; Duffy et al. 2010). Wnt signaling is necessary for the correct development of the posterior end of the planula, which transforms into the oral end of the polyp, and the presence of the localized mRNA of Wnt signaling components at the animal end is observed as early as in the egg (Plickert et al. 2006). However, *Hydractinia* embryos gastrulate by morula delamination, a non-polarized process where the outer cells of an embryo epithelialize to produce ectoderm, while the inner cells later arrange to generate endoderm. Thus, unlike in bilaterian and many cnidarian model organisms, gastrulation and body axis formation appear to be uncoupled in *Hydractinia*. The genetic underpinnings of this intriguing feature remain to be elucidated.

tion of the homology of the entocodon and the mesoderm and an evolutionary scenario of diploblastic organisms being secondarily simplified triploblasts with striated muscles as a plesiomorphic feature for Cnidaria and Bilateria (Seipel and Schmid 2005). It should be noted, however, that the entocodon is a transient structure exclusively found in hydrozoans during medusa formation, while true germ layers arise during early embryogenesis. Hence, it may well be a derived feature of the specific way hydrozoan medusae arise. Thus, the question is not resolved and requires more experimental assessment. A refined comparative expression analysis of more "mesodermal" transcription factors and muscle markers is required to answer the question of the evolutionary relationship between bilaterian mesoderm and the entocodon in hydrozoan jellyfish buds.

Podocoryne carnea

Podocoryne carnea is a close relative of *Hydractinia echinata* (described above) and in fact, according to Integrated Taxonomic Information System (http://www.itis.gov/), should be called *Hydractinia carnea*. The old name *Podocoryne carnea* is, however, widely used in the literature. Unlike *H. echinata*, *P. carnea* forms medusae (Seipel and Schmid 2005) and gastrulates by unipolar ingression (Momose and Schmid 2006). *P. carnea* was mainly used to examine the developmental origin of striated muscles in medusae and their developmental potency. In the early 1980s it was shown that the isolated striated muscle of its subumbrella are capable of transdifferentiation into various cell types upon activation by collagenase treatment in vitro (Schmid and Alder 1984). Since then, the group around Volker Schmid conducted research primarily focusing on the study of transdifferentiation and medusa bud formation. During this process, a transient structure called the entocodon detaches from the ectoderm and gives rise to the subumbrellar ectoderm including the striated muscle of the medusa. The entocodon is located between the ectoderm and endoderm and it expresses several "mesodermal" genes and generates a muscle layer (Spring et al. 2000, 2002; Seipel et al. 2004). This led to the postula-

Clytia hemisphaerica

Clytia hemisphaerica (=*Phialidium hemisphaericum*) is a metagenetic thecate hydroid belonging to the family *Campanulariidae*. It forms branching polyp colonies covered in chitinous exoskeleton called the perisarc (Fig. 6.20). The tropho- and gonozooids of *Clytia* are also protected by the funnel-like perisarc structures – the hydrothecae and the gonothecae. The medusae are small (adults up to 2.5 cm bell diameter) and, like all life stages of *Clytia*, very transparent (Fig. 6.21).

Clytia produces either male or female medusae on specialized polyp forms (Fig. 6.20). The young medusae detach and reach sexual maturity after a period of growth. Synchronized spawning can easily be induced through a light stimulus. Female *Clytia* medusae usually produce 10–20 eggs every day that may be harvested for experimental purposes. After fertilization, they develop into transparent embryos, which then develop into hollow coeloblastulae that gastrulate by unipolar ingression. The gastrula develops into a planula larva which settles and gives rise to the first polyp of the new colony (Fig. 6.22; Houliston et al. 2010).

Gene knockdown by morpholino injection and overexpression by mRNA injection have been established (Momose and Houliston 2007), making *Clytia hemisphaerica* an excellent genetic model. In a series of elegant

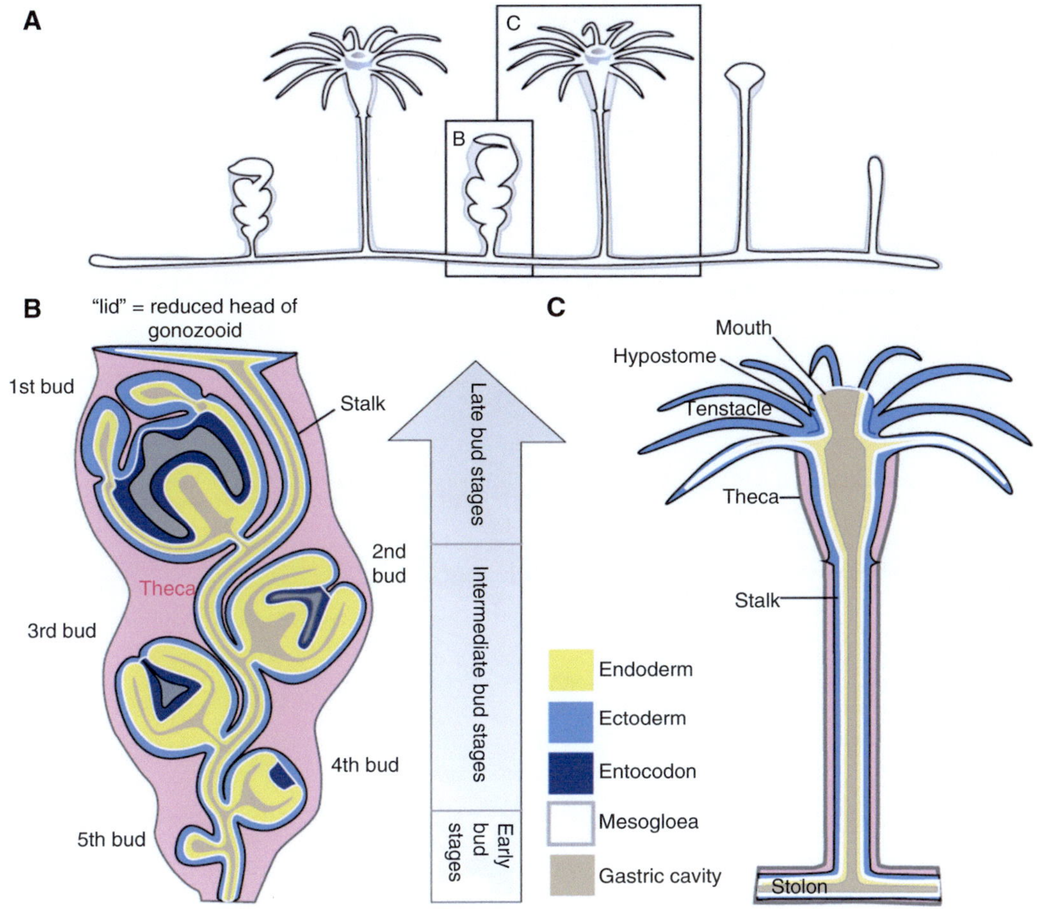

Fig. 6.20 The body plan of the dimorphic *Clytia* polyps. (**A–C**) Polyp colony with gonozooids (**B**) and trophozooids (**C**). Typically, gonozooids grow in the stolonal sections between two trophozooids. Their total number in a colony depends on the feeding regime. (**B**) Longitudinal section through a gonozooid with five medusa buds. The youngest buds are located near the stolon, while the developmentally most advanced buds are close to the "lid" of the gonozooid. (**C**) A longitudinal section through a trophozooid (© Ulrich Technau, Grigory Genikhovich, Johanna Kraus, 2015. All Rights Reserved)

experiments, Gary Freeman and colleagues demonstrated that (i) the oral-aboral axis is defined by the initiation site of the first cleavage furrow, which corresponds to the future gastrulation site and the "swimming posterior" pole of the planula larva (Freeman 1981a); (ii) *Clytia* eggs can only be fertilized at the site where the polar bodies are given off (Freeman and Miller 1982); and (iii) *Clytia* embryos bisected across the animal-vegetal axis can regulate and compensate for the lack of the missing part and generate normal planulae; however, such halves retain the original polarity and regenerate the missing positional values, i.e., vegetal halves gastrulate and develop posterior planula structures at the animal-most position available.

Moreover, if two animal or two vegetal halves are transplanted together, subsequent development continues according to the initial polarity of each half. Two combined vegetal halves initiate gastrulation at the equator of such chimeric embryos, where the animal-most positional values of both halves meet, while two animal halves form two gastrulation sites at the opposing ends of the chimera (Freeman 1981b).

Molecular studies on *Clytia hemisphaerica*, which were pioneered by the group of Evelyn Houliston, have significantly advanced our understanding of the regulation of polarity and the role of Wnt signaling. It has been shown that two Frizzled receptors are expressed at the opposing ends of the *Clytia* embryo. *CheFz1*

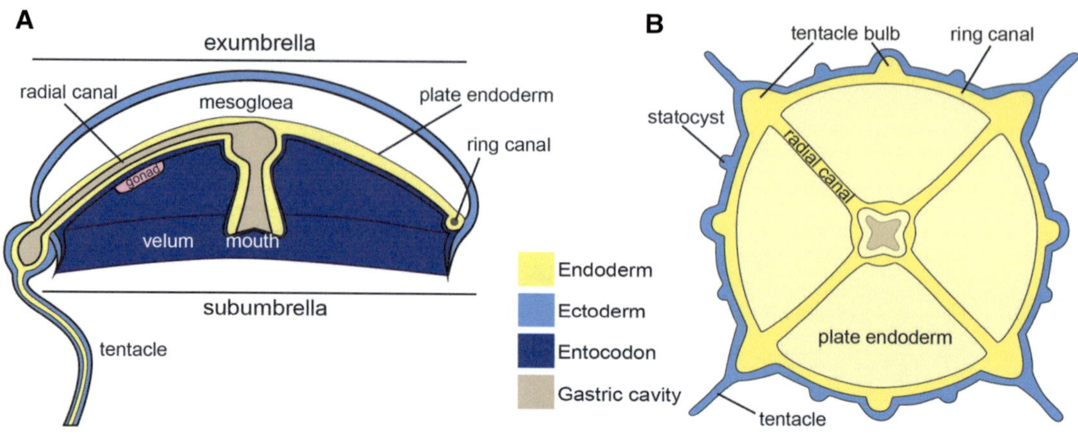

Fig. 6.21 The body plan of a *Clytia* medusa (© Ulrich Technau, Grigory Genikhovich, Johanna Kraus, 2015. All Rights Reserved). (**A**) Cross-section view. (**B**) Oral view was included

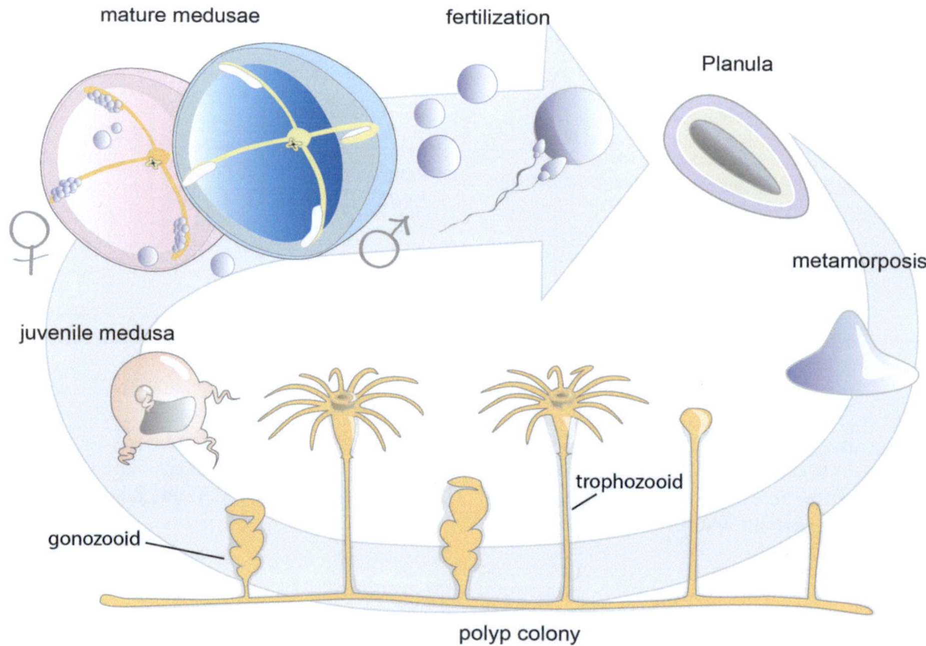

Fig. 6.22 The life cycle of *Clytia hemisphaerica* (© Ulrich Technau, Grigory Genikhovich, Johanna Kraus, 2015. All Rights Reserved)

mRNA is localized at the animal pole of the egg and the early embryo, and it is required for the development of the oral structures, while aborally localized maternal and most probably zygotic *CheFz3* antagonizes the action of the *CheFz1* (Momose and Houliston 2007). Further work revealed that it is *CheWnt3*, localized at the animal/oral pole of the eggs and early embryos, that is required for the activation of the β-catenin-mediated signaling at the gastrulation site (Momose et al. 2008). Subsequently, a link between the Wnt/β-catenin and the planar cell polarity (PCP) signaling systems has been shown. In *strabismus* morphants, apart from the disorganization of the cell polarity in the cell sheet reflected in the chaotic orientation of cilia, the embryos also failed to elongate during gastrulation, and the gastrulation site was expanded

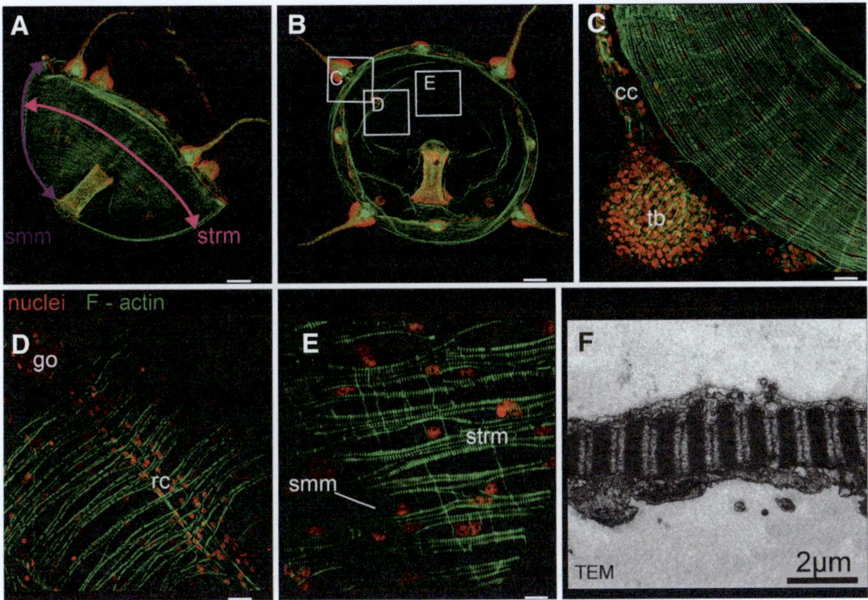

Fig. 6.23 Striated muscles in *Clytia hemisphaerica*. (**A**) Lateral view of a young medusa. *Arrows* show the orientation of smooth muscle type myonemes (smm, *purple*) and striated-type myonemes (strm, *pink*). (**B**) Oral view on a young medusa, overview. (**C**) Detail of the bell rim and velum area. *tb* tentacle bulb, *cc* circular canal. Smooth muscle type myonemes are found in the tentacle bulb, running along the circular canal and in thin rays in the velum. Striated muscle type myonemes are absent from the bell rim and around the circular canal, but are very dense in the velum (diagonal bands). (**D**) Detail of the subumbrella with a radial canal (*rc*) and a gonad (*go*). Striated-type myonemes in the subumbrella are arranged in loose circular bands, leaving large gaps in between them. Smooth muscles appear as thin rays departing from the middle of the bell running towards the bell rim. (**E**) Higher magnification of subumbrellar muscles revealing the branching pattern of the subumbrellar striated-type myonemes (strm, horizontal bands). Smooth muscles intercalate with the striated muscles in thin vertical lines (smm). (**F**) Transmission electron micrograph (*TEM*) of subumbrellar muscle showing the sarcomeric organization of the striated-type myonemes (TEM courtesy of S. Kaul-Strehlow). Scale bars: (**A**, **B**) 100 µm, (**C**) 25 µm, (**D**) 30 µm, (**E**) 8 µm (© Ulrich Technau, Grigory Genikhovich, Johanna Kraus, 2015. All Rights Reserved)

(Momose et al. 2012). A similar lack of elongation in the *CheFz1*, *CheWnt3*, and *ChDsh* (*dishevelled*) morphants suggests a link between the canonical Wnt signaling via CheWnt3, which might serve as a cue for the initial orientation of planar cell polarity (Momose et al. 2012).

Clytia also provided new insight into striated muscle evolution, with implications for the hypothesis that the entocodon is homologous to mesoderm in Bilateria. *Clytia* and *Hydractinia carnea* produce medusae by a very similar budding mechanism involving the formation of the entocodon giving rise to the striated muscles of the bell (see section "*Podocoryne carnea*").

Cnidarian and bilaterian striated muscles are strikingly similar in ultrastructural organization (Fig. 6.23; Boelsterli 1977; Squire et al. 2005), which was interpreted as evidence of a common evolutionary origin (Seipel and Schmid 2005). However, recently the evolutionary origin of 47 structural proteins with a function in vertebrate or *Drosophila* muscles has been traced by phylogenomic approaches and by expression analysis (Steinmetz et al. 2012). It turned out that some of the structural components of the striated muscle evolved much earlier than the first known striated muscles, for example, in the common ancestor of animals and plants, while others arose rather late in specific animal lineages (Steinmetz et al. 2012). Z-discs and their proteins are of particular interest, since they are likely the organizing entity of the striated muscles, generating the regularly spaced sarcomeres. Although variable in protein composition in different bilaterian clades, z-discs contain several proteins, which are shared between the striated muscle of *Drosophila* and vertebrates.

Only two of them, *mLIM* and *Ldb3*, are present in *Clytia*, yet these two genes are not expressed in the striated muscle cells of the medusa but rather in the endoderm, supporting the view that striated muscles might have evolved multiple times in animal evolution (Steinmetz et al. 2012). This unexpected result shows that even highly specialized cell types with complex arrangements of intracellular components might have arisen convergently, as crucial new proteins were added to ancestral proteins to create new structures.

Aurelia aurita and *Tripedalia* sp.

In contrast to the abovementioned hydroids with excellent embryological models, scyphozoans and cubozoans remain relatively understudied. Apart from descriptions of development, very little is known about developmental mechanisms in these cnidarian classes due to our inability (so far) to establish a complete life cycle laboratory culture of any representative of these groups. This is very unfortunate, especially if one considers the importance of gelatinous plankton in marine ecosystems and the abundance of papers in other areas of jellyfish biology, in particular in toxinology. Several lines of developmental biology research have, however, provided extremely interesting results.

Scyphozoans and cubozoans have a metagenetic development with only a few exceptions among scyphozoans. The polyp stage produces juvenile medusae by either monodisc (Cubozoa) or polydisc (most Scyphozoa) strobilation (Tardent 1978), rather than by lateral budding typical for Hydrozoa. In case of polydisc strobilation, the scyphistoma polyp subdivides into multiple juvenile medusae called ephyrae by transverse fission (Fig. 6.24). Strobilation starts at the oral end of the polyp and expands aborally until the polyp finally resembles a stack of saucers, each of which will form an ephyra (Fig. 6.25). In the wild, *Aurelia aurita* strobilates primarily in winter. Accordingly, in laboratory conditions, a decrease of temperature from 18 to 10 °C leads to induction of strobilation 3 weeks later. The molecular basis for this induction

remained elusive until very recently. In an elegant study, Fuchs and coworkers showed that strobilation is regulated by retinoic acid signaling and requires a "timer," releasing the strobilation program after a prolonged period of cold (Fuchs et al. 2014). They found a gene, *CL390*, which fulfills all the requirements for being this "timer." It encodes a secreted protein, a part of which, the peptide WSRRRWL, was shown to be an extremely potent strobilation inducer. Interestingly, structurally similar chemical compounds with two perpendicularly located indole rings, for example, indomethacin (Kuniyoshi et al. 2012) or 5-methoxy-2-methylindole (Fuchs et al. 2014), demonstrated similar potency to induce strobilation in *Aurelia*. It remains to be shown, however, whether other scyphozoans use a similar peptide to induce strobilation, as *CL390* appears to be a taxon-restricted gene.

As noted above, cubozoans form a medusa by monodisc strobilation; hence, in many cases the whole polyp transforms into a medusa, while in some cases the foot region remains attached to the substrate and regenerates a polyp (Straehler-Pohl and Jarms 2005). During this process, the polyp tentacles regress, fuse, and transform into composed sensory organs called "rhopalia" with highly developed camera eyes (Figs. 6.26 and 6.27; Stangl et al. 2002). The molecular basis of this fascinating transformation is still completely unknown.

The eyes at the bell rim are a remarkable feature of some jellyfish groups, in particular Cubomedusae but also some hydrozoan species (e.g., *Cladonema spec.*). The evolution of eye development was the one of the first EvoDevo topic addressed in jellyfish biology. Jellyfish eyes or even eye complexes are sometimes spectacularly sophisticated and are comparable to cephalopod camera eyes. In cubozoan medusae, it has been shown that these eyes play a major role in the control of directional swimming, although it is yet unclear to which extent Cubomedusae are able to see images and how the visual cues are neuronally processed. For example, the box jellyfish *Tripedalia sp.* have four rhopalia at their bell rim (Fig. 6.28c, d) that are each composed of six eyes of four different types (two lens eyes flanked by a set of pit eyes and slit eyes), besides

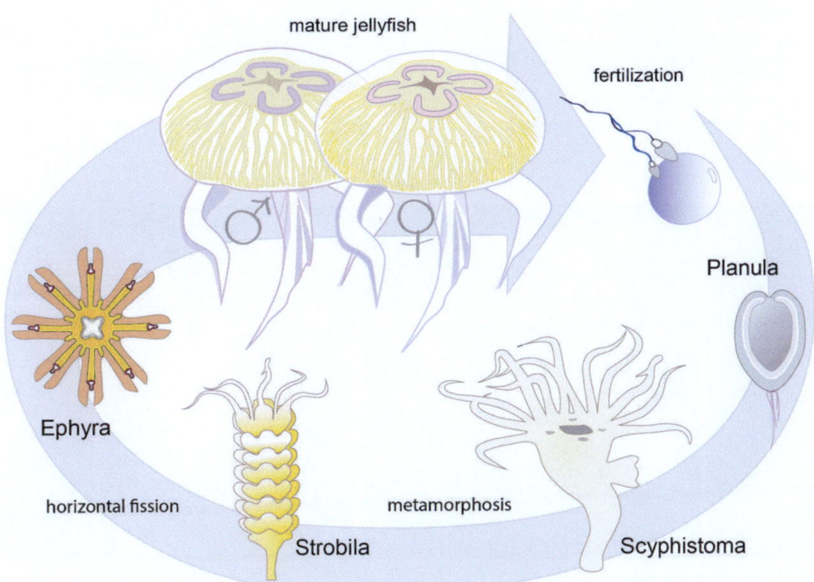

Fig. 6.24 The life cycle of a scyphozoan (*Aurelia aurita*) (© Ulrich Technau, Grigory Genikhovich, Johanna Kraus, 2015. All Rights Reserved)

Fig. 6.25 Strobilation in *Aurelia aurita* (collage). *Below*: Array of individual polyps in representative stages of the strobilation process, earliest stages on the *left*, latest on the *right*. The first visible sign of polyps undergoing strobilation is the formation of a horizontal constriction underneath the tentacles (polyp on the *left*). More and more such constrictions start to appear, until the body column of the polyp is almost entirely subdivided in a stack of discs. Each disc will differentiate into an ephyra, a juvenile jellyfish. Polyp tentacles are either gradually reduced or ectomized. The late strobila on the right displays a *reddish-brown* color due to the almost fully differentiated ephyrae in the stack. Ephyrae are one after the other liberated into the surrounding water (ephyra swimming above the strobila on the *right*) and transform into jellyfish (*top*). While the ephyrae are still being released, the foot region of the strobila regenerates into a scyphistoma. Strobilating scyphistomas are to scale large strobili being 5 mm. The young jellyfish has a diameter of 5 cm (© Ulrich Technau, Grigory Genikhovich, Johanna Kraus, 2015. All Rights Reserved)

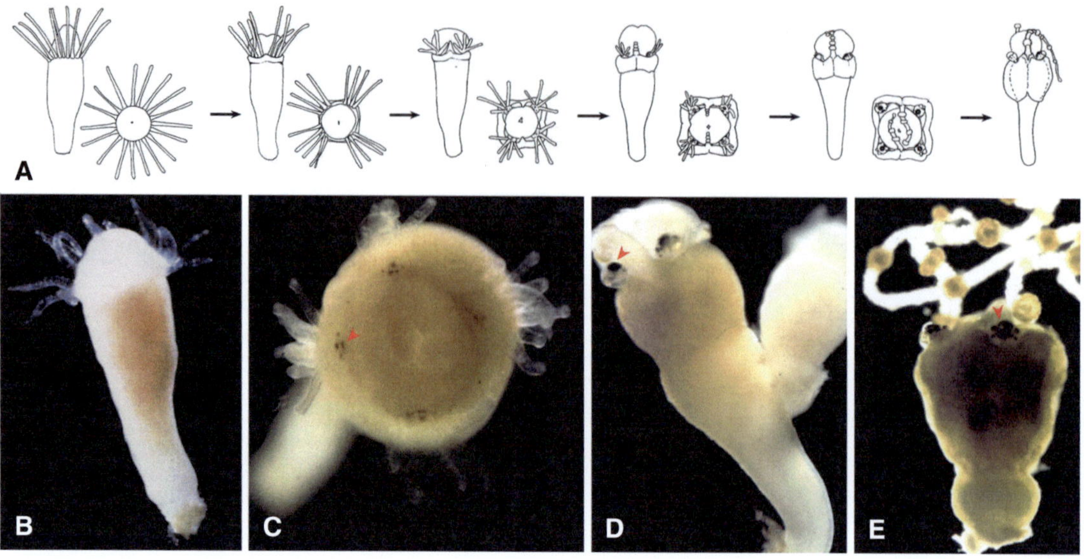

Fig. 6.26 Monodisc strobilation in the cubozoan *Carybdea spec.* (**A**) Schematic representation of the process. (**B**) *Carybdea* polyp prior to the onset of strobilation. (**C**) The tentacles fuse and degenerate to give rise to the rhopalia of the medusa. Oral view on the fusing tentacles and the forming eyes (see the pigment patches in the forming rhopalia – indicated by *red arrowheads*, the diameter of each is approximately 250 μm). (**D–E**) Later stages of medusa formation (Images reproduced with a kind permission from Stangl et al. (2002))

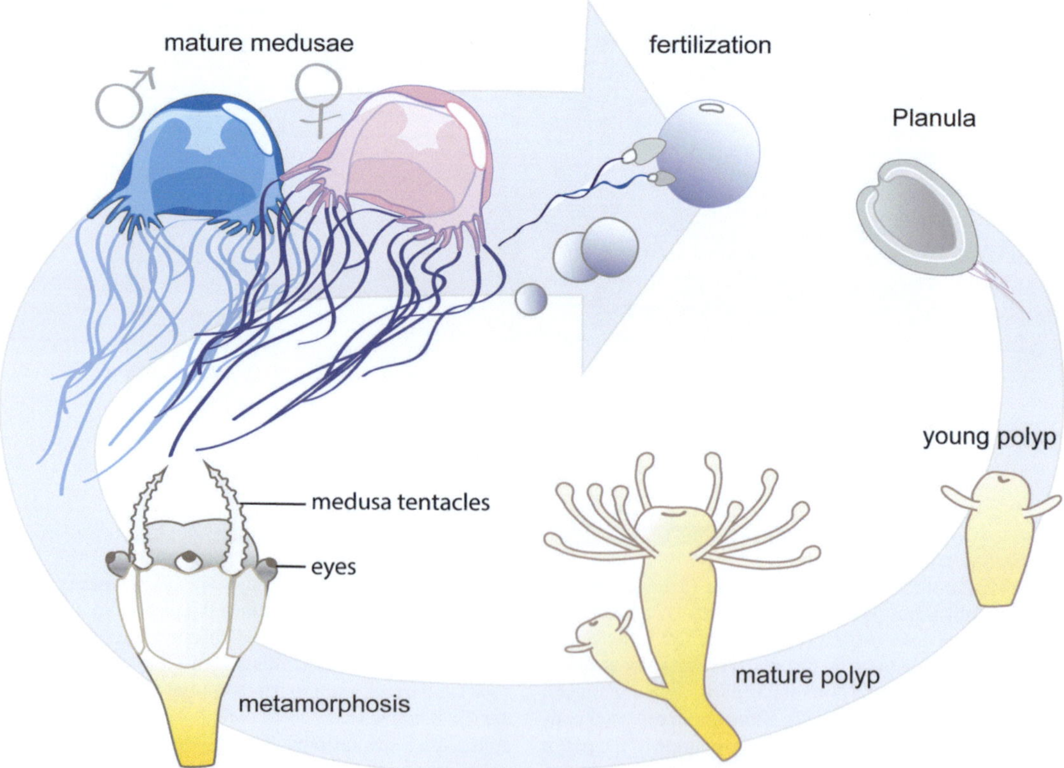

Fig. 6.27 The life cycle of a cubozoan (e.g., *Chironex fleckeri*) (© Ulrich Technau, Grigory Genikhovich, Johanna Kraus, 2015. All Rights Reserved)

a gravity-sensing statocyst. Cubozoan rhopalia are pending on thin stalks from the bell rim and carry a large statolith at the bottom acting as a stabilizer. This system keeps the rhopalia steadily orientated and more independent of the movement of the medusa bell, like a steadicam on a helicopter. One lens eye on each rhopalium is always directed towards the water surface. The other, larger lens eye is oriented downwards. The eyes allow the medusae to navigate efficiently through their mangrove habitat, to avoid predators and actively pursue prey (Garm et al. 2007a, b, 2008, 2011, 2012; Petie et al. 2011).

In *Tripedalia*, it has been shown that the camera eyes express a *PaxB* gene encoding a protein with a Pax2-like paired domain and a Pax6-like homeodomain, which is capable, if overexpressed in *Drosophila*, of inducing small ectopic eyes (Kozmik et al. 2003), similar to the *PaxB* isolated from the hydroid *Cladonema* (Suga et al. 2010). Interestingly, being sufficient to induce eye development in *Drosophila*, *Tripedalia PaxB* disrupted the development of the mouse eye (Ruzickova et al. 2009). This, together with the data on structural components of the eye (Kozmik et al. 2008; Suga et al. 2008) as well as on the formation of the rhopalia and the nervous system in *Aurelia* defined by *Otx1* and *POU* family genes (Nakanishi et al. 2010), led to two conflicting hypotheses about eye evolution: while some researchers suggest convergent development of the cnidarian and bilaterian eyes by recruitment of orthologous components (Kozmik et al. 2003, 2008), others suggest deep homology of the eyes and subfunctionalization of the Pax genes in different animal lineages (Suga et al. 2008, 2010; Graziussi et al. 2012).

Anthozoans

Acropora spec

Acropora millepora and *A. digitifera* are two of the most abundant coral reef species of the Great Barrier Reef in Australia and the Pacific Ocean around Japan, respectively. Both species have received a lot of attention in ecological studies on coral reefs. *Acropora* releases gametes only once a year in a dramatic mass spawning event.

The day of spawning is linked to the moon phase and the water temperature rise during spring and depends on the species and location. Because of these features, *Acropora* has not been developed as a lab model, but nevertheless many important insights into evolution and development have been gained, particularly from *Acropora millepora* (Ball et al. 2002). It was the first anthozoan

Cnidarian Models in EvoDevo Research
Among cnidarians, the currently most commonly used model systems are the anthozoan *Nematostella vectensis* and the hydrozoan *Hydra* (*H. vulgaris* or other *Hydra* species). *Hydra* was introduced as a model for experimental developmental biology by Abraham Trembley in 1744. Starting from the 1970s and until the year 2000, it was the dominant cnidarian system used for the study of axial pattern formation and stem cell differentiation. *Hydra* is famous for its high regenerative capacity and ability to be manipulated on the cell and tissue level, e.g., through transplantation and reaggregation experiments. The maceration of the tissue, BrdU labeling, and cell-type-specific antibodies allowed researchers to get a full quantitative account of all cell types and their differentiation kinetics. Research became refueled with the availability of the genome sequence in 2011 and the establishment of transgenic technology. In some cases, RNAi-based gene knockdown was also successful. Current research topics comprise, for instance, the role of Wnt and Notch signaling in axis formation and neuronal differentiation as well as the understanding of *Hydra* as a holobiont.

The potential of the sea anemone *Nematostella* as a lab model organism was first recognized in 1992 by Cadet Hand, who showed that the full life cycle can be maintained in the lab. Since around 2000, *Nematostella* was further established as a model organism by the Martindale and

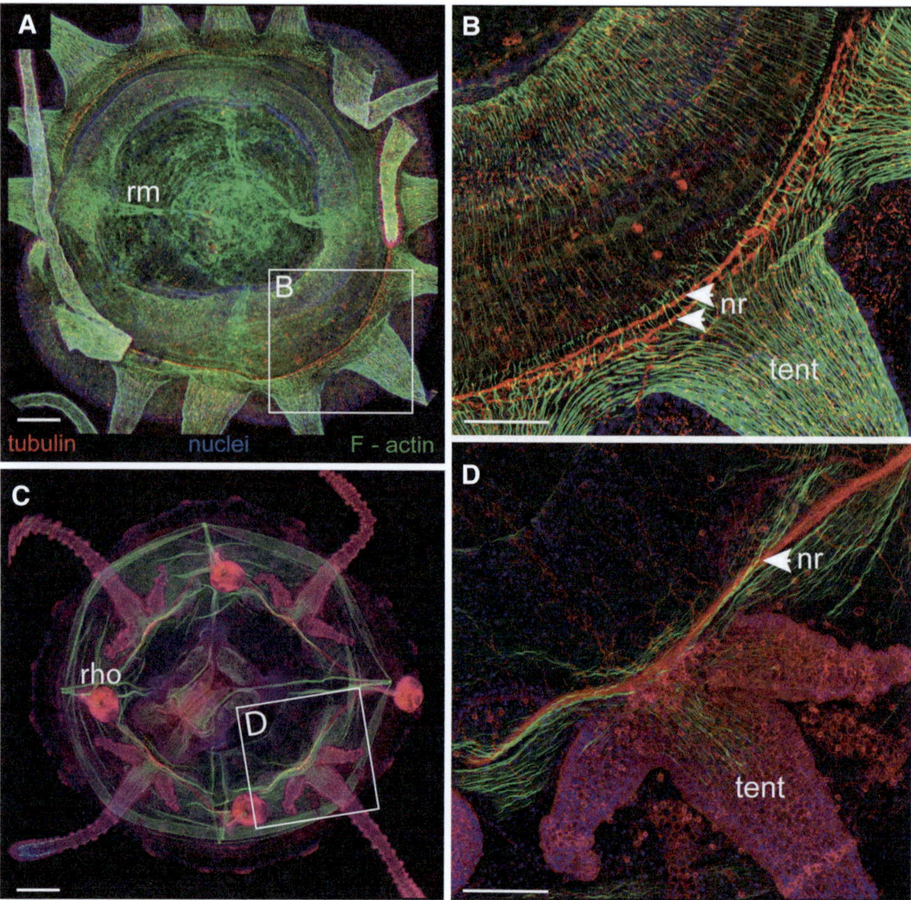

Fig. 6.28 Nerve ring in the scyphozoan *Aurelia aurita* and in the cubozoan *Tripedalia cystophora*. (**A**) Oral view of a polyp of *Aurelia aurita* stained with phalloidin for F-actin (*green*) in muscles, tyrosinated tubulin for neurons (*red*), and DAPI for nuclei (*blue*); *rm* retractor muscles. (**B**) Close-up showing the two nerve rings (*nr*) at the margin of the tentacles (*tent*). (**C**) Oral view of a young medusa of *Tripedalia cystophora* stained with phalloidin for F-actin (*green*), tyrosinated tubulin for neurons (*red*), and DAPI for nuclei (*blue*); *rho* rhopalium. (**D**) Close-up of bell rim showing numerous neurites in the velarium and the nerve ring (*nr*) connecting all rhopalia with tentacle base (*tent*). Scale bars: 200 μm (**A, C**), 100 μm (**B, D**) (© Ulrich Technau, Grigory Genikhovich, Johanna Kraus, 2015. All Rights Reserved)

Technau labs. The publication of the *Nematostella* genome in 2007 was the first reported genome sequence of a non-bilaterian animal. It corroborated the notion that anthozoans are, compared to hydrozoans, relatively slowly evolving and have maintained a surprising ancestral genetic complexity. The inducible spawning, the large numbers of embryos, the establishment of highly sensitive in situ hybridization protocols, morpholino-based gene knockdowns, overexpression by mRNA and plasmid expression, and more recently, the establishment of transgenics and CRISPR mutants have attracted many groups from other fields to this organism. As a result, the research topics have expanded enormously and now cover a wide range from early embryonic development, neuronal differentiation, functional genomics to gene regulation, regeneration, physiology, and ecology.

where an asymmetric expression of a *bmp4* homolog has been shown by in situ hybridization (Hayward et al. 2002). Before the genome of *Nematostella vectensis* was sequenced, two EST analyses corroborated the view that the transcriptome, and hence the gene repertoire, of anthozoans is surprisingly complex (Kortschak et al. 2003; Technau et al. 2005). Recently, the genome of the Japanese species, *Acropora digitifera*, was sequenced, highlighting the genomic features that might be helpful for understanding the symbiosis with the dinoflagellate *Symbiodinium* (Shinzato et al. 2011). Since the symbionts are crucial for the survival of the corals, research is directed towards understanding the uptake, interaction, and maintenance of the dinoflagellate (Davy et al. 2012; Krediet et al. 2013). Another important area of interest is to understand settlement and metamorphosis of the planula larva, since this is a decisive moment in the life cycle of any coral. Towards this goal, microarray studies have uncovered a number of candidate genes involved in settlement and calcification (Grasso et al. 2011; Hayward et al. 2011).

Nematostella vectensis

Nematostella vectensis is a brackish water sea anemone originally distributed widely at the American Atlantic coast in estuaries, but also found at the Pacific coast and in one location in England (Darling et al. 2004, 2005; Genikhovich and Technau 2009b). While *Hydra* is certainly a great model for stem cell differentiation and regeneration, its embryonic development is quite atypical and experimentally inaccessible. By comparison, *Nematostella* has an inducible and accessible embryogenesis (Hand and Uhlinger 1992; Fritzenwanker and Technau 2002), which offers the possibility to molecularly dissect its embryonic development and compare it with the development of bilaterians.

Molecular Tools and Genomic Features

A simple induction protocol for spawning based on light and temperature shift allows to control the life cycle and obtain thousands of embryos reproducibly on a daily basis (Figs. 6.29 and 6.30; Fritzenwanker and Technau 2002). This

facilitated studies of the cellular, molecular, and genetic basis of embryogenesis and allowed researchers to compare it with embryogenesis of bilaterians. Within a few years, two main labs established and improved an in situ hybridization protocol (Scholz and Technau 2003; Martindale et al. 2004; Genikhovich and Technau 2009a), generated ESTs (Technau et al. 2005), a BAC library (Chourrout et al. 2006), and the first non-bilaterian animal genome sequence (Putnam et al. 2007). Importantly, in order to dissect gene functions, morpholino-mediated gene knockdown (Magie et al. 2007; Rentzsch et al. 2008), over-expression by mRNA injection (Wikramanayake et al. 2003), and transgenesis with germ line transmission (Renfer et al. 2010) have been established alongside routine gene expression studies by in situ hybridization (Figs. 6.31, 6.32, and 6.33).

The combination of these resources and tools allowed for the rapid cloning and accumulation of gene expression data, which are – at least in part – collected in the expression database Kahi Kai (Ormestad et al. 2011) and also enabled functional analysis of many genes. Several families of transcription factors have been systematically assayed by in situ hybridization (e.g., Magie et al. 2005; Matus et al. 2007).

The genome of *Nematostella* was the first one sequenced of a non-bilaterian animal, and it offered many important insights into the evolution of metazoan genomes (Putnam et al. 2007). The sequence analysis of the genome revealed a surprising complexity in the gene repertoire. Furthermore, the high number of conserved exon-intron boundaries proved that the intron-rich gene structure of vertebrates is ancestral and the intron-poor gene structure of *Drosophila melanogaster* and *Caenorhabditis elegans* is derived. On the chromosome level, *Nematostella* also showed a surprising degree of synteny conservation to humans, suggesting that the chromosome structure in these two lineages is slowly evolving (Putnam et al. 2007). In line with this, recent genome-wide mapping of enhancers and promoters based on histone modifications revealed that the principle of complex cis-regulation is ancient and can be traced back to at

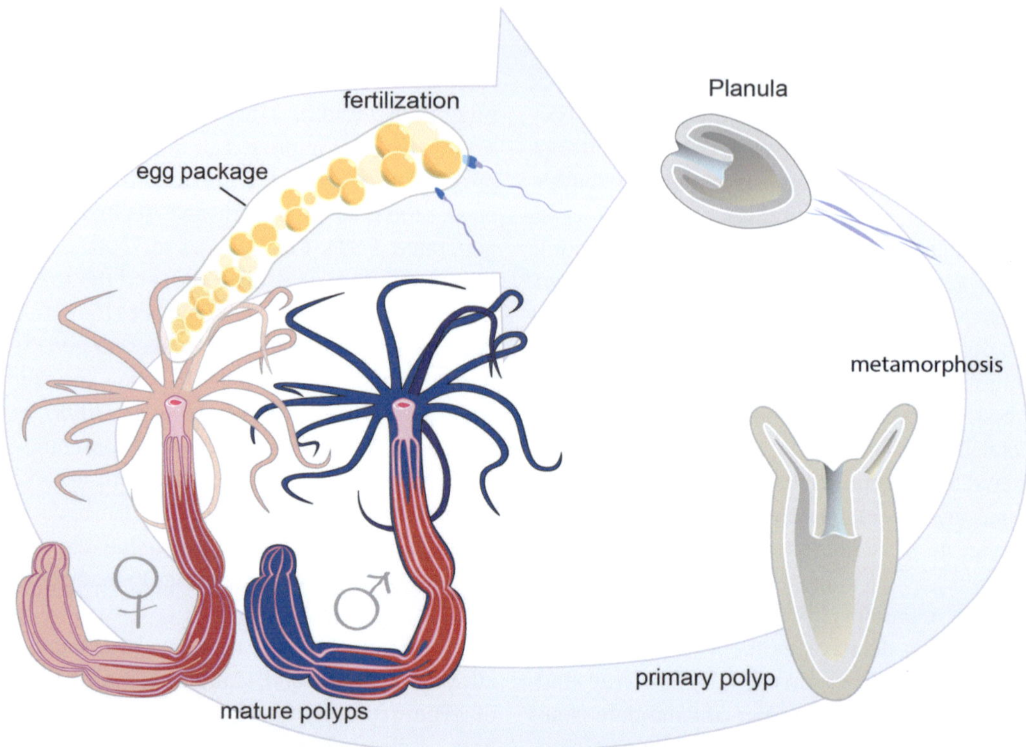

Fig. 6.29 The life cycle of *Nematostella vectensis* (© Ulrich Technau, Grigory Genikhovich, Johanna Kraus, 2015. All Rights Reserved)

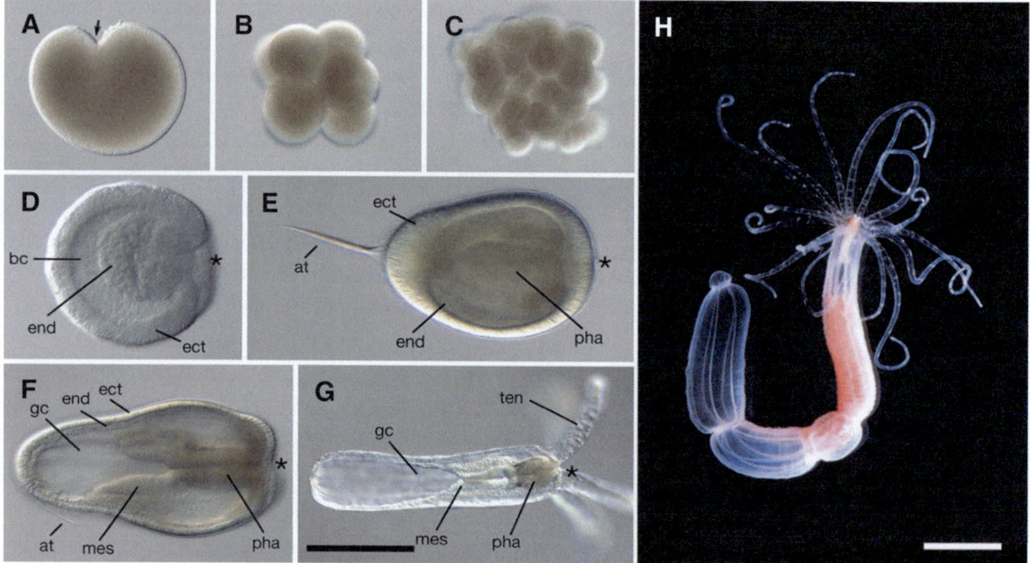

Fig. 6.30 Overview of the embryonic and larval development of *Nematostella vectensis*. (**A**) First cleavage. *Arrow* points at the animal pole. (**B**) Eight-cell stage embryo. (**C**) Morula stage. (**D**) Gastrula stage. (**E**) Planula stage. (**F**) Metamorphosing late planula. (**G**) Primary polyp. (**H**) Adult polyp. Abbreviations: *ect* ectoderm, *end* endoderm, *pha* pharynx, *at* apical tuft, *bc* blastocoel, *mes* mesentery, *gc* gastric cavity, *ten* tentacle. *Asterisk* indicates blastopore. Scale bar in (**G**) represents 200 μm and is valid for (**A–G**). Scale bar in (**H**) represents 1 cm (Images taken from Genikhovich and Technau (2009b): Emerging model organisms. Cold Spring Harbor Press, with permission)

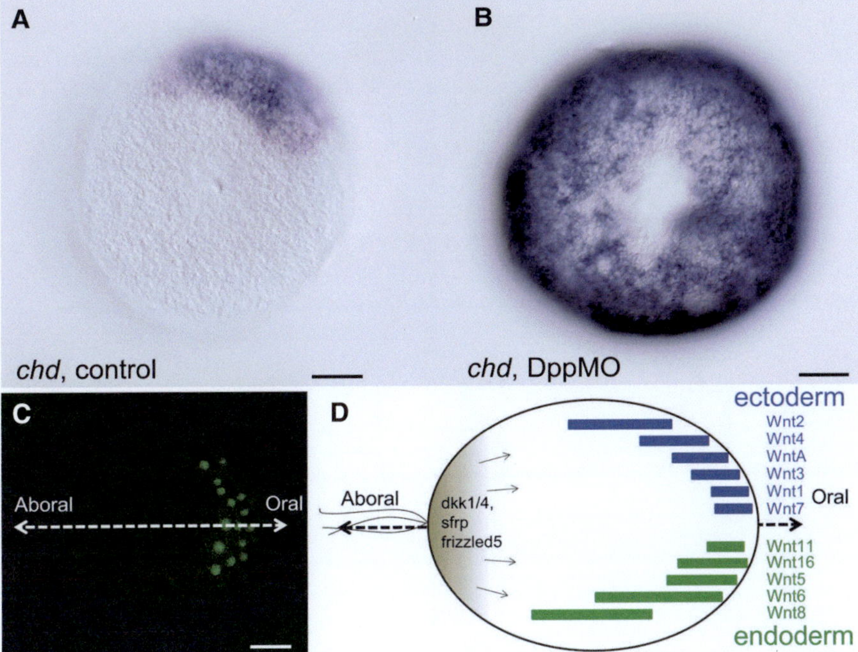

Fig. 6.31 Wnt and BMP signaling regulate the body axes in *Nematostella vectensis*. *Chordin* expression, which is normally confined to one side of the blastopore (**A**), is radialized (**B**) upon a morpholino-mediated knockdown of *NvDpp* (*bmp4* homolog). β-catenin fused to the fluorescent protein Venus is observed in the nuclei on the future oral side of the blastula stage embryo (**C**), which has been injected with β-catenin: Venus mRNA at the zygote stage (the β-catenin::Venus construct was kindly provided by T. Momose, Villefranche-sur-Mer). (**D**) Different Wnt genes are expressed in the domains staggered along the oral-aboral axis in the ectoderm and the endoderm of the *Nematostella* planula. Aborally, both, Wnt receptor *frizzled5* and secreted Wnt antagonists *dkk1/4* and *sfrp* are expressed. Scale bars: 40 μm (© Ulrich Technau, Grigory Genikhovich, Johanna Kraus, 2015. All Rights Reserved)

least the common ancestor of bilaterians and cnidarians (Schwaiger et al. 2014). Surprisingly, post-transcriptional regulation by miRNAs differs drastically from that of bilaterians. Unlike bilaterian miRNAs, which bind their mRNA targets via the seed sequence and inhibit translation, *Nematostella* miRNAs bind to their target mRNAs with almost full complementarity and lead to cleavage of the target mRNA (Moran et al. 2014). There is evidence that this slicing mechanism is also at work in *Hydra* and possibly also in sponges. This mode of action is reminiscent of that of plant miRNAs or siRNAs and may represent the ancestral mechanism of miRNA-mediated silencing. Notably, all non-bilaterian animals possess crucial miRNA biogenesis factors (e.g., HYL-1) that were previously thought to be plant-specific, but they also have the full animal biogenesis machinery (Moran et al. 2013). These findings suggest that transcriptional and post-transcriptional regulation of expression evolved independently in animals and that seed-based mechanisms of miRNA-mediated silencing only became dominant in bilaterians.

Normal Development

Each *Nematostella* female produces several hundred eggs embedded into a gelatinous egg package, which is formed in the gastric cavity and released through the mouth (Fig. 6.29). Cleavage starts 2–3 h after fertilization; however, the first two rounds of nuclear divisions are not followed by complete cell division. The two-cell stage is rarely observed, and the cells remain connected by cytoplasmic bridges at the four cell stage. The full separation of blastomeres is achieved at the eight-cell stage. At 18 °C, cells divide every 50 min and the

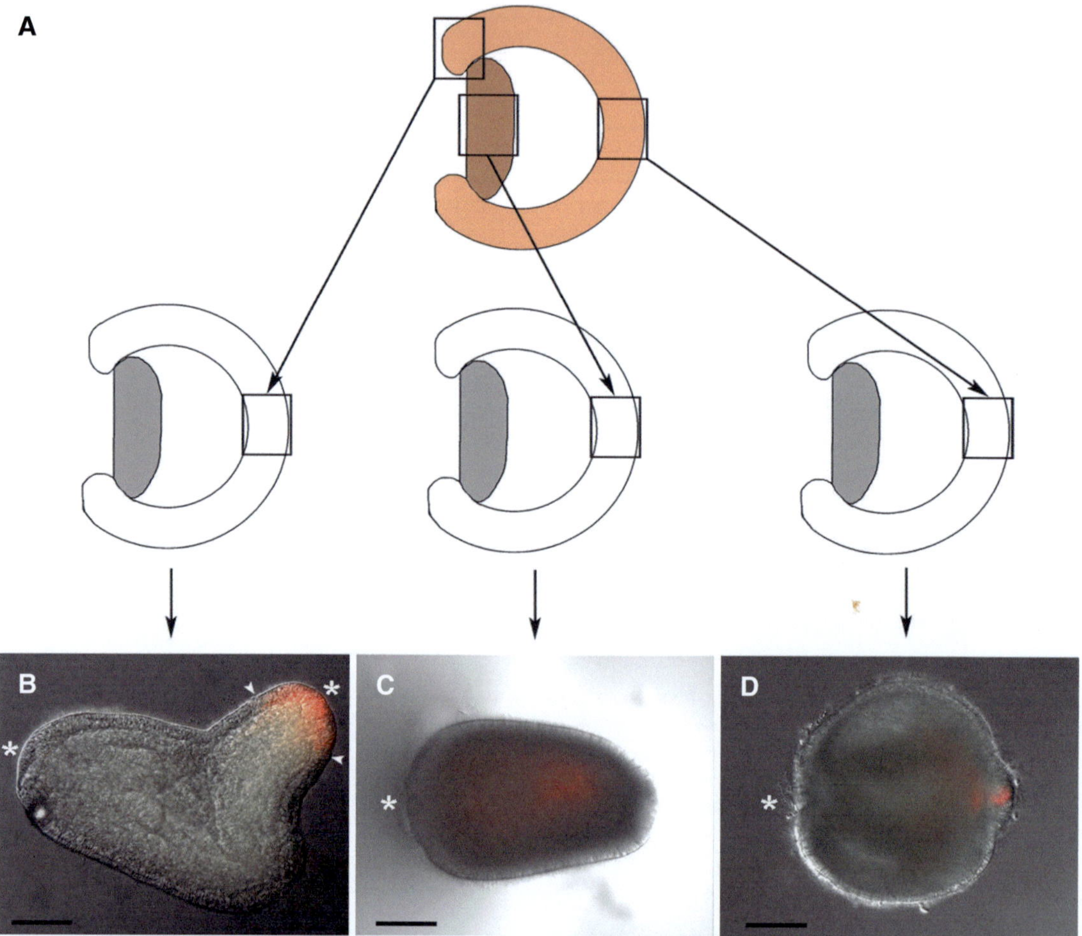

Fig. 6.32 Organizer activity of the blastopore lip of a *Nematostella* gastrula. (**A**) The transplantation scheme is exemplified at the *top* with the donor tissue marked in *red*. (**B**) Transplantation of the blastopore lip leads to the efficient formation of a secondary body axis (*arrowheads*). Transplanted pre-endodermal plate cells become internalized and intermingled with the host tissue. (**C**) Aboral pole tissue remains at an aboral pole position in the ectoderm without altering the body axis of the host. (**D**) *Asterisks* indicate the oral pole of the embryos. Transplantations were performed 24 h post-fertilization. Images were taken at 96 h post-fertilization. Scale bars: 50 μm (Reproduced with permission from Kraus et al. (2007). Copyright Elsevier 2007)

blastocoel becomes visible already at a 16–32 cell stage (Fig. 6.30; Fritzenwanker et al. 2007). When the blastula stage is reached, marked by the formation of a single cell layer of blastomeres and a blastocoel, the embryos undergo four to five pulsating invaginations each linked to a synchronized cell division. Inhibitor experiments have shown that these pulsating invaginations depend on actin, microtubuli, and DNA synthesis. The pulsations stop when the cell divisions become asynchronous at a mid-blastula stage (Fritzenwanker et al. 2007).

Since the polar bodies are usually lost as the egg squeezes through the endoderm when leaving the gonad, there is no landmark to tell the animal pole from the vegetal pole. It therefore was at first also unclear where gastrulation occurred and whether the later oral opening corresponds to the animal or vegetal pole. However, labeling techniques have demonstrated that it is the animal pole where cleavage starts and where gastrulation occurs, giving rise later to the oral opening (Fritzenwanker et al. 2007; Lee et al. 2007).

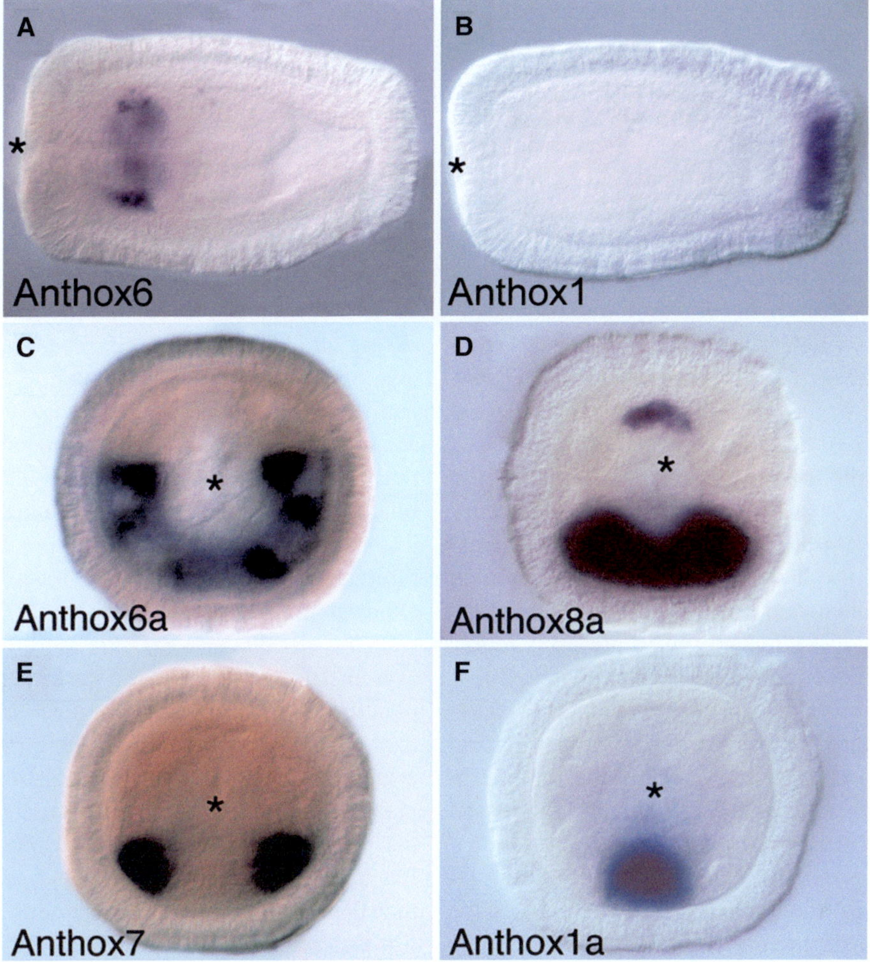

Fig. 6.33 Hox gene expression in *Nematostella*. (**A**) *Anthox6* (=*HoxA*) is expressed in the pre-pharyngeal endoderm in a radially symmetric domain. (**B**) *Anthox1* (=*HoxF*) is expressed in the aboral ectoderm in a radially symmetric domain. (**C–F**) Another four Hox genes are expressed in staggered endodermal domains along the directive axis. (**C**) *Anthox6a* (=*HoxB*) is expressed in five mesenterial chambers. (**D**) *Anthox8a* (=*HoxDa*) is expressed in three mesenterial chambers. (**E**) *Anthox7* is expressed in two mesenterial chambers to the *left* and to the *right* of the one where *Anthox1a* (=*HoxE*) (**F**) is expressed. The length of the embryo along the oral-aboral axis is approximately 250 μm; *asterisks* mark the blasto-pore (The images are reproduced from Ryan et al. (2007))

If cultured at 18 °C, gastrulation starts at around 18–20 h of development and begins with the formation of a pre-endodermal plate, which is marked by a flattening on the animal side of the embryo. Prior to invagination, presumptive endodermal cells show hallmarks of bottle cells and undergo an incomplete epithelial-mesenchymal transition (EMT): they form apical constrictions and loosen their apical adherens junctions, their nucleus migrates to a more basal position, and the basal surface of the cells forms numerous filopodia (Kraus and Technau 2006). However, the cells never seem to completely detach from each other during the invagination process (Magie et al. 2007). After gastrulation, the embryo develops into a planula which leaves the egg package, swims for several days, and gradually transforms into a primary polyp with four tentacles (Figs. 6.29 and 6.30). The polyp starts to feed and reaches sexual maturity after 4–6 months depending on the intensity of feeding.

Germ Layer Formation and the Evolution of the Mesoderm

Cnidarians are diploblastic and since they are the sister group to the triploblastic bilaterians, they are very informative for understanding the evolution of the mesoderm. To this end, research has been directed at revealing the repertoire and role of "mesodermal" genes in cnidarians. Many genes known to play a conserved role in mesendoderm formation and differentiation in bilaterians were isolated from *Nematostella*. These include *Brachyury*, *foxA*/*forkhead*, *mox*, *gata*, *snail*, *twist*, and *mef2*. Virtually all of these genes are expressed either in the whole endoderm, in part of the endoderm, or at the blastopore lip: *Brachyury* and *forkhead* are expressed at the ectodermal margin of the blastopore (Scholz and Technau 2003; Fritzenwanker et al. 2004; Martindale et al. 2004); *snailA* and *snailB* are early markers of the endoderm (Fritzenwanker et al. 2004; Martindale et al. 2004; Magie et al. 2007); *mox* and *twist* are expressed only after gastrulation in a circumpharyngeal ring of the endoderm (Martindale et al. 2004); *gata* is first expressed in ectodermal single cells, presumably neuronal precursors, and then in broad longitudinal domains of the endoderm, presumably also neurons along the parietal muscles (Martindale et al. 2004); *mef2* has several splice variants of which one class, coding for short proteins, is expressed in single cells of the ectoderm, whereas the other class, coding for long proteins, is expressed in the whole endoderm and weakly in the ectoderm (Genikhovich and Technau 2011). Notably, for most of these genes, the function is not yet clear. Knockdown of *snail* paralogs surprisingly failed to show a phenotype (Magie et al. 2007). In the case of *mef2*, morpholino knockdown of either all or selected splice variants showed that different splice variants are involved in either nematocyte differentiation or endoderm differentiation after gastrulation (Genikhovich and Technau 2011). Thus, more functional work needs to be done to clarify the specific functions of these "mesodermal" genes in a diploblastic organism.

Axis Formation: The Role of Wnt and BMP Signaling

Cnidarians have only one apparent body axis and have been often categorized as "Radiata" in older textbooks. However, morphologists have long since recognized that anthozoan polyps display an internal asymmetry in the arrangement of the retractor muscles in the mesenteries and the position of the siphonoglyph in the pharynx, making these animals clearly bilaterally or, in some cases, biradially symmetric. The second body axis running orthogonally to the oral-aboral axis has been named the directive axis. In *Nematostella*, for example, the location of the retractor muscles shows that this sea anemone is bilaterally symmetric. Long before the mesenteries or the siphonoglyph form, the presence of the directive axis can be observed at the level of gene expression. Homologs of *dpp* (*bmp2/4*), *bmp5-8*, *chordin*, *gremlin*, and *gdf5-like* were found to be asymmetrically expressed in gastrulae and early planula larvae, providing a molecular correlate for the directive axis (Finnerty et al. 2004; Matus et al. 2006a, b; Rentzsch et al. 2006). Interestingly, *dpp* and *chordin* expression starts radially around the blastopore and then shifts to one side of it (Rentzsch et al. 2006). Functional studies have then demonstrated that BMP signaling is required for the symmetry break of *dpp* and *chordin* expression and the establishment of the directive axis at the molecular level (Fig. 6.30; Saina et al. 2009; Genikhovich et al. 2015).

While the formation of the directive axis obviously depends on BMP signaling, the oral-aboral axis requires Wnt signaling. A total of 13 Wnt ligands belonging to 11 subfamilies have been isolated from *Nematostella* (Fig. 6.31; Kusserow et al. 2005). Since only a Wnt9 is lacking, it is clear that the common ancestor of cnidarians and bilaterians possessed almost the complete set of Wnt subfamilies, and subsequently, in some animal lineages like *Drosophila* or, even more drastically, in *Caenorhabditis elegans*, several Wnt subfamilies were secondarily lost. All of these 13 Wnt subfamilies are expressed in wider or narrower rings around the blastopore, either in the endoderm or in the ectoderm, forming a staggered expression pattern (Kusserow et al. 2005;

Guder et al. 2006a; Lee et al. 2006). In line with this, nuclear β-catenin accumulates only in the oral half of the early embryo (Wikramanayake et al. 2003), where dishevelled protein is localized (Lee et al. 2007).

The presence of Wnt/β-catenin signaling in the animal (oral) half of the embryo is correlated with the developmental potential of these cells: when eight-cell embryos or early gastrula stage embryos are divided into animal and vegetal halves, only the animal half can regenerate a small but normal primary polyp, whereas the vegetal half is forming only a "dauerblastula," unable to restore the full pattern (Fritzenwanker et al. 2007; Lee et al. 2007). This indicates that the animal/oral half has organizing activity which is lacking in the vegetal/aboral half. Similar results were obtained in the hydrozoan *Podocoryne carnea* (Momose and Schmid 2006), but not in *Clytia gregaria* (Freeman 1981b). Moreover, the instructive capacity can clearly be localized to the blastopore lip of the mid-gastrula, as only the transplantation of the fragment of the blastopore lip, but not of the pre-endodermal plate or blastocoel roof, was able to induce a secondary body axis (Kraus et al. 2007). This is reminiscent of the inductive capacities of the dorsal blastopore lip of frogs (the "Spemann organizer"), indicating that the principle of a blastoporal organizer may be an ancestral feature of eumetazoan development.

Ectopic activation of the Wnt pathway by LiCl or azakenpaullone treatment induces an expansion of nuclear β-catenin to the vegetal side and concomitantly to an expansion of the endodermal tissue. This has been interpreted such that β-catenin plays a role in the establishment of the mesendoderm, similar to sea urchins (Wikramanayake et al. 2003). Supporting this view, knockdown of β-catenin appeared to block endoderm differentiation, but not gastrulation, whereas knockdown of Strabismus, a Frizzled co-receptor in the planar cell polarity pathway of non-canonical Wnt signaling, blocked gastrulation. However, analysis of downstream targets of β-catenin by microarray revealed that also numerous ectodermal genes are activated by β-catenin and ectopic activation of β-catenin by azaken-

paullone can lead to expansion of oral marker gene expression and formation of ectopic heads (Rottinger et al. 2012; Marlow et al. 2013). Thus, Wnt/β-catenin signaling might be involved both in axis formation and germ layer differentiation. Interestingly, the aboral pole expresses both a Frizzled receptor (*frz5*) and *sfrp*, a Wnt antagonist (Kumburegama et al. 2011). Functional analysis has yet to clarify their role in the regulation of the oral-aboral body axis.

In summary, Wnt/β-catenin signaling seems to be crucial for establishing an oral-aboral polarity and is likely to be involved in the organizing capacity of the blastopore lip, whereas BMP signaling is responsible for the symmetry break and establishment of the directive axis, perpendicular to the oral-aboral axis.

Hox Genes

Since in many bilaterians canonical Wnt signaling is involved in posterior development and BMP signaling forms a morphogen gradient along the dorsoventral axis, it is tempting to speculate that these oral-aboral and directive axes of the anthozoan are homologous to the anterior-posterior and dorsoventral axis of bilaterians, respectively. A famous example of the conservation of developmental processes is the staggered, colinear expression of Hox genes along the anterior-posterior axis of bilaterians (review in Gehring et al. 2009). In vertebrates and insects, Hox genes have been shown to have a homeotic role in the segmental identity along the anterior-posterior axis. As putative sister group of the Bilateria, Cnidaria were of obvious interest for assaying when the role of Hox genes in axis patterning arose. Indeed, early expression analyses showed that one anterior Hox gene is expressed in the pharyngeal endoderm and that another Hox gene that has been interpreted as a posterior Hox gene is expressed in the aboral ectoderm, while three others were expressed in the middle of the planula larva of *Nematostella* (Fig. 6.33; Finnerty et al. 2004).

On the basis of these patterns, the authors concluded that the oral-aboral axis of *Nematostella* corresponds to the anterior-posterior axis of Bilateria. However, the orthology of the non-anterior Hox genes has been much disputed and it is therefore not clear whether they are homologous to posterior or central Hox genes or whether they arose independently in Cnidaria (Chourrout et al. 2006; Ryan et al. 2007; Thomas-Chollier et al. 2010). At least, the interactions of the Hox proteins with the ancient TALE transcription factors Meis and Pbx are conserved and most probably were in place in the common ancestor of Cnidaria and Bilateria (Hudry et al. 2014). Importantly, however, the majority of Hox genes in *Nematostella* display a staggered expression pattern in the endoderm along the directive axis in planulae, on the opposite side of *bmp4/dpp* (Finnerty et al. 2004; Ryan et al. 2007). Although at present functional data are lacking, these patterns are suggestive of a role of the Hox genes along the directive axis, possibly by defining the positions of mesenteries, endodermal folds harboring retractor muscles and gonads. Irrespectively, the Hox cluster has been broken down in *Nematostella* except for few lineage-specific linkages (Chourrout et al. 2006). In *Acropora digitifera*, however, at least the linkage between the anterior genes and one of the non-anterior/posterior Hox genes has been maintained (DuBuc et al. 2012). Interestingly, both in *Acropora* and *Nematostella*, *evx*, a non-Hox gene usually flanking the Hox cluster in bilaterians on the posterior Hox genes side (Minguillon and Garcia-Fernandez 2003), is located between two anterior Hox genes, suggesting an ancient genomic rearrangement in the anthozoan or cnidarian lineage, disrupting a contiguous cluster organization. Together with the findings of Wnt and BMP signaling, it remains therefore open whether or how the two body axes of anthozoans can be homologized with the two body axes of Bilateria.

The Development of the Nervous System

In most Bilateria, the dorsoventral axis is linked to the formation of a central nervous system at the side of BMP repression through *chordin*. Despite the localized *chordin* expression in *Nematostella*, its nervous system, as in other cnidarians, is diffuse and so far no signs of centralization beyond the ring-like accumulation of certain subpopulations of neurons in oral or aboral body regions (Marlow et al. 2009) as well

as the paired nerve cord-like structures on either side of the eight mesenteries have been shown (Fig. 6.34; Nakanishi et al. 2012).

A number of conserved neuronal markers and transcription factors are expressed in neurons or neuronal precursors in *Nematostella*. This includes the basic helix-loop-helix (bHLH) proneural gene *achaete-scute*, of which four paralogs have been identified. At least one of them, *NvAshA*, is expressed in single cells, presumably neuronal precursor cells, in the ectodermal aboral half of the embryo (Layden et al. 2012). Functional manipulation showed that this *achaete-scute* homolog drives neuronal differentiation (Layden et al. 2012). Furthermore, several intracellular transport, synaptic, and transmitter systems of Bilateria have been also identified in *Nematostella*, suggesting a common origin of cnidarian and bilaterian nervous systems (Marlow et al. 2009; Nakanishi et al. 2012). *Nematostella* also uses conserved ion channels for neuronal signaling, both in neurons as well as in nematocytes (Mahoney et al. 2011). Strikingly though, the comparison of amino acid composition of the ion selection filter shows that *Nematostella* voltage-gated Na^+ channels have evolved from voltage-gated Ca^{2+} channels independently from Bilateria-specific Na^+ channels (Gur Barzilai et al. 2012). Notch signaling, which is involved in singling out neuronal precursors in the neuroectoderm of bilaterians, is also involved in neuronal and possibly also nematocyte differentiation (Marlow et al. 2012; Layden et al. 2014). On the other hand, genes regulating fast neurotransmission using GABA, monoamines, and acetylcholine surprisingly appear to be not restricted to neurons but rather to specific parts of the endoderm (Oren et al. 2014).

The establishment of transgenesis (Renfer et al. 2010) now allows researchers to study neurogenesis in vivo. Using an ELAV-promoter-driven transgenic line, Nakanishi and colleagues showed that the first neurons are born in the late blastula/early gastrula, hence before both germ layers are formed (Nakanishi et al. 2012). It also revealed that neurons are formed in both endoderm and ectoderm independently and do not require migration of precursor cells between the layers (Nakanishi et al. 2012). Furthermore,

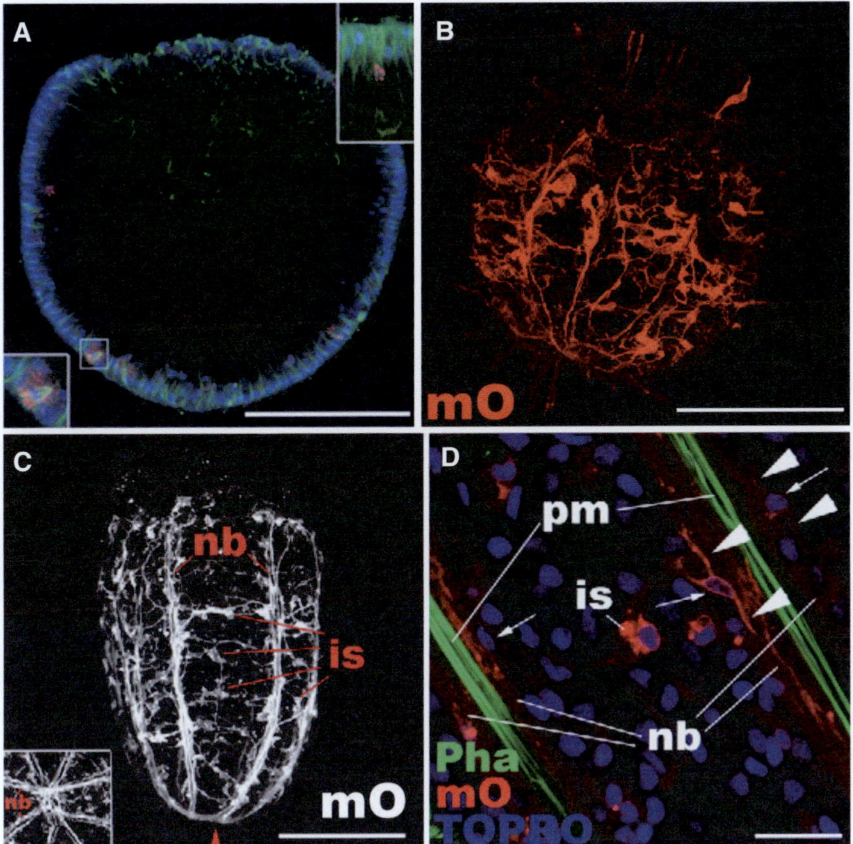

Fig. 6.34 Formation of the ELAV-positive nervous system monitored in NvElav 1::mOrange transgenics. (**A**) Elav -positive cells (*red*) neurons differentiate already at an early gastrula stage in the aboral half from epithelial cells. (*blue* nuclei, *green* acetylated tubulin), (**B**) Late planula expressing NvELAV1::mOrange with fluorescent neurons in the ectoderm and endoderm. (**C**) Endodermal nervous system in the transgenic early primary polyp shows the presence of neurite bundles on either side of each mesentery merging at the aboral pole (*red arrowhead*, inset). (**D**) Close-up view of the neurite bundles of the transgenic polyp (*red*, mO) on either side of the phalloidin-positive parietal muscle (*green*, Pha) at the base of the mesentery. The nuclei are stained with TOPRO (*blue*). *pm* parietal muscle, *nb* neurite bundle, *is* isolated ELAV::mOrange positive neuron. Scale bars: (**A–C**) 100 μm, (**D**) 10 μm (Reproduced with permission from Nakanishi et al. (2012)

ELAV neurons are born directly from epithelial cells, as in bilaterians, and not from a population of interstitial stem cells, as in *Hydra* and other hydrozoans (see above). This raises the question whether the interstitial stem cell system in hydrozoans is a derived feature of this cnidarian class. In summary, the molecular underpinning of the nervous system of *Nematostella* shows many developmental and structural components also known from bilaterian nervous systems, but also a number of unexpected features. Thus, while many advances have been made in recent years, many questions remain to be addressed (see below).

BEYOND DEVELOPMENT: CNIDARIANS AND THE ENVIRONMENT

Recently, researchers in many fields have realized the importance of symbiotic interactions between a host organism and bacteria or other eucaryotes for physiology, immunity, and development. Rosenberg has coined the term "holobiont" to express the notion that an organism is in fact an assemblage of several organisms (Rosenberg et al. 2007, 2010; Bosch 2012b). Indeed, the role of symbiotic algae on the physiology of corals and sea anemones has long been recognized. More recently, distinct bacterial associations with cnidarians, in particular in *Hydra*, have been identified and their role is being revealed (reviewed in Bosch 2012a). These close microbial associations are often species-specific and are shaped actively by the host (Fraune and Bosch 2007; Franzenburg et al. 2012, 2013) and may influence specific traits of the development and of the physiology of these animals.

CONCLUDING REMARKS: CNIDARIAN AND BILATERIAN EVODEVO

Cnidaria is a diverse, evolutionarily and ecologically successful phylum and as such this group is worth being investigated in its own right. However, as the putative sister group to the bilaterians, cnidarians are of crucial importance for understanding the evolution of key bilaterian traits: the evolution of bilaterality, the evolution of mesoderm and its derivatives, the evolution of a (central) nervous system, the evolution of pluripotent and multipotent stem cells, and regeneration.

Much can be learnt by comparing cnidarian and bilaterian genomes in order to understand evolutionary trajectories. For instance, genes in *Drosophila* and *Caenorhabditis elegans* are intron-poor, whereas in vertebrates they are intron-rich. The genome analysis of the sea anemone *Nematostella vectensis* revealed that intron-rich gene structures is the ancestral state, at least for the last common ancestor of cnidarians and bilaterians. In fact, humans and the sea anemone have both retained about 80 % of the ancestral intron positions, whereas only about 20 % are retained in ecdysozoans (Putnam et al. 2007; Miller and Ball 2008). Likewise, *Drosophila melanogaster* has six different Wnt subfamilies, *Caenorhabditis elegans* has only three, while vertebrates have 12 subfamilies. Strikingly, in *Nematostella* also 12 subfamilies could be identified, one of which (WntA) is shared exclusively with protostomes (Kusserow et al. 2005; Lee et al. 2006). Only Wnt9 is missing. This clearly demonstrates that the common ancestor of cnidarians possessed already the full complement of

Wnt subfamilies, while a considerable part of this genetic complexity has been lost in the ecdysozoan model organisms *Drosophila melanogaster* and *Caenorhabditis elegans* (Guder et al. 2006a). These are only two out of many examples that show that cnidarians and vertebrates have retained more ancestral traits on the molecular level than the model ecdysozoans.

Cnidarian genes are often more similar to their vertebrate than to their ecdysozoan homologs and cnidarians tend to share more genes exclusively with vertebrates than with ecdysozoans, reflecting the different rates of evolution (Technau et al. 2005). In terms of gene repertoire, cnidarians are much more similar to vertebrates than to other non-bilaterians, such as placozoans, sponges, and ctenophores, which, for instance, lack Hox genes altogether (see Chapters 4, 5, and 8; Srivastava et al. 2008, 2010; Ryan et al. 2013). Yet, surprisingly, cnidarians exhibit a mode of post-transcriptional regulation by miRNAs that is much more plant-like than bilaterian-like (Moran et al. 2014). Hence, some parts of information processing are likely to have evolved de novo in bilaterians.

Cnidarians are diploblastic, but contain most of the conserved important transcription factors involved in mesoderm formation and differentiation, yet they also lack a few crucial ones, such as MyoD. Similarly, most muscle protein-coding genes are present in diploblasts (and even in non-metazoan organisms), but a few crucial ones, such as titin and troponins, are lacking (Steinmetz et al. 2012). Several important conserved neuronal determinants appear to have a role in neurogenesis in cnidarians as well, but NeuroD appears to be a bilaterian novelty. Perhaps most strikingly, in spite of the high regenerative capacity of cnidarians, which is based on the multipotent stem cells and their properties of self-renewal and differentiation, no bona fide orthologs of *Sox2*, *Oct4*, and *Nanog* have been found in cnidarian genomes. It is likely, however, that other related proteins fulfill this function in cnidarians (Millane et al. 2011) and cnidarians may therefore serve as a model for how to generate and maintain stem cells in long-living animals.

OPEN QUESTIONS

- Do different Wnt ligands have distinct roles in axial development?
- Are the oral-aboral and the directive body axes of cnidarians homologous to any of the bilaterian body axes?
- How is the nervous system formed and maintained in the various cnidarian subgroups?
- How does the diffuse nervous system function in such a dynamic tissue?
- What is the function of "mesodermal" transcription factors in cnidarians and how did it change in the transition to triploblasts?
- How did functions of mesoderm and endoderm tissues and cells segregate from the presumably ancestral diploblasty as found today in Cnidaria?
- How is the longevity or potential immortality of cnidarians maintained and regulated?

Acknowledgements We thank the members of the Technau lab for continuous fruitful discussions. We especially thank Andy Aman for critically reading and correcting the manuscript. Also, we wish to acknowledge the Core Facility Cell Imaging and Ultrastructure Research for support in confocal imaging. Work in the Technau lab is supported by grants of the Austrian Science Fund FWF to UT and GG (P24858; P22618; P26962).

References

Ambrosone A, Marchesano V, Tino A, Hobmayer B, Tortiglione C (2012) Hymyc1 downregulation promotes stem cell proliferation in *Hydra* vulgaris. PLoS One 7(1):e30660

Augustin R, Franke A, Khalturin K, Kiko R, Siebert S, Hemmrich G, Bosch TC (2006) Dickkopf related genes are components of the positional value gradient in *Hydra*. Dev Biol 296(1):62–70

Ball EE, Hayward DC, Reece-Hoyes JS, Hislop NR, Samuel G, Saint R, Harrison PL, Miller DJ (2002) Coral development: from classical embryology to molecular control. Int J Dev Biol 46(4):671–678

Bielen H, Oberleitner S, Marcellini S, Gee L, Lemaire P, Bode HR, Rupp R, Technau U (2007) Divergent functions of two ancient *Hydra* brachyury paralogues suggest specific roles for their C-terminal domains in tissue fate induction. Development 134(23):4187–4197

Bode HR (1992) Continuous conversion of neuron phenotype in *Hydra*. Trends Genet 8(8):279–284

Bode PM, Bode HR (1980) Formation of pattern in regenerating tissue pieces of *Hydra* attenuata. I. Head-body proportion regulation. Dev Biol 78(2):484–496

Bode HR, Heimfeld S, Chow MA, Huang LW (1987) Gland cells arise by differentiation from interstitial cells in *Hydra* attenuata. Dev Biol 122(2):577–585

Bode PM, Awad TA, Koizumi O, Nakashima Y, Grimmelikhuijzen CJ, Bode HR (1988) Development of the two-part pattern during regeneration of the head in *Hydra*. Development 102(1):223–235

Boehm AM, Khalturin K, Anton-Erxleben F, Hemmrich G, Klostermeier UC, Lopez-Quintero JA, Oberg HH, Puchert M, Rosenstiel P, Wittlieb J, Bosch TC (2012) FoxO is a critical regulator of stem cell maintenance in immortal *Hydra*. Proc Natl Acad Sci U S A 109(48):19697–19702

Boelsterli U (1977) An electron microscopic study of early developmental stages, myogenesis, oogenesis, and cnidogenesis in the anthomedusa, *Podocoryne carnea* M. Sars. J Morphol 154:259–289

Bosch TC (2012a) Understanding complex host-microbe interactions in *Hydra*. Gut Microbes 3(4):345–351

Bosch TC (2012b) What *Hydra* has to say about the role and origin of symbiotic interactions. Biol Bull 223(1):78–84

Broun M, Bode HR (2002) Characterization of the head organizer in *Hydra*. Development 129(4):875–884

Broun M, Gee L, Reinhardt B, Bode HR (2005) Formation of the head organizer in *Hydra* involves the canonical Wnt pathway. Development 132(12):2907–2916

Browne E (1909) The production of new *Hydra*nts by the insertion of small grafts. J Exp Zool 7:1–37

Byrum CA, Martindale MQ (2004) Gastrulation in the Cnidaria and Ctenophora. In: Stern CD (ed) Gastrulation. From cells to embryos. Cold Spring Harbor Laboratory Press, New York, pp 33–50

Campbell RD (1976) Elimination by *Hydra* interstitial and nerve cells by means of colchicine. J Cell Sci 21(1):1–13

Chapman JA, Kirkness EF, Simakov O, Hampson SE, Mitros T, Weinmaier T, Rattei T, Balasubramanian PG, Borman J, Busam D, Disbennett K, Pfannkoch C, Sumin N, Sutton GG, Viswanathan LD, Walenz B, Goodstein DM, Hellsten U, Kawashima T, Prochnik SE, Putnam NH, Shu S, Blumberg B, Dana CE, Gee L, Kibler DF, Law L, Lindgens D, Martinez DE, Peng J, Wigge PA, Bertulat B, Guder C, Nakamura Y, Ozbek S, Watanabe H, Khalturin K, Hemmrich G, Franke A, Augustin R, Fraune S, Hayakawa E, Hayakawa S, Hirose M, Hwang JS, Ikeo K, Nishimiya-Fujisawa C, Ogura A, Takahashi T, Steinmetz PR, Zhang X, Aufschnaiter R, Eder MK, Gorny AK, Salvenmoser W, Heimberg AM, Wheeler BM, Peterson KJ, Bottger A, Tischler P, Wolf A, Gojobori T, Remington KA, Strausberg RL, Venter JC, Technau U, Hobmayer B, Bosch TC, Holstein TW, Fujisawa T, Bode HR, David CN, Rokhsar DS, Steele RE (2010) The dynamic genome of *Hydra*. Nature 464(7288):592–596

Chera S, Ghila L, Dobretz K, Wenger Y, Bauer C, Buzgariu W, Martinou JC, Galliot B (2009) Apoptotic cells provide an unexpected source of Wnt3 signaling to drive *Hydra* head regeneration. Dev Cell 17(2):279–289

Chourrout D, Delsuc F, Chourrout P, Edvardsen RB, Rentzsch F, Renfer E, Jensen MF, Zhu B, de Jong P, Steele RE, Technau U (2006) Minimal ProtoHox cluster inferred from bilaterian and cnidarian Hox complements. Nature 442(7103):684–687

Collins, AG (2002) Phylogeny of Medusozoa and the evolution of cnidarian life cycles. J Evol Biol 15:418–432

Collins AG, Schuchert P, Marques AC, Jankowski T, Medina M, Schierwater B (2006) Medusozoan Phylogeny and Character Evolution Clarified by New Large and Small Subunit rDNA Data and an Assessment of the Utility of Phylogenetic Mixture Models. Syst Biol 55(1):97–115

Darling JA, Reitzel AM, Finnerty JR (2004) Regional population structure of a widely introduced estuarine invertebrate: *nematostella vectensis* Stephenson in New England. Mol Ecol 13(10):2969–2981

Darling JA, Reitzel AR, Burton PM, Mazza ME, Ryan JF, Sullivan JC, Finnerty JR (2005) Rising starlet: the starlet sea anemone, *Nematostella vectensis*. Bioessays 27(2):211–221

David CN (1973) A quantitative method for maceration of *Hydra* tissue. Wilhelm Roux Arch 171:259–268

David CN (2012) Interstitial stem cells in *Hydra*: multipotency and decision-making. Int J Dev Biol 56(6–8):489–497

David CN, Campbell RD (1972) Cell cycle kinetics and development of *Hydra* attenuata. I. Epithelial cells. J Cell Sci 11(2):557–568

David CN, MacWilliams H (1978) Regulation of the self-renewal probability in *Hydra* stem cell clones. Proc Natl Acad Sci U S A 75(2):886–890

David CN, Murphy S (1977) Characterization of interstitial stem cells in *Hydra* by cloning. Dev Biol 58(2):372–383

David CN, Plotnick I (1980) Distribution of interstitial stem cells in *Hydra*. Dev Biol 76(1):175–184

Davy SK, Allemand D, Weis VM (2012) Cell biology of cnidarian-dinoflagellate symbiosis. Microbiol Mol Biol Rev 76(2):229–261

DuBuc TQ, Ryan JF, Shinzato C, Satoh N, Martindale MQ (2012) Coral comparative genomics reveal expanded Hox cluster in the cnidarian-bilaterian ancestor. Integr Comp Biol 52(6):835–841

Duffy DJ, Plickert G, Kuenzel T, Tilmann W, Frank U (2010) Wnt signaling promotes oral but suppresses aboral structures in *Hydractinia* metamorphosis and regeneration. Development 137(18):3057–3066

Elms P, Siggers P, Napper D, Greenfield A, Arkell R (2003) Zic2 is required for neural crest formation and hindbrain patterning during mouse development. Dev Biol 264(2):391–406

Engel U, Ozbek S, Streitwolf-Engel R, Petri B, Lottspeich F, Holstein TW (2002) Nowa, a novel protein with mini-collagen Cys-rich domains, is involved in nematocyst formation in *Hydra*. J Cell Sci 115(Pt 20):3923–3934

Finnerty JR, Pang K, Burton P, Paulson D, Martindale MQ (2004) Origins of bilateral symmetry: hox and dpp expression in a sea anemone. Science 304(5675):1335–1337

Frank U, Leitz T, Muller WA (2001) The hydroid *Hydractinia*: a versatile, informative cnidarian representative. Bioessays 23(10):963–971

Franzenburg S, Fraune S, Kunzel S, Baines JF, Domazet-Loso T, Bosch TC (2012) MyD88-deficient *Hydra* reveal an ancient function of TLR signaling in sensing bacterial colonizers. Proc Natl Acad Sci U S A 109(47):19374–19379

Franzenburg S, Walter J, Kunzel S, Wang J, Baines JF, Bosch TC, Fraune S (2013) Distinct antimicrobial peptide expression determines host species-specific bacterial associations. Proc Natl Acad Sci U S A 110(39):E3730–E3738

Fraune S, Bosch TC (2007) Long-term maintenance of species-specific bacterial microbiota in the basal metazoan *Hydra*. Proc Natl Acad Sci U S A 104(32):13146–13151

Freeman G (1981a) The cleavage initiation site establishes the posterior pole of the hydrozoan embryo. Roux's Arch Dev Biol 190:123–125

Freeman G (1981b) The role of polarity in the development of the hydrozoan planula larva. Roux's Arch Dev Biol 190:168–184

Freeman G, Miller RL (1982) Hydrozoan eggs can only be fertilized at the site of the polar body formation. Dev Biol 94:142–152

Fritzenwanker JH, Technau U (2002) Induction of gametogenesis in the basal cnidarian *Nematostella vectensis*(Anthozoa). Dev Genes Evol 212(2):99–103

Fritzenwanker JH, Saina M, Technau U (2004) Analysis of forkhead and snail expression reveals epithelial-mesenchymal transitions during embryonic and larval development of *Nematostella vectensis*. Dev Biol 275(2):389–402

Fritzenwanker JH, Genikhovich G, Kraus Y, Technau U (2007) Early development and axis specification in the sea anemone *Nematostella vectensis*. Dev Biol 310(2):264–279

Fuchs B, Wang W, Graspeuntner S, Li Y, Insua S, Herbst EM, Dirksen P, Bohm AM, Hemmrich G, Sommer F, Domazet-Loso T, Klostermeier UC, Anton-Erxleben F, Rosenstiel P, Bosch TC, Khalturin K (2014) Regulation of polyp-to-jellyfish transition in *Aurelia aurita*. Curr Biol 24(3):263–273

Galliot B, Quiquand M (2011) A two-step process in the emergence of neurogenesis. Eur J Neurosci 34(6):847–862

Galliot B, Quiquand M, Ghila L, de Rosa R, Miljkovic-Licina M, Chera S (2009) Origins of neurogenesis, a cnidarian view. Dev Biol 332(1):2–24

Garm A, Coates MM, Gad R, Seymour J, Nilsson DE (2007a) The lens eyes of the box jellyfish *Tripedalia cystophora* and *Chiropsalmus* sp. are slow and color-blind. J Comp Physiol A Neuroethol Sens Neural Behav Physiol 193(5):547–557

Garm A, O'Connor M, Parkefelt L, Nilsson DE (2007b) Visually guided obstacle avoidance in the box jellyfish *Tripedalia cystophora* and *Chiropsella bronzie*. J Exp Biol 210(Pt 20):3616–3623

Garm A, Andersson F, Nilsson DE (2008) Unique structure and optics of the lesser eyes of the box jellyfish *Tripedalia cystophora*. Vision Res 48(8):1061–1073

Garm A, Oskarsson M, Nilsson DE (2011) Box jellyfish use terrestrial visual cues for navigation. Curr Biol 21(9):798–803

Garm A, Bielecki J, Petie R, Nilsson DE (2012) Opposite patterns of diurnal activity in the box jellyfish *Tripedalia cystophora* and *Copula sivickisi*. Biol Bull 222(1):35–45

Gauchat D, Kreger S, Holstein T, Galliot B (1998) prdl-a, a gene marker for *Hydra* apical differentiation related to triploblastic paired-like head-specific genes. Development 125(9):1637–1645

Gauchat D, Escriva H, Miljkovic-Licina M, Chera S, Langlois MC, Begue A, Laudet V, Galliot B (2004) The orphan COUP-TF nuclear receptors are markers for neurogenesis from cnidarians to vertebrates. Dev Biol 275(1):104–123

Gee L, Hartig J, Law L, Wittlieb J, Khalturin K, Bosch TC, Bode HR (2010) beta-catenin plays a central role in setting up the head organizer in *Hydra*. Dev Biol 340(1):116–124

Gehring WJ, Kloter U, Suga H (2009) Evolution of the Hox gene complex from an evolutionary ground state. Curr Top Dev Biol 88:35–61

Genikhovich G, Technau U (2009a) In situ hybridization of starlet sea anemone (*Nematostella vectensis*) embryos, larvae, and polyps. CSH Protoc 2009(9):pdb prot5282

Genikhovich G, Technau U (2009b) The starlet sea anemone *Nematostella vectensis*: an anthozoan model organism for studies in comparative genomics and functional evolutionary developmental biology. CSH Protoc 2009(9):pdb emo129

Genikhovich G, Technau U (2011) Complex functions of Mef2 splice variants in the differentiation of endoderm and of a neuronal cell type in a sea anemone. Development 138(22):4911–4919

Genikhovich G, Fried P, Prünster MM, Schinko JB, Gilles AF, Fredman D, Meier K, Iber D, Technau U (2015) Axis patterning by BMPs: cnidarian network reveals evolutionary constraints. Cell Rep pii: S2211–1247(15)00181–00183

Genikhovich G, Kurn U, Hemmrich G, Bosch TC (2006) Discovery of genes expressed in *Hydra* embryogenesis. Dev Biol 289(2):466–481

Gierer A, Meinhardt H (1972) A theory of biological pattern formation. Kybernetik 12(1):30–39

Gierer A, Berking S, Bode H, David CN, Flick K, Hansmann G, Schaller H, Trenkner E (1972) Regeneration of *Hydra* from reaggregated cells. Nat New Biol 239(91):98–101

Grasso LC, Negri AP, Foret S, Saint R, Hayward DC, Miller DJ, Ball EE (2011) The biology of coral metamorphosis: molecular responses of larvae to inducers of settlement and metamorphosis. Dev Biol 353(2):411–419

Graziussi DF, Suga H, Schmid V, Gehring WJ (2012) The "eyes absent" (eya) gene in the eye-bearing hydrozoan jellyfish *Cladonema radiatum*: conservation of the retinal determination network. J Exp Zool B Mol Dev Evol 318(4):257–267

Grens A, Mason E, Marsh JL, Bode HR (1995) Evolutionary conservation of a cell fate specification gene: the *Hydra achaete-scute* homolog has proneural activity in *Drosophila*. Development 121(12):4027–4035

Guder C, Philipp I, Lengfeld T, Watanabe H, Hobmayer B, Holstein TW (2006a) The Wnt code: cnidarians signal the way. Oncogene 25(57):7450–7460

Guder C, Pinho S, Nacak TG, Schmidt HA, Hobmayer B, Niehrs C, Holstein TW (2006b) An ancient Wnt-Dickkopf antagonism in *Hydra*. Development 133(5):901–911

Gur Barzilai M, Reitzel AM, Kraus JE, Gordon D, Technau U, Gurevitz M, Moran Y (2012) Convergent evolution of sodium ion selectivity in metazoan neuronal signaling. Cell Rep 2(2):242–248

Hager G, David CN (1997) Pattern of differentiated nerve cells in *Hydra* is determined by precursor migration. Development 124(2):569–576

Hanaoka K (1934) Notes to the early development of a stalked medusa. Proc Imp Acad Jpn 10:117–120

Hand C, Uhlinger KR (1992) The culture, sexual and asexual reproduction, and growth of the sea anemone *Nematostella vectensis*. Biol Bull 182: 169–176

Hartl M, Mitterstiller AM, Valovka T, Breuker K, Hobmayer B, Bister K (2010) Stem cell-specific activation of an ancestral myc protooncogene with conserved basic functions in the early metazoan *Hydra*. Proc Natl Acad Sci U S A 107(9):4051–4056

Hassel M, Bieller A (1996) Stepwise transfer from high to low lithium concentrations increases the head-forming potential in *Hydra* vulgaris and possibly activates the PI cycle. Dev Biol 177(2):439–448

Hassel M, Albert K, Hofheinz S (1993) Pattern formation in *Hydra* vulgaris is controlled by lithium-sensitive processes. Dev Biol 156(2):362–371

Hayakawa E, Fujisawa C, Fujisawa T (2004) Involvement of *Hydra* achaete-scute gene CnASH in the differentiation pathway of sensory neurons in the tentacles. Dev Genes Evol 214(10):486–492

Hayward DC, Samuel G, Pontynen PC, Catmull J, Saint R, Miller DJ, Ball EE (2002) Localized expression of a dpp/BMP2/4 ortholog in a coral embryo. Proc Natl Acad Sci U S A 99(12):8106–8111

Hayward DC, Hetherington S, Behm CA, Grasso LC, Forêt S, Miller DJ, Ball EE (2011) Differential gene expression at coral settlement and metamorphosis – a subtractive hybridization study. PLoS One 6(10):e26411

Hemmrich G, Anokhin B, Zacharias H, Bosch TC (2007) Molecular phylogenetics in *Hydra*, a classical model in evolutionary developmental biology. Mol Phylogenet Evol 44(1):281–290

Hemmrich G, Khalturin K, Boehm AM, Puchert M, Anton-Erxleben F, Wittlieb J, Klostermeier UC, Rosenstiel P, Oberg HH, Domazet-Loso T, Sugimoto T, Niwa H, Bosch TC (2012) Molecular signatures of the three stem cell lineages in *Hydra* and the emergence of stem cell function at the base of multicellularity. Mol Biol Evol 29(11):3267–3280

Hobmayer E, Holstein TW, David CN (1990) Tentacle morphogenesis in *Hydra*. II. Formation of a complex between a sensory nerve cell and a battery cell. Development 109:897–904

Hobmayer B, Holstein TW, David CN (1997) Stimulation of tentacle and bud formation by the neuropeptide head activator in *Hydra* magnipapillata. Dev Biol 183(1):1–8

Hobmayer B, Rentzsch F, Kuhn K, Happel CM, von Laue CC, Snyder P, Rothbacher U, Holstein TW (2000) WNT signalling molecules act in axis formation in the diploblastic metazoan *Hydra*. Nature 407(6801):186–189

Holstein TW (2008) Wnt signaling in cnidarians. Methods Mol Biol 469:47–54

Holstein TW (2012) The evolution of the Wnt pathway. Cold Spring Harb Perspect Biol 4(7):a007922

Holstein T, Tardent P (1984) An ultrahigh-speed analysis of exocytosis: nematocyst discharge. Science 223(4638):830–833

Holstein TW, Hobmayer E, David CN (1991) Pattern of epithelial cell cycling in *Hydra*. Dev Biol 148(2): 602–611

Holstein TW, Benoit M, Herder GV, David CN, Wanner G, Gaub HE (1994) Fibrous mini-collagens in *Hydra* nematocysts. Science 265(5170):402–404

Holstein TW, Hobmayer E, Technau U (2003) Cnidarians: an evolutionarily conserved model system for regeneration? Dev Dyn 226(2):257–267

Houliston E, Momose T, Manuel M (2010) *Clytia hemisphaerica*: a jellyfish cousin joins the laboratory. Trends Genet 26(4):159–167

Hudry B, Thomas-Chollier M, Volovik Y, Duffraisse M, Dard A, Frank D, Technau U, Merabet S (2014) Molecular insights into the origin of the Hox-TALE patterning system. eLife 3:e01939

Hwang JS, Takaku Y, Momose T, Adamczyk P, Ozbek S, Ikeo K, Khalturin K, Hemmrich G, Bosch TC, Holstein TW, David CN, Gojobori T (2010) Nematogalectin, a nematocyst protein with GlyXY and galectin domains, demonstrates nematocyte-specific alternative splicing in *Hydra*. Proc Natl Acad Sci U S A 107(43):18539–18544

Juliano CE, Reich A, Liu N, Gotzfried J, Zhong M, Uman S, Reenan RA, Wessel GM, Steele RE, Lin H (2014) PIWI proteins and PIWI-interacting RNAs function in *Hydra* somatic stem cells. Proc Natl Acad Sci U S A 111(1):337–342

Kayal E, Roure B, Philippe H, Collins AG, Lavrov DV (2013) Cnidarian phylogenetic relationships as revealed by mitogenomics. BMC Evol Biol 13:5

Khalturin K, Anton-Erxleben F, Milde S, Plotz C, Wittlieb J, Hemmrich G, Bosch TC (2007) Transgenic stem cells in *Hydra* reveal an early evolutionary origin for key elements controlling self-renewal and differentiation. Dev Biol 309(1):32–44

Koch AW, Holstein TW, Mala C, Kurz E, Engel J, David CN (1998) Spinalin, a new glycine- and histidine-rich protein in spines of *Hydra* nematocysts. J Cell Sci 111(Pt 11):1545–1554

Koizumi O, Bode HR (1991) Plasticity in the nervous system of adult *Hydra*. III. Conversion of neurons to expression of a vasopressin-like immunoreactivity depends on axial location. J Neurosci 11(7): 2011–2020

Kortschak RD, Samuel G, Saint R, Miller DJ (2003) EST analysis of the cnidarian *Acropora millepora* reveals extensive gene loss and rapid sequence divergence in the model invertebrates. Curr Biol 13(24):2190–2195

Kowalewsky A (1884) Zur Entwicklung der Lucernaria (vorläufige Mittelung). Zool Anz 7:712–717

Kozmik Z, Daube M, Frei E, Norman B, Kos L, Dishaw LJ, Noll M, Piatigorsky J (2003) Role of Pax genes in eye evolution: a cnidarian *PaxB* gene uniting *Pax2* and *Pax6* functions. Dev Cell 5(5):773–785

Kozmik Z, Ruzickova J, Jonasova K, Matsumoto Y, Vopalensky P, Kozmikova I, Strnad H, Kawamura S, Piatigorsky J, Paces V, Vlcek C (2008) Assembly of the cnidarian camera-type eye from vertebrate-like components. Proc Natl Acad Sci U S A 105(26):8989–8993

Kraus Y, Technau U (2006) Gastrulation in the sea anemone *Nematostella vectensis* occurs by invagination and immigration: an ultrastructural study. Dev Genes Evol 216(3):119–132

Kraus Y, Fritzenwanker JH, Genikhovich G, Technau U (2007) The blastoporal organiser of a sea anemone. Curr Biol 17(20):R874–R876

Krediet CJ, Ritchie KB, Paul VJ, Teplitski M (2013) Coral-associated micro-organisms and their roles in promoting coral health and thwarting diseases. Proc Biol Sci 280(1755):20122328

Kumburegama S, Wijesena N, Xu R, Wikramanayake AH (2011) Strabismus-mediated primary archenteron invagination is uncoupled from Wnt/beta-catenin-dependent endoderm cell fate specification in *Nematostella vectensis* (Anthozoa, Cnidaria): implications for the evolution of gastrulation. EvoDevo 2(1):2

Kuniyoshi H, Okumura I, Kuroda R, Tsujita N, Arakawa K, Shoji J, Saito T, Osada H (2012) Indomethacin induction of metamorphosis from the asexual stage to sexual stage in the moon jellyfish, *Aurelia aurita*. Biosci Biotechnol Biochem 76(7):1397–1400

Kunzel T, Heiermann R, Frank U, Muller W, Tilmann W, Bause M, Nonn A, Helling M, Schwarz RS, Plickert G (2010) Migration and differentiation potential of stem cells in the cnidarian *Hydractinia* analysed in eGFP-transgenic animals and chimeras. Dev Biol 348(1):120–129

Kurz EM, Holstein TW, Petri BM, Engel J, David CN (1991) Mini-collagens in *Hydra* nematocytes. J Cell Biol 115(4):1159–1169

Kusserow A, Pang K, Sturm C, Hrouda M, Lentfer J, Schmidt HA, Technau U, von Haeseler A, Hobmayer B, Martindale MQ, Holstein TW (2005) Unexpected complexity of the Wnt gene family in a sea anemone. Nature 433(7022):156–160

Layden MJ, Martindale, MQ (2014) Non-canonical Notch signaling represents an ancestral mechanism to regulate neural differentiation. Evodevo 5:30

Layden MJ, Boekhout M, Martindale MQ (2012) *Nematostella vectensis* achaete-scute homolog NvashA regulates embryonic ectodermal neurogenesis and represents an ancient component of the metazoan neural specification pathway. Development 139(5):1013–1022

Lee PN, Pang K, Matus DQ, Martindale MQ (2006) A WNT of things to come: evolution of Wnt signaling and polarity in cnidarians. Semin Cell Dev Biol 17(2):157–167

Lee PN, Kumburegama S, Marlow HQ, Martindale MQ, Wikramanayake AH (2007) Asymmetric developmental potential along the animal-vegetal axis in the anthozoan cnidarian, *Nematostella vectensis*, is mediated by Dishevelled. Dev Biol 310(1): 169–186

Lengfeld T, Watanabe H, Simakov O, Lindgens D, Gee L, Law L, Schmidt HA, Ozbek S, Bode H, Holstein TW (2009) Multiple Wnts are involved in *Hydra* organizer formation and regeneration. Dev Biol 330(1):186–199

Lindgens D, Holstein TW, Technau U (2004) Hyzic, the *Hydra* homolog of the *zic/odd*-paired gene, is involved in the early specification of the sensory nematocytes. Development 131(1):191–201

Littlefield CL, Bode HR (1986) Germ cells in *Hydra* oligactis males. II. Evidence for a subpopulation of interstitial stem cells whose differentiation is limited to sperm production. Dev Biol 116(2):381–386

Littlefield CL, Dunne JF, Bode HR (1985) Spermatogenesis in *Hydra oligactis*. I. Morphological description and characterization using a monoclonal antibody specific for cells of the spermatogenic pathway. Dev Biol 110(2):308–320

Littlefield CL, Finkemeier C, Bode HR (1991) Spermatogenesis in *Hydra* oligactis. II. How temperature controls the reciprocity of sexual and asexual reproduction. Dev Biol 146(2):292–300

MacWilliams HK (1982) Numerical simulations of *Hydra* head regeneration using a proportion-regulating version of the Gierer-Meinhardt model. J Theor Biol 99(4):681–703

MacWilliams HK (1983a) *Hydra* transplantation phenomena and the mechanism of *Hydra* head regeneration. I. Properties of the head inhibition. Dev Biol 96(1):217–238

MacWilliams HK (1983b) *Hydra* transplantation phenomena and the mechanism of *Hydra* head regeneration. II. Properties of the head activation. Dev Biol 96(1):239–257

Magie CR, Pang K, Martindale MQ (2005) Genomic inventory and expression of Sox and Fox genes in the cnidarian *Nematostella vectensis*. Dev Genes Evol 215(12):618–630

Magie CR, Daly M, Martindale MQ (2007) Gastrulation in the cnidarian *Nematostella vectensis* occurs via invagination not ingression. Dev Biol 305(2):483–497

Mahoney JL, Graugnard EM, Mire P, Watson GM (2011) Evidence for involvement of TRPA1 in the detection of vibrations by hair bundle mechanoreceptors in sea anemones. J Comp Physiol A Neuroethol Sens Neural Behav Physiol 197(7):729–742

Marcum BA, Campbell RD (1978) Development of *Hydra* lacking nerve and interstitial cells. J Cell Sci 29:17–33

Marlow HQ, Srivastava M, Matus DQ, Rokhsar D, Martindale MQ (2009) Anatomy and development of the nervous system of *Nematostella vectensis*, an anthozoan cnidarian. Dev Neurobiol 69(4):235–254

Marlow H, Roettinger E, Boekhout M, Martindale MQ (2012) Functional roles of Notch signaling in the cnidarian *Nematostella vectensis*. Dev Biol 362(2):295–308

Marlow H, Matus DQ, Martindale MQ (2013) Ectopic activation of the canonical wnt signaling pathway affects ectodermal patterning along the primary axis during larval development in the anthozoan *Nematostella vectensis*. Dev Biol 380(2):324–334

Martin VJ, Littlefield CL, Archer WE, Bode HR (1997) Embryogenesis in *Hydra*. Biol Bull 192(3):345–363

Martindale MQ, Pang K, Finnerty JR (2004) Investigating the origins of triploblasty: "mesodermal" gene expression in a diploblastic animal, the sea anemone *Nematostella vectensis* (phylum, Cnidaria; class, Anthozoa). Development 131(10):2463–2474

Marques AC, Collins AG (2004) Cladistic analysis of Medusozoa and cnidarian evolution. Invert Biol 123: 23–42

Matus DQ, Pang K, Marlow H, Dunn CW, Thomsen GH, Martindale MQ (2006a) Molecular evidence for deep evolutionary roots of bilaterality in animal development. Proc Natl Acad Sci U S A 103(30):11195–11200

Matus DQ, Thomsen GH, Martindale MQ (2006b) Dorso/ventral genes are asymmetrically expressed and involved in germ-layer demarcation during cnidarian gastrulation. Curr Biol 16(5):499–505

Matus DQ, Pang K, Daly M, Martindale MQ (2007) Expression of Pax gene family members in the anthozoan cnidarian, *Nematostella vectensis*. Evol Dev 9(1):25–38

Meinhardt H (1993) A model for pattern formation of hypostome, tentacles, and foot in *Hydra*: how to form structures close to each other, how to form them at a distance. Dev Biol 157(2):321–333

Meinhardt H (2012) Modeling pattern formation in *Hydra*: a route to understanding essential steps in development. Int J Dev Biol 56(6–8):447–462

Meinhardt H, Gierer A (2000) Pattern formation by local self-activation and lateral inhibition. Bioessays 22(8):753–760

Miljkovic-Licina M, Chera S, Ghila L, Galliot B (2007) Head regeneration in wild-type *Hydra* requires de novo neurogenesis. Development 134(6):1191–1201

Millane RC, Kanska J, Duffy DJ, Seoighe C, Cunningham S, Plickert G, Frank U (2011) Induced stem cell neoplasia in a cnidarian by ectopic expression of a POU domain transcription factor. Development 138(12):2429–2439

Miller DJ, Ball EE (2008) Cryptic complexity captured: the *Nematostella* genome reveals its secrets. Trends Genet 24(1):1–4

Miller MA, Technau U, Smith KM, Steele RE (2000) Oocyte development in *Hydra* involves selection from competent precursor cells. Dev Biol 224(2):326–338

Minguillon C, Garcia-Fernandez J (2003) Genesis and evolution of the *Evx* and *Mox* genes and the extended Hox and ParaHox gene clusters. Genome Biol 4(2):R12

Mokady O, Buss LW (1996) Transmission genetics of allorecognition in *Hydractinia symbiolongicarpus* (Cnidaria:Hydrozoa). Genetics 143(2):823–827

Momose T, Houliston E (2007) Two oppositely localised frizzled RNAs as axis determinants in a cnidarian embryo. PLoS Biol 5(4):e70

Momose T, Schmid V (2006) Animal pole determinants define oral-aboral axis polarity and endodermal cell-fate in hydrozoan jellyfish *Podocoryne carnea*. Dev Biol 292(2):371–380

Momose T, Derelle R, Houliston E (2008) A maternally localised Wnt ligand required for axial patterning in the cnidarian *Clytia hemisphaerica*. Development 135(12):2105–2113

Momose T, Kraus Y, Houliston E (2012) A conserved function for Strabismus in establishing planar cell polarity in the ciliated ectoderm during cnidarian larval development. Development 139(23): 4374–4382

Moran Y, Praher D, Fredman D, Technau U (2013) The evolution of MicroRNA pathway protein components in Cnidaria. Mol Biol Evol 30(12):2541–2552

Moran Y, Fredman D, Praher D, Li XZ, Wee LM, Rentzsch F, Zamore PD, Technau U, Seitz H (2014) Cnidarian microRNAs frequently regulate targets by cleavage. Genome Res 24(4):651–663

Müller WA (1964) Experimental investigations on colony development, polyp differentiation and sexual chimeras in *Hydractinia echinata* Wilhelm Roux'. Arch Entwicklungsmechanik 155:181–268

Müller WA (1982) Intercalation and pattern regulation in hydroids. Differentiation 22:141–150

Muller WA (1990) Ectopic head and foot formation in *Hydra*: diacylglycerol-induced increase in positional value and assistance of the head in foot formation. Differentiation 42(3):131–143

Muller WA, Teo R, Frank U (2004) Totipotent migratory stem cells in a hydroid. Dev Biol 275(1):215–224

Muller W, Frank U, Teo R, Mokady O, Guette C, Plickert G (2007) Wnt signaling in hydroid development: ectopic heads and giant buds induced by GSK-3beta inhibitors. Int J Dev Biol 51(3):211–220

Nakanishi N, Yuan D, Hartenstein V, Jacobs DK (2010) Evolutionary origin of rhopalia: insights from cellular-level analyses of *Otx* and *POU* expression patterns in the developing rhopalial nervous system. Evol Dev 12(4):404–415

Nakanishi N, Renfer E, Technau U, Rentzsch F (2012) Nervous systems of the sea anemone *Nematostella vectensis* are generated by ectoderm and endoderm and shaped by distinct mechanisms. Development 139(2):347–357

Nicotra ML, Powell AE, Rosengarten RD, Moreno M, Grimwood J, Lakkis FG, Dellaporta SL, Buss LW (2009) A hypervariable invertebrate allodeterminant. Curr Biol 19(7):583–589

Nishimiya-Fujisawa C, Kobayashi S (2012) Germline stem cells and sex determination in *Hydra*. Int J Dev Biol 56(6–8):499–508

Nüchter T, Benoit M, Engel U, Ozbek S, Holstein TW (2006) Nanosecond-scale kinetics of nematocyst discharge. Curr Biol 16(9):R316–R318

Oliver D, Brinkmann M, Sieger T, Thurm U (2008) Hydrozoan nematocytes send and receive synaptic signals induced by mechano-chemical stimuli. J Exp Biol 211(Pt 17):2876–2888

Oren M, Brikner I, Appelbaum L, Levy O (2014) Fast neurotransmission related genes are expressed in non nervous endoderm in the sea anemone *Nematostella vectensis*. PLoS One 9(4):e93832

Ormestad M, Martindale MQ, Rottinger E (2011) A comparative gene expression database for invertebrates. EvoDevo 2:17

Otto JJ (1976) Early development and planula movement in *Haliclystus* (Scyphozoa, Stauromedusae). In: Mackie GO (ed) Coelenterate ecology and behavior. Plenum Press, New York, pp 319–329

Petie R, Garm A, Nilsson DE (2011) Visual control of steering in the box jellyfish *Tripedalia cystophora*. J Exp Biol 214(Pt 17):2809–2815

Philipp I, Aufschnaiter R, Ozbek S, Pontasch S, Jenewein M, Watanabe H, Rentzsch F, Holstein TW, Hobmayer B (2009) Wnt/beta-catenin and noncanonical Wnt signaling interact in tissue evagination in the simple eumetazoan *Hydra*. Proc Natl Acad Sci U S A 106(11):4290–4295

Philippe H, Derelle R, Lopez P, Pick K, Borchiellini C, Boury-Esnault N, Vacelet J, Renard E, Houliston E, Queinnec E, Da Silva C, Wincker P, Le Guyader H, Leys S, Jackson DJ, Schreiber F, Erpenbeck D, Morgenstern B, Worheide G, Manuel M (2009) Phylogenomics revives traditional views on deep animal relationships. Curr Biol 19(8):706–712

Pick KS, Philippe H, Schreiber F, Erpenbeck D, Jackson DJ, Wrede P, Wiens M, Alie A, Morgenstern B, Manuel M, Worheide G (2010) Improved phylogenomic taxon sampling noticeably affects nonbilaterian relationships. Mol Biol Evol 27(9):1983–1987

Plickert G, Jacoby V, Frank U, Muller WA, Mokady O (2006) Wnt signaling in hydroid development: formation of the primary body axis in embryogenesis and its subsequent patterning. Dev Biol 298(2):368–378

Plickert G, Frank U, Muller WA (2012) *Hydractinia*, a pioneering model for stem cell biology and reprogramming somatic cells to pluripotency. Int J Dev Biol 56(6–8):519–534

Poudyal M, Rosa S, Powell AE, Moreno M, Dellaporta SL, Buss LW, Lakkis FG (2007) Embryonic chimerism does not induce tolerance in an invertebrate model organism. Proc Natl Acad Sci U S A 104(11):4559–4564

Putnam NH, Srivastava M, Hellsten U, Dirks B, Chapman J, Salamov A, Terry A, Shapiro H, Lindquist E, Kapitonov VV, Jurka J, Genikhovich G, Grigoriev IV, Lucas SM, Steele RE, Finnerty JR, Technau U, Martindale MQ, Rokhsar DS (2007) Sea anemone genome reveals ancestral eumetazoan gene repertoire and genomic organization. Science 317(5834):86–94

Raikova EV (1994) Life cycle, cytology and morphology on Polypodium hydriforme, a coelenterate parasite of the eggs of acipenseriform fishes. J Parasitol 80(1):1–22

Rebscher N, Volk C, Teo R, Plickert G (2008) The germ plasm component vasa allows tracing of the interstitial stem cells in the cnidarian *Hydractinia echinata*. Dev Dyn 237(6):1736–1745

Renfer E, Amon-Hassenzahl A, Steinmetz PR, Technau U (2010) A muscle-specific transgenic reporter line of the sea anemone, *Nematostella vectensis*. Proc Natl Acad Sci U S A 107(1):104–108

Rentzsch F, Anton R, Saina M, Hammerschmidt M, Holstein TW, Technau U (2006) Asymmetric expression of the BMP antagonists chordin and gremlin in the sea anemone *Nematostella vectensis*: implications for the evolution of axial patterning. Dev Biol 296(2):375–387

Rentzsch F, Fritzenwanker JH, Scholz CB, Technau U (2008) FGF signalling controls formation of the apical sensory organ in the cnidarian *Nematostella vectensis*. Development 135(10):1761–1769

Rosa SF, Powell AE, Rosengarten RD, Nicotra ML, Moreno MA, Grimwood J, Lakkis FG, Dellaporta SL, Buss LW (2010) *Hydractinia* allodeterminant alr1 resides in an immunoglobulin superfamily-like gene complex. Curr Biol 20(12):1122–1127

Rosenberg E, Koren O, Reshef L, Efrony R, Zilber-Rosenberg I (2007) The role of microorganisms in coral health, disease and evolution. Nat Rev Microbiol 5(5):355–362

Rosenberg E, Sharon G, Atad I, Zilber-Rosenberg I (2010) The evolution of animals and plants via symbiosis with microorganisms. Environ Microbiol Rep 2(4):500–506

Rosengarten RD, Moreno MA, Lakkis FG, Buss LW, Dellaporta SL (2011) Genetic diversity of the allodeterminant alr2 in *Hydractinia symbiolongicarpus*. Mol Biol Evol 28(2):933–947

Rottinger E, Dahlin P, Martindale MQ (2012) A framework for the establishment of a cnidarian gene regulatory network for "endomesoderm" specification: the inputs of ss-catenin/TCF signaling. PLoS Genet 8(12):e1003164

Ruzickova J, Piatigorsky J, Kozmik Z (2009) Eye-specific expression of an ancestral jellyfish *PaxB* gene interferes with *Pax6* function despite its conserved *Pax6/Pax2* characteristics. Int J Dev Biol 53(4):469–482

Ryan JF, Mazza ME, Pang K, Matus DQ, Baxevanis AD, Martindale MQ, Finnerty JR (2007) Pre-bilaterian origins of the Hox cluster and the Hox code: evidence from the sea anemone, *Nematostella vectensis*. PLoS One 2(1):e153

Ryan JF, Pang K, Schnitzler CE, Nguyen AD, Moreland RT, Simmons DK, Koch BJ, Francis WR, Havlak P, Program NCS, Smith SA, Putnam NH, Haddock SH, Dunn CW, Wolfsberg TG, Mullikin JC, Martindale MQ, Baxevanis AD (2013) The genome of the ctenophore *Mnemiopsis leidyi* and its implications for cell type evolution. Science 342(6164):1242592

Saina M, Genikhovich G, Renfer E, Technau U (2009) BMPs and chordin regulate patterning of the directive axis in a sea anemone. Proc Natl Acad Sci U S A 106(44):18592–18597

Sarras MP Jr (2012) Components, structure, biogenesis and function of the *Hydra* extracellular matrix in regeneration, pattern formation and cell differentiation. Int J Dev Biol 56(6–8):567–576

Sato M, Bode HR, Sawada Y (1990) Patterning processes in aggregates of *Hydra* cells visualized with the monoclonal antibody, TS19. Dev Biol 141(2):412–420

Schaller H, Gierer A (1973) Distribution of the head-activating substance in *Hydra* and its localization in membranous particles in nerve cells. J Embryol Exp Morphol 29(1):39–52

Schaller HC, Hofmann M, Javois LC (1990) Effect of head activator on proliferation, head-specific determination and differentiation of epithelial cells in *Hydra*. Differentiation 43(3):157–164

Schmich J, Trepel S, Leitz T (1998) The role of GLWamides in metamorphosis of *Hydractinia echinata*. Dev Genes Evol 208(5):267–273

Schmid V, Alder H (1984) Isolated, mononucleated, striated muscle can undergo pluripotent transdifferentiation and form a complex regenerate. Cell 38(3):801–809

Schmidt T, David CN (1986) Gland cells in *Hydra*: cell cycle kinetics and development. J Cell Sci 85:197–215

Scholz CB, Technau U (2003) The ancestral role of *Brachyury*: expression of *NemBra1* in the basal cnidarian *Nematostella vectensis* (Anthozoa). Dev Genes Evol 212(12):563–570

Schwaiger M, Schonauer A, Rendeiro AF, Pribitzer C, Schauer A, Gilles AF, Schinko JB, Renfer E, Fredman D, Technau U (2014) Evolutionary conservation of the eumetazoan gene regulatory landscape. Genome Res 24(4):639–650

Seipel K, Schmid V (2005) Evolution of striated muscle: jellyfish and the origin of triploblasty. Dev Biol 282(1):14–26

Seipel K, Yanze N, Schmid V (2004) Developmental and evolutionary aspects of the basic helix-loop-helix transcription factors Atonal-like 1 and Achaete-scute homolog 2 in the jellyfish. Dev Biol 269(2):331–345

Seipp S, Schmich J, Kehrwald T, Leitz T (2007) Metamorphosis of Hydractinia echinata–natural versus artificial induction and developmental plasticity. Dev Genes Evol 217(5):385–394

Seipp S, Schmich J, Will B, Schetter E, Plickert G, Leitz T (2010) Neuronal cell death during metamorphosis of *Hydractina echinata* (Cnidaria, Hydrozoa). Invert Neurosci 10(2):77–91

Shimizu H (2012) Transplantation analysis of developmental mechanisms in *Hydra*. Int J Dev Biol 56(6–8):463–472

Shinzato C, Shoguchi E, Kawashima T, Hamada M, Hisata K, Tanaka M, Fujie M, Fujiwara M, Koyanagi R, Ikuta T, Fujiyama A, Miller DJ, Satoh N (2011) Using the *Acropora digitifera* genome to understand coral responses to environmental change. Nature 476(7360):320–323

Spring J, Yanze N, Middel AM, Stierwald M, Groger H, Schmid V (2000) The mesoderm specification factor *twist* in the life cycle of jellyfish. Dev Biol 228(2):363–375

Spring J, Yanze N, Josch C, Middel AM, Winninger B, Schmid V (2002) Conservation of *Brachyury*, *Mef2*, and *Snail* in the myogenic lineage of jellyfish: a connection to the mesoderm of Bilateria. Dev Biol 244(2):372–384

Squire JM, Al-Khayat HA, Knupp C, Luther PK (2005) Molecular architecture in muscle contractile assemblies. Adv Protein Chem 71:17–87. doi:10.1016/S0065-3233(04)71002-5

Srivastava M, Begovic E, Chapman J, Putnam NH, Hellsten U, Kawashima T, Kuo A, Mitros T, Salamov A, Carpenter ML, Signorovitch AY, Moreno MA, Kamm K, Grimwood J, Schmutz J, Shapiro H, Grigoriev IV, Buss LW, Schierwater B, Dellaporta SL, Rokhsar DS (2008) The *Trichoplax* genome and the nature of placozoans. Nature 454(7207):955–960

Srivastava M, Simakov O, Chapman J, Fahey B, Gauthier ME, Mitros T, Richards GS, Conaco C, Dacre M, Hellsten U, Larroux C, Putnam NH, Stanke M, Adamska M, Darling A, Degnan SM, Oakley TH, Plachetzki DC, Zhai Y, Adamski M, Calcino A, Cummins SF, Goodstein DM, Harris C, Jackson DJ, Leys SP, Shu S, Woodcroft BJ, Vervoort M, Kosik KS, Manning G, Degnan BM, Rokhsar DS (2010) The *Amphimedon queenslandica* genome and the evolution of animal complexity. Nature 466(7307):720–726

Stangl K, Salvini-Plawen L, Holstein TW (2002) Staging and induction of medusa metamorphosis in *Carybdea marsupialis* (Cnidaria, Cubozoa). Vie Mileu 52:131–140

Steinmetz PR, Kraus JE, Larroux C, Hammel JU, Amon-Hassenzahl A, Houliston E, Worheide G, Nickel M, Degnan BM, Technau U (2012) Independent evolution of striated muscles in cnidarians and bilaterians. Nature 487(7406):231–234

Straehler-Pohl I, Jarms G (2005) Life cycle of Carybdea marsupialis Linnaeus, 1758 (Cubozoa, Carybdeidae) reveals metamorphosis to be a modified strobilation. Mar Biol 147:1271–1277

Suga H, Schmid V, Gehring WJ (2008) Evolution and functional diversity of jellyfish opsins. Curr Biol 18(1):51–55

Suga H, Tschopp P, Graziussi DF, Stierwald M, Schmid V, Gehring WJ (2010) Flexibly deployed Pax genes in eye development at the early evolution of animals demonstrated by studies on a hydrozoan jellyfish. Proc Natl Acad Sci U S A 107(32):14263–14268

Takahashi T (2013) Neuropeptides and epitheliopeptides: structural and functional diversity in an ancestral metazoan *Hydra*. Protein Pept Lett 20(6):671–680

Takahashi T, Fujisawa T (2009) Important roles for epithelial cell peptides in *Hydra* development. Bioessays 31(6):610–619

Takahashi K, Yamanaka S (2006) Induction of pluripotent stem cells from mouse embryonic and adult fibroblast cultures by defined factors. Cell 126(4):663–676

Tardent P (1978) Coelenterata, Cnidaria. Morphogenese der Tiere. Gustav Fischer Verlag, Jena

Technau U (2001) *Brachyury*, the blastopore and the evolution of the mesoderm. Bioessays 23(9):788–794

Technau U, Bode HR (1999) *HyBra1*, a *Brachyury* homologue, acts during head formation in *Hydra*. Development 126(5):999–1010

Technau U, Holstein TW (1992) Cell sorting during the regeneration of *Hydra* from reaggregated cells. Dev Biol 151(1):117–127

Technau U, Holstein TW (1995a) Boundary cells of endodermal origin define the mouth of *Hydra vulgaris* (Cnidaria). Cell Tissue Res 280:235–242

Technau U, Holstein TW (1995b) Head formation is different at apical and basal levels. Development 121:1273–1282

Technau U, Holstein TW (1996) Phenotypic maturation of neurons and continuous precursor migration in the formation of the peduncle nerve net in *Hydra*. Dev Biol 177(2):599–615

Technau U, Steele RE (2011) Evolutionary crossroads in developmental biology: cnidaria. Development 138(8):1447–1458

Technau U, Cramer von Laue C, Rentzsch F, Luft S, Hobmayer B, Bode HR, Holstein TW (2000) Parameters of self-organization in *Hydra* aggregates. Proc Natl Acad Sci U S A 97(22):12127–12131

Technau U, Miller MA, Bridge D, Steele RE (2003) Arrested apoptosis of nurse cells during *Hydra* oogenesis and embryogenesis. Dev Biol 260(1):191–206

Technau U, Rudd S, Maxwell P, Gordon PM, Saina M, Grasso LC, Hayward DC, Sensen CW, Saint R, Holstein TW, Ball EE, Miller DJ (2005) Maintenance of ancestral complexity and non-metazoan genes in two basal cnidarians. Trends Genet 21(12):633–639

Teo R, Mohrlen F, Plickert G, Muller WA, Frank U (2006) An evolutionary conserved role of Wnt signaling in stem cell fate decision. Dev Biol 289(1):91–99

Thomas-Chollier M, Ledent V, Leyns L, Vervoort M (2010) A non-tree-based comprehensive study of metazoan Hox and ParaHox genes prompts new insights into their origin and evolution. BMC Evol Biol 10:73

Turing A (1952) The chemical basis for morphogenesis. Phil Trans R Soc B 237:37–72

Watanabe H, Hoang VT, Mattner R, Holstein TW (2009) Immortality and the base of multicellular life: lessons from cnidarian stem cells. Semin Cell Dev Biol 20(9):1114–1125

Weber J (1990) Poly(gamma-glutamic acid)s are the major constituents of nematocysts in *Hydra* (Hydrozoa, Cnidaria). J Biol Chem 265(17):9664–9669

Weill R (1934) Contribution à l'étude des cnidaires et leurs nematocystes. Trav Stat Zool Wimereux 10–11:1–700

Wietrzykowski W (1910) Sur le développement des Lucernaridés (note préliminaire). Arch Zool Exp 2:10–27

Wietrzykowski W (1912) Recherches sur le développement des Lucernaires. Arch Zool Exp Gen 10:1–95

Wikramanayake AH, Hong M, Lee PN, Pang K, Byrum CA, Bince JM, Xu R, Martindale MQ (2003) An ancient role for nuclear beta-catenin in the evolution of axial polarity and germ layer segregation. Nature 426(6965):446–450

Wittlieb J, Khalturin K, Lohmann JU, Anton-Erxleben F, Bosch TC (2006) Transgenic *Hydra* allow in vivo tracking of individual stem cells during morphogenesis. Proc Natl Acad Sci U S A 103(16):6208–6211

Yamaguchi TP, Takada S, Yoshikawa Y, Wu N, McMahon AP (1999) T (*Brachyury*) is a direct target of Wnt3a during paraxial mesoderm specification. Genes Dev 13(24):3185–3190

Zenkert C, Takahashi T, Diesner MO, Ozbek S (2011) Morphological and molecular analysis of the *Nematostella vectensis* cnidom. PLoS One 6(7): e22725

Myxozoa

Alexander Gruhl

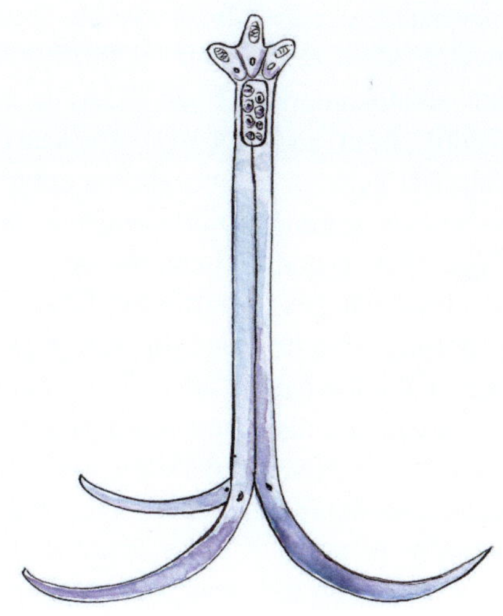

Although recognised as a subtaxon of Cnidaria by most recent phylogenetic analyses, the Myxozoa, with their highly aberrant life cycles, are covered separately herein.

Chapter vignette artwork by Brigitte Baldrian.
© Brigitte Baldrian and Andreas Wanninger.

A. Gruhl
Natural History Museum,
Cromwell Road, London SW7 5BD, UK
e-mail: a.gruhl@nhm.ac.uk

A. Wanninger (ed.), *Evolutionary Developmental Biology of Invertebrates 1:
Introduction, Non-Bilateria, Acoelomorpha, Xenoturbellida, Chaetognatha*
DOI 10.1007/978-3-7091-1862-7_7, © Springer-Verlag Wien 2015

INTRODUCTION

Myxozoa are endoparasitic animals exhibiting complex life cycles that in most known cases involve two hosts: a vertebrate (usually fish, but also rarely amphibian, avian, or mammalian) intermediate host and an invertebrate, mostly annelid or freshwater ectoproct (bryozoan), definitive host. Direct fish-to-fish transmission has been demonstrated in only one species (Diamant 1997). About 2,200 species are known, but only about 100 life cycles have been completely resolved. Myxozoans occur in both marine and freshwater habitats; only few exclusively terrestrial life cycles are suspected. For general reviews on Myxozoa, see, e.g. Kent et al. (2001), Canning and Okamura (2004), Feist and Longshaw (2006), Lom and Dyková (2006), and Okamura et al. (2015).

Transmission between the two hosts is accomplished by microscopic spores (Figs. 7.1A–D and 7.2B, C). The spores can have diverse shapes, but the principal morphology is uniform: one or two sporoplasms, constituting the actual infective agent, are encased by a layer of flattened cells called valve cells, which can secrete protective surface coatings and form elaborate floatation appendages. Integrated into the layer of valve cells are two to four (in rare cases one or up to 15) specialised capsulogenic cells, each of which bears one polar capsule, an extrudable organelle that is used for host recognition, contact, and entry. Capsulogenic cells and valve cells are connected by cell junctions (septate and adherens junctions), thus forming a sealing epithelium.

With the revelation of the first complete life cycles (Markiw and Wolf 1983; Wolf and Markiw 1984), older classification schemes, which, based on spore morphology, differentiated actinosporeans (Fig. 7.1D), and myxosporeans (Fig. 7.1C), had to be revised. Instead of representing different taxa, these turned out to be different stages of the same life cycle: actinosporeans are the spores released from the definitive invertebrate host and myxosporeans from the intermediate vertebrate host (Fig. 7.3). In turn, only Myxosporea was continued as a taxon name (Kent et al. 1994); however, "actinosporeans" and "myxosporeans" are still used as technical terms for the invertebrate and vertebrate phases of Myxosporea, respectively. Recently, a further subtaxon was erected and shown to represent the sister group to Myxosporea: the Malacosporea, consisting of only few species so far, including *Tetracapsuloides bryosalmonae,* the causative agent of salmonid proliferative kidney disease (PKD) (Canning et al. 2000), the enigmatic worm *Buddenbrockia plumatellae* (Okamura et al. 2002), and few further lineages. However, malacosporean diversity appears highly underestimated (Hartikainen et al. 2014).

All known malacosporeans exclusively infect fresh water ectoprocts (Phylactolaemata) as their definitive hosts. Malacosporean spores (Figs. 7.1A, B and 7.2B, C) differ from those of myxosporeans (Fig. 7.1C, D) in the fact that they are uncuticularised with more or less similar morphology of both transmission phases. Spore characters are traditionally used for taxonomic purposes; however, molecular data increasingly demonstrate high levels of homoplasy in these traits. Thus, many traditional myxozoan genera and families are polyphyletic (Fiala 2006; Fiala and Bartosová 2010; Bartošová and Fiala 2011; Bartošová et al. 2013).

As opposed to the dormant spores, all myxozoan trophic stages (Fig. 7.1E–J) are exclusively found, either inter- or intracellularly, within the hosts, and exhibit extremely simple morphologies, lacking any form of gut, gonads, clearly recognisable gametes, and even nervous system and sensory organs. In fact, even some general metazoan cytological features such as cilia and centrioles are absent. In general, the trophic stages in invertebrate hosts (Fig. 7.1E, F, and H) are more complex, being delimited by at least one outer epithelial tissue layer, whereas stages in vertebrate hosts are syncytial plasmodia (large multinucleate cells, Fig. 7.1J) or pseudoplasmodia (large uninucleate cells, Fig. 7.1G, I). Uptake of nutrients from the host is in all cases facilitated by endocytosis via the external membrane. A characteristic phenomenon in many stages of the myxozoan life cycle is endogeny, where one cell (the primary cell) completely surrounds another cell (the secondary cell). Sometimes, even tertiary cells occur.

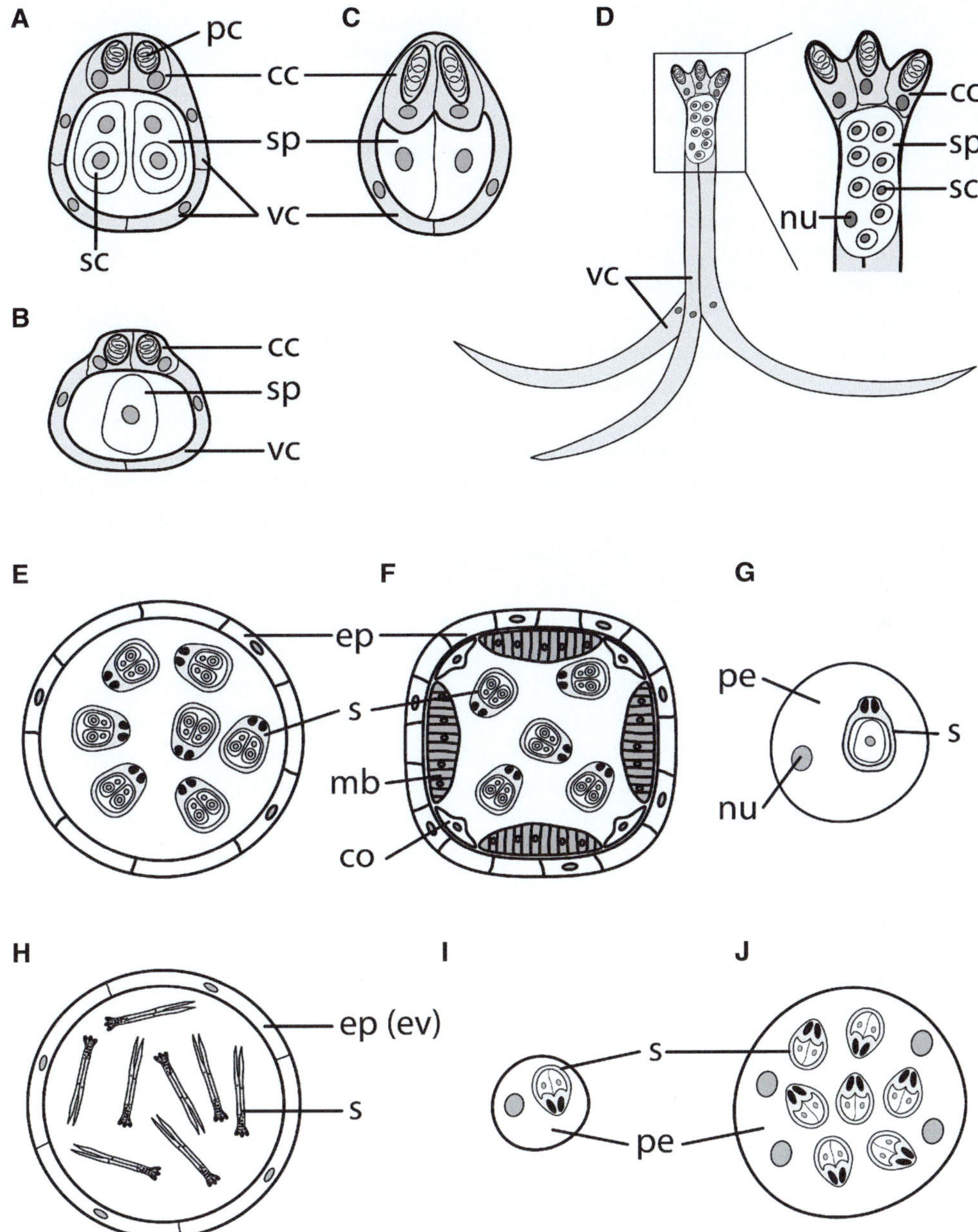

Fig. 7.1 Morphology of myxozoan life cycle stages. (**A–D**) Spores. (**A**) Malacospore (malacosporean spore produced in ectoproct host). The spore wall is formed by capsulogenic cells and flattened valve cells. These spores typically contain two sporoplasms, each of which encloses one secondary cell. (**B**) Fishmalacospore containing one uninucleate sporoplasm. (**C**) Myxospore (myxosporean spore produced in fish host) containing two uninucleate or one binucleate sporoplasm(s). (**D**) Actinospores (myxosporean spore produced in annelid host) are usually triradiate with three valves and three capsulogenic cells. Sporoplasms are large uninucleate cells harbouring numerous secondary (germ) cells. (**E–J**) Trophic stages, schematic cross sections. (**E**) Saclike malacosporean (e.g. *Tetracapsuloides bryosalmonae*) with epidermis and spores. (**F**) Wormlike malacosporeans (*Buddenbrockia*) with four longitudinal muscle blocks and four rows of connecting cells. (**G**) Malacosporean pseudoplasmodium producing one single spore. (**H**) Myxosporean pansporocyst with eight envelope cells and eight actinospores (the valve cells gain turgescence upon release and acquire the shape depicted in (**D**)). (**I**) Myxosporean pseudoplasmodium producing one single spore. (**J**) Myxosporean plasmodium with multinuclear pericyte producing multiple spores. *cc* capsulogenic cell, *co* connecting cell, *ep* epidermis, *ev* envelope cell, *mb* muscle block, *nu* nucleus, *pc* polar capsule, *pe* pericyte, *s* spore, *sc* secondary cell, *sp* sporoplasm, *vc* valve cell (© Alexander Gruhl, 2015. All Rights Reserved)

Pansporocysts (Fig. 7.1H), the intra-invertebrate stages of myxosporeans, represent simple, more or less spherical sacs delimited by two to eight cells, the envelope cells, which are flat and interconnected by cell junctions, thus constituting an epithelial layer. In some species, surface extensions of the envelope cells' apical and/or basal membranes occur, indicating transcytotic uptake of nutrients and potential secretion of waste products.

Malacosporean intra-invertebrate trophic stages occur inside the fluid-filled body cavity of fresh water ectoprocts, either as large sacs (*T. bryosalmonae, Buddenbrockia allmani,* Figs. 7.1E and 7.2A) or as worms (*Buddenbrockia plumatellae,* Figs. 7.1F and 7.2Q). Both sacs and worms are lined by an outer epithelium. In *Buddenbrockia* species, a further, internal epithelium exists in juveniles. *Buddenbrockia* worms (*B. plumatella* and a few further, undescribed lineages) exhibit, between the inner and outer epithelium, four longitudinal muscle blocks which span the entire length of the worm (Fig. 7.2K) and are used to facilitate helicoidal movements of the worm. The muscle blocks consist of individual, obliquely arranged elongate muscle cells (Fig. 7.2U). At least two cell types are discernible in the inner epithelium: connecting cells, which reside in a single line between the muscle blocks, and

remaining cells, which are considered sporogonic. When the worm matures (Fig. 7.2Q–V), the inner epithelium disintegrates, with the connecting cells remaining in place between the muscle blocks and the sporogonic cells detaching from each other and beginning to float freely in the internal cavity to form spores, a process similar to that in saclike malacosporeans.

Vertebrate stages of both malacosporeans and myxosporeans are plasmodia or pseudoplasmodia bearing internal proliferative or sporogonic cells within them. Plasmodia can become large and often show differentiation into an outer and inner layer. The former is specialised in secretion, endocytotic uptake of nutrients, and defence against host attacks. It differs in cytoplasmic composition from the inner compartment which bears the nuclei, most of the other organelles and the endogenous cells.

For a long time, myxozoans had been considered as protists, due to their simple morphologies. However, with the advent of better microscopic techniques, their metazoan nature became obvious. Connections to both Bilateria and Cnidaria were suggested; recent molecular phylogenetic and phylogenomic analyses along with morphological and protein data now provide convincing evidence that Myxozoa are an ingroup of Cnidaria. Potentially due to their

Fig. 7.2 Malacosporean stages. (**A**) *Tetracapsuloides bryosalmonae* spore sac. Light micrograph. Scale bar: 30 μm. (**B**) *T. bryosalmonae* spore. Light micrograph. Scale bar: 10 μm. (**C**) Scanning electron micrograph of a *T. bryosalmonae* spore. Scale bar: 5 μm. (**D**) Dissected ectoproct gut with attached early stages of *Buddenbrockia* (*arrowheads*). Scale bar: 150 μm. (**E–J**) Confocal images (optical sections) of early *Buddenbrockia* developmental stages. *Green* – F-actin, *red* – nuclei. (**E**) Attached spherical stage with beginning segregation into outer epidermis and inner compact tissue. Scale bar: 30 μm. (**F**) Attached elongate stage, developing cavity and inner epithelium at distal end. Scale bar: 50 μm. (**G**) Detached early worm. Scale bar: 50 μm. (**H–J**) Optical cross sections through (**G**), from proximal to distal. Scale bar: 10 μm. (**K–P**) Immature *Buddenbrockia* worm with developing muscle blocks, confocal images. (**K**) Whole mount specimen. Scale bar: 200 μm. (**L**) Proximal tip with cluster of undifferentiated cells (*arrowhead*). (**M**) Cross section through

central region. (**N**) Horizontal optical section of central region. (**O**) Maximum intensity projection of central region. (**P**) Horizontal optical section of distal tip. Scale bar for l-p: 30 μm. (**Q–V**) Light and confocal micrographs of a mature *Buddenbrockia* worm. The inner epithelium has disintegrated, and spores are fully developed. (**Q**) Whole-mount specimen. Scale bar: 200 μm. (**R**) Proximal tip with cluster of undifferentiated cells (*arrowhead*). (**S**) Partial optical cross section through central region showing muscle block. (**T**) Optical horizontal section showing inner cavity filled with spores. (**U**) Maximum intensity projection of the central region showing arrangement of muscle fibres. (**V**) Optical horizontal section of distal tip. Scale bar for **R–V**: 30 μm. *cc* capsulogenic cell, *cl* central lumen, *co* connecting cell, *dt* distal tip, *ep* epidermis, *ev* envelope cell, *gu* ectoproct gut, *ic* inner compact cells, *ie* inner epithelium, *mb* muscle block, *nu* nucleus, *pc* polar capsule, *pt* proximal tip, *s* spore, *sp* sporoplasm, *vc* valve cell (Images (**D–V**) from Gruhl and Okamura (2012))

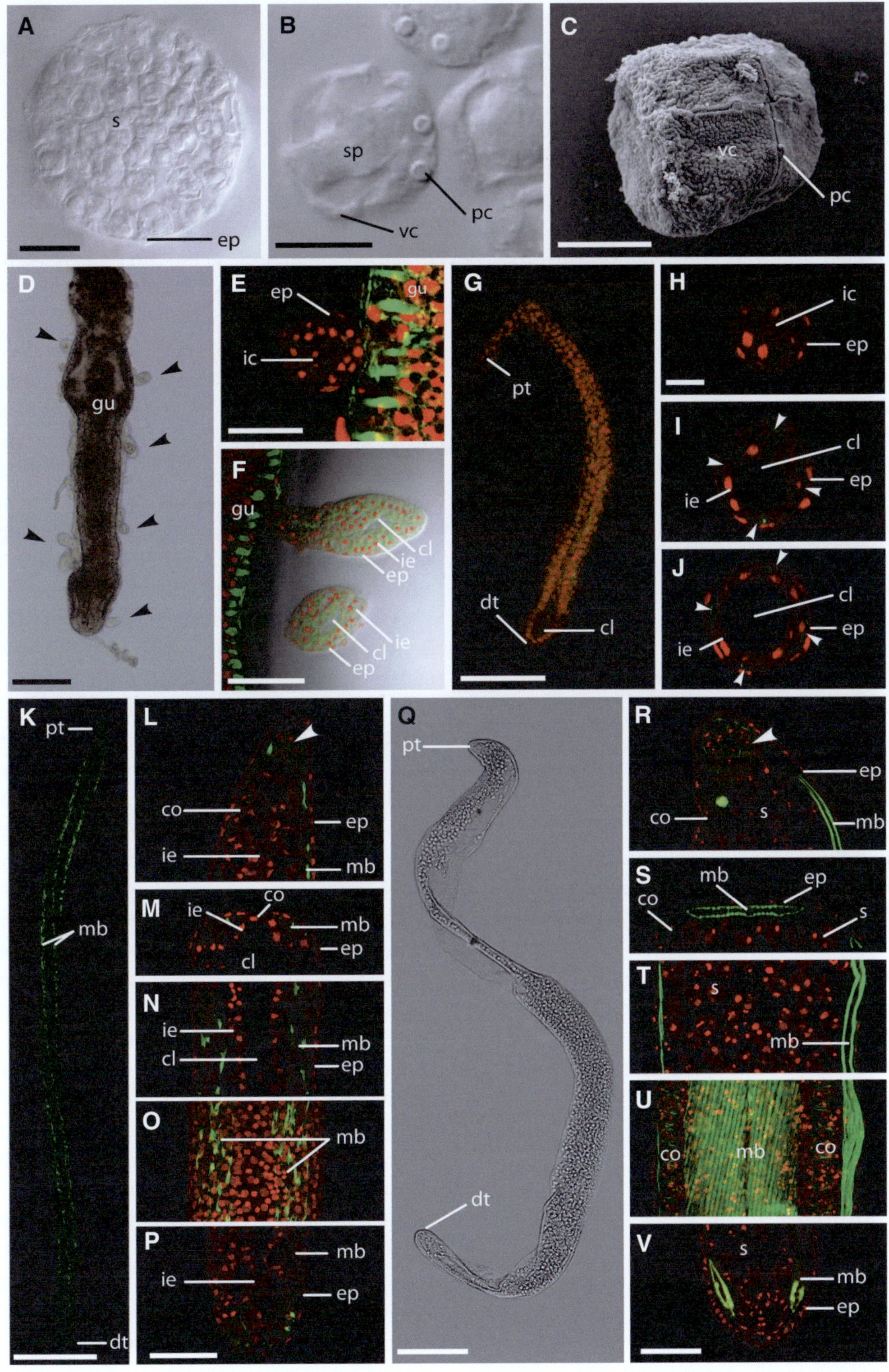

aberrant morphology and high rates of sequence evolution, the exact position within the Cnidaria is not well resolved, the most parsimonious assumption being a sister group relationship to Medusozoa (Staurozoa, Cubozoa, Hydrozoa, Scyphozoa) (Nesnidal et al. 2013; Chapter 6). However, rDNA data still support alternative positions at the base of Bilateria with a sister group relationship to *Polypodium hydriforme,* a further parasitic cnidarian that, however, is likely to be related to leptomedusans (Evans et al. 2008, 2010). Many morphological characters support the metazoan nature of myxozoans, but only few correspondences to Cnidaria exist. Myxozoan polar capsules exhibit striking resemblance to cnidarian nematocysts (Weill 1938; Siddall et al. 1995) that, apart from ultrastructure and genesis, includes molecular architecture, with the presence of cnidarian-specific proteins (minicollagens) (Holland et al. 2011). Further evidence comes from the myoarchitecture of *Buddenbrockia plumatellae,* which shows tetraradial symmetry, a pattern unique to medusozoans (Gruhl and Okamura 2012).

DEVELOPMENT

Due to the complex life cycle and the still unclear state of sexual reproduction, several phases of development can be identified that might correspond to either embryological stages or forms of asexual development found in related cnidarians. Most studies of developmental stages do not have a high temporal resolution, thus, many inferences about developmental processes are based on observations of a few stages only and have to be considered with care.

Development in the Invertebrate Host

Malacosporea

Usually referred to as "sacculogenesis", the development of malacosporeans in the invertebrate host (Fig. 7.3A–G) has been described ultrastructurally in a few publications (Canning et al. 1996, 2000, 2007, 2008; Okamura et al. 2002; McGurk et al. 2006; Morris and Adams 2007a, b; Gruhl and Okamura 2012). The development differs between the species.

The simplest form of development is found in *Tetracapsuloides bryosalmonae,* where unicellular stages firstly occur in the body cavity or attached to the apical surface of the peritoneum of the ectoproct host. These can remain relatively inactive inside the host, facilitating a long-term cryptic infection or, potentially triggered by host condition, begin to proliferate rapidly by division initiating an acute overt infection. Early multicellular stages are clusters of several cells, similar in their ultrastructure to the unicellular stages. These clusters are believed to arise either by aggregation (thus potentially being chimaeras of different genotypes) or by mitosis. Subsequent stages have developed an outer epithelial layer embracing a compact mass of inner cells. Vegetative growth leads to sacs of up to 300 μm in diameter. At some stage, the inner cells undergo sporogony, which, in contrast to the growth, happens rather synchronously. Thus, usually sacs of different sizes but at a similar stage of sporogony are found within one host.

The early development in sac-forming *Buddenbrockia* is essentially similar; however, the inner compact mass differentiates into a transitory epithelium, encompassing a central lumen. Prior to sporogony, this epithelium disintegrates, and cells detach and float freely inside the sac.

The most complex trophic stages are found in the wormlike *Buddenbrockia.* In contrast to the other malacosporeans, early unicellular stages are not free in the host coelom, but within the extracellular matrix (ECM), usually between gut epithelium and peritoneum, but also, more rarely, between epidermis and peritoneum. These stages proliferate by division and form clusters of cells which seem rather unstructured, invading large portions of the ECM. At some point, parts of the multicellular masses penetrate the peritoneum towards the coelom (Fig. 7.2D–F). The part that lies in the coelom becomes bilayered (similar to early stages in the sac-forming species), with an outer epithelial layer and an inner compact mass of cells.

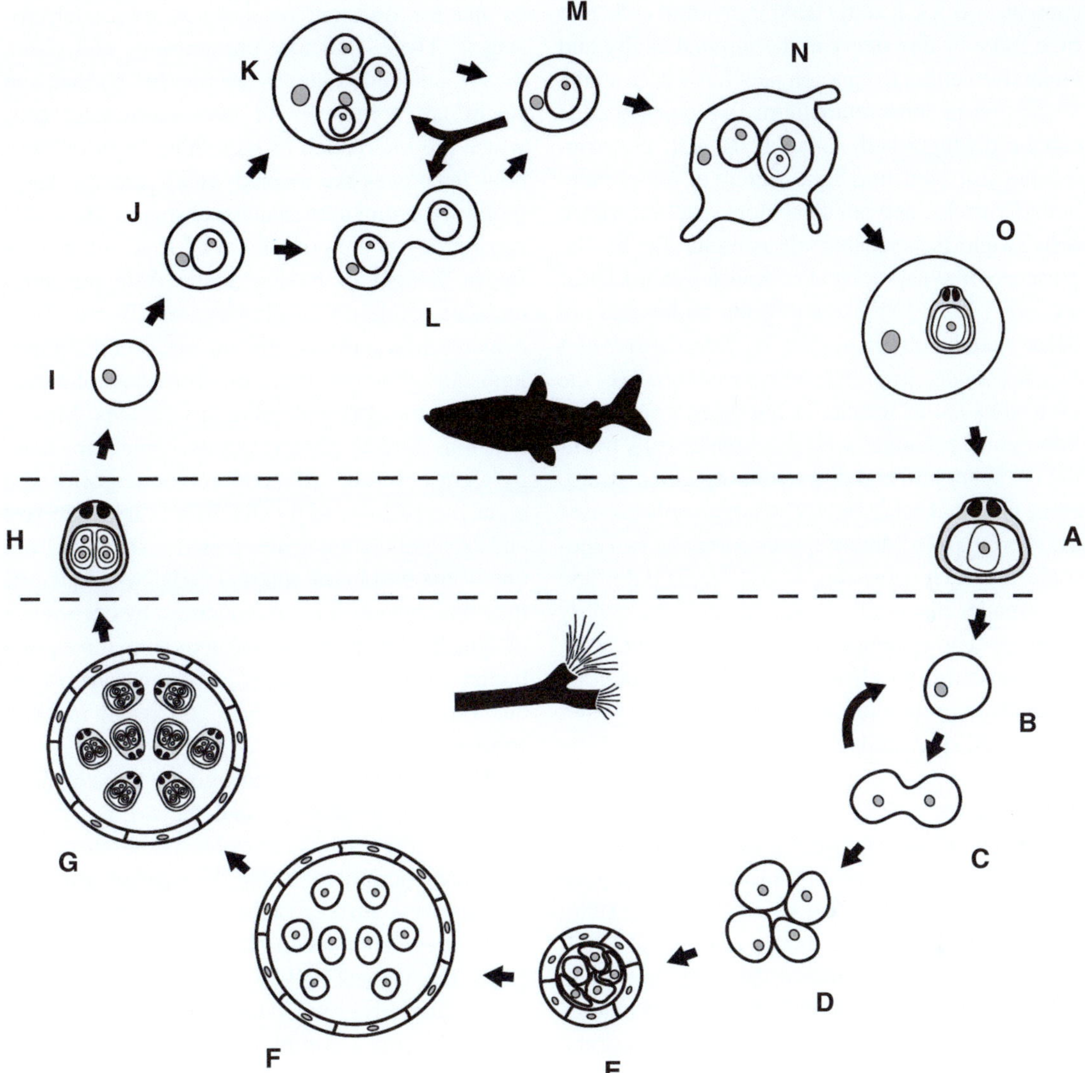

Fig. 7.3 Malacosporean life cycle (exemplified by *Tetracapsuloides bryosalmonae*; see main text for deviations in other malacosporeans). (**A**) Fishmalacospore. (**B**) Sporoplasm enters the ectoproct host via gut epithelium and epidermis. (**C**) Proliferation in the host coelom via mitoses. (**D**) Early cell cluster. (**E**) Early compact bilayered stage. (**F**) Immature sac, sporogonic cells floating freely in inner cavity. (**G**) Mature sac, filled with spores. (**H**) Malacospore. (**I**) Sporoplasm germ enters the fish host via epidermis. (**J**) Endogenic stage with secondary cell. (**K**) Proliferation by mitotic division and release of secondary cells. (**L**) Proliferation by mitotic division of primary and secondary cells. (**M**) Endogenic stage with secondary cell. (**N**) Sporogonic pseudoplasmodium. (**O**) Mature sporogonic pseudoplasmodium (© Alexander Gruhl, 2015. All Rights Reserved)

The inner cells differentiate further into an inner epithelium and muscle precursor cells, which reside between the two epithelial tissue layers (Fig. 7.2G–J). At some stage, the worms detach from the cell mass in the ECM and swim freely in the host coelom, retaining their initial body polarity. Further growth leads to elongation, differentiation of the musculature, and differentiation of the inner epithelium into connecting cells and sporogonic cells (Fig. 7.2K–P).

Sporogony has not been described in sufficient detail, but so far seems to be similar in all malacosporean species. It starts from individual cells that float more or less freely in the internal cavity and are derived either from a compact mass or from the disintegrating inner epithelium. Two types of cells can be distinguished: larger cells with electron-lucent cytoplasm, and high content of membrane-bound vesicles and smaller, more electron-lucent cells. Meiosis is visible ultrastructurally by the presence of synaptonemal complexes in nuclei of the larger cells. Expulsion of polar bodies has not been directly observed, but in *Tetracapsuloides bryosalmonae*, tiny cells remain within the sac even after spore formation has been completed. Sporogony continues with the smaller cells forming the spore hull and differentiating into capsulogenic cells and valve cells. The larger cells become the sporoplasms. Mature spores comprise two sporoplasms, each consisting of one larger outer and one smaller inner cell (endogenic cell, secondary cell), four capsulogenic cells and eight valve cells. It is currently not clear if fusion of the meiotic products happens during sporogony, i.e. whether some or all spore cells are haploid.

Myxosporea

The development of pansporocysts (Fig. 7.4A–H), especially of the early stages, has been followed in detail only in very few cases (Marques 1986; Lom and Dyková 1992, 1997; Lom et al. 1997; El-Matbouli and Hoffmann 1998; Hallett et al. 1998; Oumouna et al. 2002). Most other studies focus on spore formation; however, occasional reports of early developmental stages are given (e.g. Bartholomew et al. 1997; El-Mansy et al. 1998; Hallett and Lester 1999; Özer and Wootten 2001; Alvarez-Pellitero et al. 2002; Meaders and Hendrickson 2009; Rangel et al. 2009; Morris 2010, 2012a; Marton and Eszterbauer 2012), adding bitwise information to the whole picture.

The most comprehensive study so far is by El-Matbouli and Hoffmann (1998) on *Myxobolus cerebralis* development in the oligochaete *Tubifex tubifex*. Tubificid worms get infected by ingestion of myxospores from decaying fish. Upon contact with the gut lining, the polar capsules discharge, open, and release amoeboid binucleate sporoplasms, which penetrate the gut epithelium and undergo presporogonic proliferation (schizogony)

by nuclear division, resulting in multinucleated stages. These undergo plasmotomy and, later, mostly unicellular stages are present. Subsequent stages are complexes of two uninucleate cells which are interpreted as beginning to fuse. Next, binucleate cells are formed which undergo karyotomy, resulting in tetranucleate stages. The initial pansporocysts are complexes of four uninucleate stages thought of having arisen from the tetranucleate condition by plasmotomy. Two of these cells are in a more peripheral position and envelope the other two cells by formation of membrane protrusions and cell-cell junctions. Mitotic divisions lead to pansporocysts, which are lined by eight envelope cells and contain eight α- and eight β-cells. α- and β-cells differ slightly in size and each population is interpreted as deriving from one of the two initial internal cells. Subsequently, they undergo meiosis (as evidenced by occurrence of synaptonemal complexes), expelling three polar bodies each. Eight complexes of each one α- and one β-cell form which fuse and result in eight zygotes.

The next phase is sporogony: the envelope cells remain unchanged, and the zygotes divide twice to form clusters (sporoblasts) of one central and three peripheral cells. The peripheral cells undergo one further division, and of these six cells, three become capsulogenic and three valve cells. The central cell is the prospective sporoplasm and divides asymmetrically (endogenically) to form a complex of one outer and one inner (generative cell). The inner cell gives rise to up to 64 germ cells. Spores are released by rupture into the intestinal tract of the tubificid worm.

Deviations from the above pattern have been reported. In many species, only two or four cells constitute the sporoplasm envelope. Early stages have been described differently. For an *Aurantiactinomyxon*, Lom et al. (1997) described no tetranucleate or four cell stages, but the pansporocyst starting from an endogenic (primary/secondary) cell. However, tetranucleate stages have been described for *Aurantiactinomyxon* and *Raabeia* (Oumouna et al. 2002). Hallett and Lester (1999) described the genus *Tetraspora* in which only four spores develop in one pansporocyst.

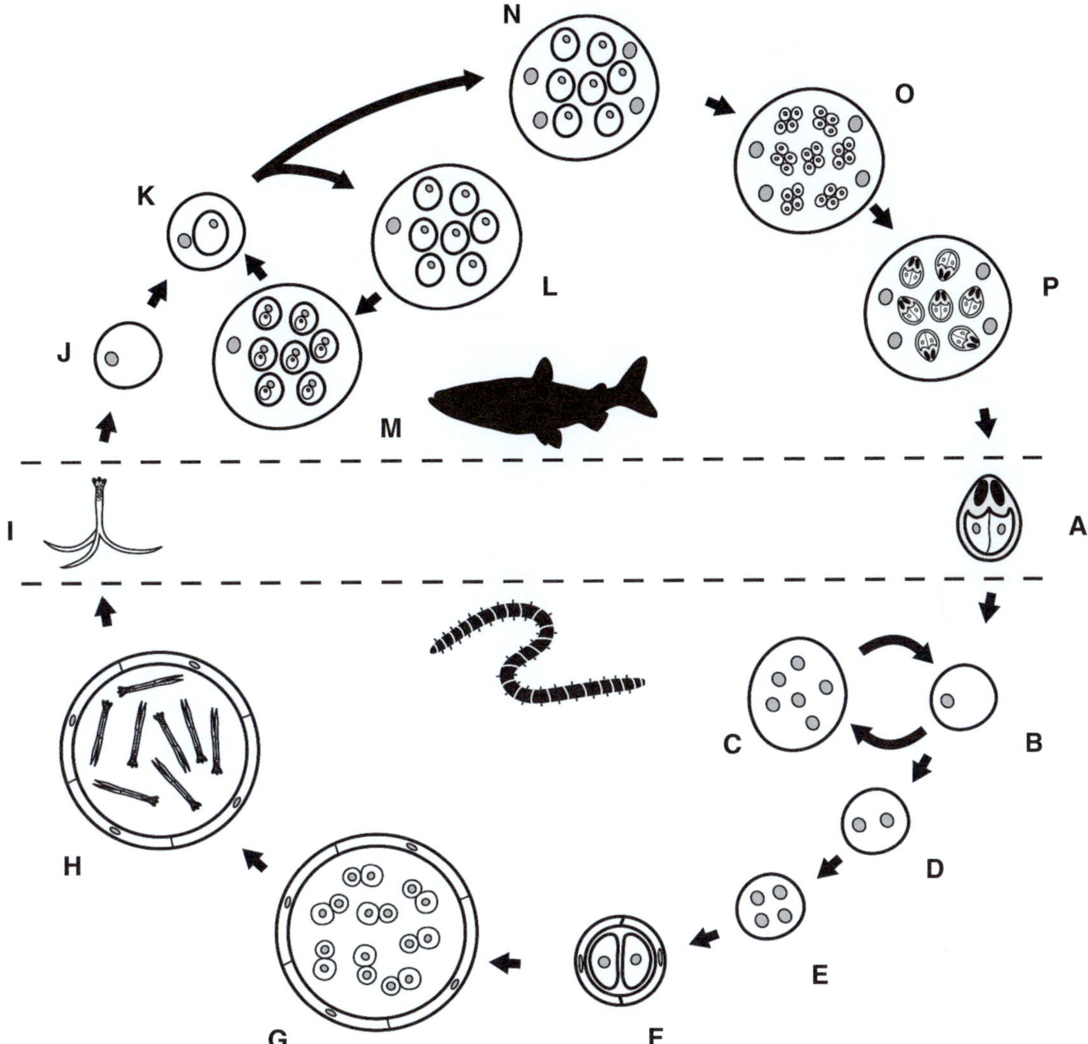

Fig. 7.4 Myxosporean life cycle (mostly adopted from *Myxobolus cerebralis*; see main text for deviations in other myxosporeans). (**A**) Myxospore. (**B**) Sporoplasm enters the annelid host via gut epithelium or epidermis. (**C**) Extrasporogonic proliferation by nuclear divisions followed by plasmotomy. (**D**) Binucleate cell. (**E**) Tetranucleate cell. (**F**) Early pansporocyst consisting of two internal and two envelope cells. (**G**) Pansporocyst with fusing α- and β-cells, resulting in eight zygotes. (**H**) Mature pansporocyst contain-ing eight actinospores. (**I**) Actinospore. (**J**) Sporoplasm germ cell released from sporoplasm after penetration of host epidermis. (**K**) Endogenic (primary/secondary) stage. (**L, M**) Extrasporogonic proliferation by division of secondary cells and formation and release of secondary/tertiary cell doublets. (**N**) Sporogonic multinucleate plasmodium containing several secondary cells. (**O**) Formation of sporoblasts by division of secondary cells. (**P**) Mature sporogonic plasmodium (© Alexander Gruhl, 2015. All Rights Reserved)

Development in the Vertebrate Host

Malacosporea

Malacosporean development within the vertebrate (fish) host (Fig. 7.3H–O) has only been studied in *Tetracapsuloides bryosalmonae* (Kent and Hedrick 1986; Morris and Adams 2008).

The entry portals seem to be thin parts of the epidermis and mucus cells (Grabner and El-Matbouli 2010). The earliest stages visible are unicellular stages, most likely representing the inner (secondary) cells of the sporoplasms. Shortly after infection, typical cell doublets consisting of one primary and one internalised secondary cell occur. Multiple extrasporogonic proliferation cycles prior to spore formation seem to be the norm and take place in the kidney interstitium with the parasite stages engulfed by host phagocytes. Proliferation involves division of secondary cells followed by division of primary cells. The maximum number of secondary cells found in these stages is three (Morris and Adams 2008), with rare findings of tertiary cells.

Sporogonic stages migrate into the kidney tubules, where larger so-called pseudoplasmodia are formed. These are essentially large pericytes, which harbour several secondary cells and secondary/tertiary cell doublets. The cytoplasm of the outer primary cell differs from that of the extrasporogonic stage in the absence of sporoplasmosomes, secretory vesicles characteristic for many myxozoan cells. Sporogony commences by the secondary cells developing into sporoplasms and the secondary/tertiary cell complexes differentiating into capsulogenic cells and valve cells. Each pseudoplasmodium produces a single so-called fishmalacospore consisting of one sporoplasm cell, two capsulogenic cells and four valve cells. The spores are released from the fish via the urine.

Myxosporea

Early development and sporogony of myxosporeans (Fig. 7.4J–P) have been described extensively especially in economically important species. However, many studies are based on pathological results, and exact developmental sequences are sometimes difficult to infer.

The fish host is infected by the actinospore attaching to the host and the sporoplasm penetrating the epidermis. In most cases, the outer cell disintegrates soon to release the sporoplasm (germ) cells, which then spread into the species-specific target tissues. In most cases, extrasporogonic proliferation cycles precede spore formation. These can happen in a range of different ways: one extreme would be represented by the way as described for malacosporeans (i.e. by mitotic division of secondary cells, followed by division of the primary cells). The other extreme would be the formation of

large complex stages which harbour large numbers of secondary cells and secondary/tertiary cell doublets which are either released successively or by breakdown of the large pericyte. Many species undergo several extrasporogonic proliferative cycles, each in a different host tissue.

Sporogony takes place in small pseudoplasmodia that produce only one or few spores, or in larger multinucleate plasmodia. In the latter, two pathways are common. In the first, the spores are produced individually within so-called pansporoblasts, which are essentially cell doublets where the secondary cell undergoes several mitotic divisions to form sporoplasm, valve, and capsulogenic cells. Alternatively, a large population of secondary cells is present, which separately develop into sporoplasms, valve cells, or capsulogenic cells. The differentiated cells then aggregate to form the spores.

EXPERIMENTAL AND GENE EXPRESSION STUDIES

In situ hybridization studies focussing on developmental gene expression are lacking so far for myxozoans. However, a survey of genes expressed during spore activation has been conducted (Eszterbauer et al. 2009), the first myxozoan genome has been published (Yang et al. 2014), and several transcriptome and a few genome sequencing projects are currently on the way. Light and electron microscopic techniques are well established and other procedures, such as confocal microscopy, fluorescence staining, and antibody labeling, have been successfully applied (e.g. Alama-Bermejo et al. 2012; Gruhl and Okamura 2012). At least for a few species, partial replication of the life cycle in the laboratory is possible (Tops and Okamura 2003; Tops et al. 2004; Hartikainen et al. 2009; Kumar et al. 2013), and attempts for in vitro culture have been made (Morris 2012b). Thus, although myxozoans are comparatively difficult to access, possibilities to tackle more EvoDevo questions using experimental and molecular techniques are within reach.

OPEN QUESTIONS

- How is the myxozoan life cycle related to that of free-living cnidarians/medusozoans?
- Where and how does sexual reproduction and outcrossing happen in the myxozoan life cycle?
- What are the mechanisms that have caused extreme body simplification and loss of cytological features in myxozoans?
- How do mechanisms of tissue specification differ within myxozoans and between myxozoans and other cnidarians?
- Is myxozoan body plan diversity underestimated?
- When and where are key developmental genes such as Hox, ParaHox, and other homeobox genes expressed in the various myxozoan life cycles, and what are their roles?

References

Alama-Bermejo G, Bron JE, Raga JA, Holzer AS (2012) 3D morphology, ultrastructure and development of *Ceratomyxa puntazzi* stages: first insights into the mechanisms of motility and budding in the Myxozoa. PLoS One 7:e32679

Alvarez-Pellitero P, Molnár K, Sitjà-Bobadilla A, Székely C (2002) Comparative ultrastructure of the actinosporean stages of *Myxobolus bramae* and *M. pseudodispar* (Myxozoa). Parasitol Res 88:198–207

Bartholomew J, Whipple M, Stevens D, Fryer JL (1997) The life cycle of *Ceratomyxa shasta*, a myxosporean parasite of salmonids, requires a freshwater polychaete as an alternate host. J Parasitol 83:859–868

Bartošová P, Fiala I (2011) Molecular evidence for the existence of cryptic species assemblages of several myxosporeans (Myxozoa). Parasitol Res 108:573–583

Bartošová P, Fiala I, Jirků M, Cinková M, Caffara M, Fioravanti ML, Atkinson SD, Bartholomew JL, Holzer AS (2013) *Sphaerospora* sensu stricto: taxonomy, diversity and evolution of a unique lineage of myxosporeans (Myxozoa). Mol Phylogenet Evol 68:93–105

Canning EU, Okamura B (2004) Biodiversity and evolution of the Myxozoa. Adv Parasitol 56:43–131

Canning EU, Okamura B, Curry A (1996) Development of a myxozoan parasite *Tetracapsula bryozoides* gen. n. et sp. n. in *Cristatella mucedo* (Bryozoa: Phylactolaemata). Folia Parasitol (Praha) 43:259–261

Canning EU, Curry A, Feist SW, Longshaw M, Okamura B (2000) A new class and order of myxozoans to accommodate parasites of bryozoans with ultrastructural observations on *Tetracapsula bryosalmonae* (PKX organism). J Eukaryot Microbiol 47:456–468

Canning EU, Curry A, Hill SLL, Okamura B (2007) Ultrastructure of *Buddenbrockia allmani* n. sp. (Myxozoa, Malacosporea), a parasite of *Lophopus crystallinus* (Bryozoa, Phylactolaemata). J Eukaryot Microbiol 54:247–262

Canning EU, Curry A, Okamura B (2008) Early development of the myxozoan *Buddenbrockia plumatellae* in the bryozoans *Hyalinella punctata* and *Plumatella fungosa*, with comments on taxonomy and systematics of the Myxozoa. Folia Parasitol (Praha) 45:241–255

Diamant A (1997) Fish-to-fish transmission of a marine myxosporean. Dis Aquat Organ 30:99–105

El-Mansy A, Molnár K, Székely C (1998) Development of *Myxobolus portucalensis* Saraiva & Molnár, 1990 (Myxosporea: Myxobolidae) in the oligochaete *Tubifex tubifex* (Müller). Syst Parasitol 41:95–103

El-Matbouli M, Hoffmann RW (1998) Light and electron microscopic studies on the chronological development of *Myxobolus cerebralis* to the actinosporean stage in *Tubifex tubifex*. Int J Parasitol 28:195–217

Eszterbauer E, Kallert DM, Grabner D, El-Matbouli M (2009) Differentially expressed parasite genes involved in host recognition and invasion of the triactinomyxon stage of *Myxobolus cerebralis* (Myxozoa). Parasitology 136:367–377

Evans N, Lindner A, Raikova EV, Collins AG, Cartwright P (2008) Phylogenetic placement of the enigmatic parasite, *Polypodium hydriforme*, within the phylum Cnidaria. BMC Evol Biol 8:139

Evans NM, Holder MT, Barbeitos MS, Okamura B, Cartwright P (2010) The phylogenetic position of Myxozoa: exploring conflicting signals in phylogenomic and ribosomal data sets. Mol Biol Evol 27:2733–2746. doi:10.1093/molbev/msq159

Feist SW, Longshaw M (2006) Phylum Myxozoa. In: Woo PTK (ed) Fish diseases and disorders, vol 1, 2nd edn, Protozoan and metazoan infections. Cab Intl, Wallingford, pp 230–296

Fiala I (2006) The phylogeny of Myxosporea (Myxozoa) based on small subunit ribosomal RNA gene analysis. Int J Parasitol 36:1521–1534

Fiala I, Bartosová P (2010) History of myxozoan character evolution on the basis of rDNA and EF-2 data. BMC Evol Biol 10:228

Grabner D, El-Matbouli M (2010) *Tetracapsuloides bryosalmonae* (Myxozoa: Malacosporea) portal of entry into the fish host. Dis Aquat Organ 90:197–206

Gruhl A, Okamura B (2012) Development and myogenesis of the vermiform *Buddenbrockia* (Myxozoa) and implications for cnidarian body plan evolution. EvoDevo 3:10

Hallett SL, Lester RJ (1999) Actinosporeans (Myxozoa) with four developing spores within a pansporocyst: *Tetraspora discoidea* n.g. n.sp. and *Tetraspora rotundum* n.sp. Int J Parasitol 29:419–427

Hallett SL, O'Donoghue PJ, Lester RJG (1998) Structure and development of a marine actinosporean, *Sphaeractinomyxon ersei* n. sp. (Myxozoa). J Eukaryot Microbiol 45:142–150

Hartikainen H, Johnes P, Moncrieff C, Okamura B (2009) Bryozoan populations reflect nutrient enrichment and productivity gradients in rivers. Freshw Biol 54:2320–2334

Hartikainen H, Gruhl A, Okamura B (2014) Diversification and repeated morphological transitions in endoparasitic cnidarians (Myxozoa: Malacosporea). Mol Phylogenet Evol 76:261–269

Holland JW, Okamura B, Hartikainen H, Secombes CJ (2011) A novel minicollagen gene links cnidarians and myxozoans. Proc Biol Sci 278:546–553

Kent ML, Hedrick R (1986) Development of the PKX myxosporean in rainbow trout *Salmo gairdneri*. Dis Aquat Organ 1:169–182

Kent ML, Margolis L, Corliss JO (1994) The demise of a class of protists: taxonomic and nomenclatural revisions proposed for the protist phylum Myxozoa Grassé, 1970. Can J Zool 72:932–937

Kent ML, Andree KB, Bartholomew JL, El-Matbouli M, Desser SS, Devlin RH, Feist SW, Hedrick RP, Hoffmann RW, Khattra J, Hallett SL, Lester RJG, Longshaw M, Palenzeula O, Siddall ME, Xiao CX (2001) Recent advances in our knowledge of the Myxozoa. J Eukaryot Microbiol 48:395–413

Kumar G, Abd-Elfattah A, Soliman H, El-Matbouli M (2013) Establishment of medium for laboratory cultivation and maintenance of Fredericella sultana for in vivo experiments with Tetracapsuloides bryosalmonae (Myxozoa). J Fish Dis 36:81–88

Lom J, Dyková I (1992) Fine structure of *Triactinomyxon* early stages and sporogony: myxosporean and actinosporean features compared. J Eukaryot Microbiol 39:16–27

Lom J, Dyková I (1997) Ultrastructural features of the actinosporean phase of Myxosporea (Phylum Myxozoa): a comparative study. Acta Protozool 36:83–103

Lom J, Dyková I (2006) Myxozoan genera: definition and notes on taxonomy, life-cycle terminology and pathogenic species. Folia Parasitol (Praha) 53:1–36

Lom J, Yokoyama H, Dykova I (1997) Comparative ultrastructure of *Aurantiactinomyxon* and *Raabeia*, actinosporean stages of myxozoan life cycles. Arch Protistenkd 148:173–189

Markiw ME, Wolf K (1983) *Myxosoma cerebralis* (Myxozoa: Myxosporea) etiologic agent of salmonid whirling disease requires tubificid worm (Annelida: Oligochaeta) in its life cycle. J Protozool 30:561–564

Marques A (1986) La sexualite chez les Actinomyxidies: etude chez *Neoactinomyxon eiseniellae* (Ormieres et Frezil, 1969), Actinosporea, Noble, 1980; Myxozoa, Grasse, 1970. Ann Sci Nat Zool Paris Be ser 8:81–101

Marton S, Eszterbauer E (2012) The susceptibility of diverse species of cultured oligochaetes to the fish parasite *Myxobolus pseudodispar* Gorbunova (Myxozoa). J Fish Dis 35:303–314

McGurk C, Morris DJ, Adams A (2006) Sequential development of *Buddenbrockia plumatellae* (Myxozoa:

Malacosporea) within *Plumatella repens* (Bryozoa: Phylactolaemata). Dis Aquat Organ 73:159–169

Meaders MD, Hendrickson GL (2009) Chronological development of *Ceratomyxa shasta* in the polychaete host, *Manayunkia speciosa*. J Parasitol 95:1

Morris DJ (2010) Cell formation by myxozoan species is not explained by dogma. Proc Biol Sci 277:2565–2570

Morris D (2012a) Towards an in vitro culture method for the rainbow trout pathogen *Tetracapsuloides bryosalmonae*. J Fish Dis 35:941–944

Morris DJ (2012b) A new model for myxosporean (Myxozoa) development explains the endogenous budding phenomenon, the nature of cell within cell life stages and evolution of parasitism from a cnidarian ancestor. Int J Parasitol 42:829–840

Morris DJ, Adams A (2007a) Sacculogenesis of *Buddenbrockia plumatellae* (Myxozoa) within the invertebrate host *Plumatella repens* (Bryozoa) with comments on the evolutionary relationships of the Myxozoa. Int J Parasitol 37:1163–1171

Morris DJ, Adams A (2007b) Sacculogenesis and sporogony of *Tetracapsuloides bryosalmonae* (Myxozoa: Malacosporea) within the bryozoan host *Fredericella sultana* (Bryozoa: Phylactolaemata). Parasitol Res 100:983–992

Morris DJ, Adams A (2008) Sporogony of *Tetracapsuloides bryosalmonae* in the brown trout *Salmo trutta* and the role of the tertiary cell during the vertebrate phase of myxozoan life cycles. Parasitology 135:1075–1092

Nesnidal MP, Helmkampf M, Bruchhaus I, El-Matbouli M, Hausdorf B (2013) Agent of whirling disease meets orphan worm: phylogenomic analyses firmly place Myxozoa in Cnidaria. PLoS One 8:e54576

Okamura B, Curry A, Wood TS, Canning EU (2002) Ultrastructure of *Buddenbrockia* identifies it as a myxozoan and verifies the bilaterian origin of the Myxozoa. Parasitology 124:215–223

Okamura B, Gruhl A, Bartholomew J (eds) (2015) Myxozoan evolution, ecology and development. Springer International Publishing. ISBN 978-3-319-14752-9

Oumouna M, Hallett SL, Hoffmann RW, El-Matbouli M (2002) Early developmental stages of two actinosporeans, *Raabeia* and *Aurantiactinomyxon* (Myxozoa), as detected by light and electron microscopy. J Invertebr Pathol 79:17–26

Özer A, Wootten R (2001) Ultrastructural observations on the development of some actinosporean types within their oligochaete hosts. Turk J Zool 25:199–216

Rangel LF, Santos MJ, Cech G, Székely C (2009) Morphology, molecular data, and development of *Zschokkella mugilis* (Myxosporea, Bivalvulida) in a polychaete alternate host, *Nereis diversicolor*. J Parasitol 95:561–569

Siddall ME, Martin DS, Bridge D, Desser SS, Cone DK (1995) The demise of a phylum of protists: phylogeny of Myxozoa and other parasitic Cnidaria. J Parasitol 81:961–967

Tops S, Okamura B (2003) Infection of bryozoans by *Tetracapsuloides bryosalmonae* at sites endemic for salmonid proliferative kidney disease. Dis Aquat Organ 57:221–226

Tops S, Baxa DV, McDowell TS, Hedrick RP, Okamura B (2004) Evaluation of malacosporean life cycles through transmission studies. Dis Aquat Organ 60:109–121

Weill R (1938) L'interpretation des Cnidosporidies et la valeur taxonomique de leur cnidome. Leur cycle comparé à la phase larvaire des Narcomeduses cuninides. Trav Stn Zool Wimereaux 13:727–744

Wolf K, Markiw ME (1984) Biology contravenes taxonomy in the Myxozoa: new discoveries show alternation of invertebrate and vertebrate hosts. Science 225:1449–1452

Yang Y, Xiong J, Zhou Z, Huo F, Miao W, Ran C, Liu Y, Zhang J, Feng J, Wang M, Wang L, Yao B (2014) The genome of the myxosporean *Thelohanellus kitauei* shows adaptations to nutrient acquisition within its fish host. Genome Biol Evol 6:3182–3198

Ctenophora

8

Mark Q. Martindale and Jonathan Q. Henry

Chapter vignette artwork by Brigitte Baldrian.
© Brigitte Baldrian and Andreas Wanninger.

M.Q. Martindale (✉)
Whitney Lab for Marine Bioscience,
University of Florida, 9505 Oceanshore Blvd,
St. Augustine, FL 32080, USA
e-mail: mqmartin@whitney.ufl.edu

J.Q. Henry
Department of Cell and Developmental Biology,
University of Illinois, 601 S. Goodwin Ave.,
Urbana, IL 61801, USA
e-mail: j-henry4@illinois.edu

A. Wanninger (ed.), *Evolutionary Developmental Biology of Invertebrates 1:
Introduction, Non-Bilateria, Acoelomorpha, Xenoturbellida, Chaetognatha*
DOI 10.1007/978-3-7091-1862-7_8, © Springer-Verlag Wien 2015

INTRODUCTION

Ctenophores produce exquisite embryological material for both descriptive and experimental manipulation (e.g., Martindale 1997a, b, c). The fact that they are free-spawned and optically clear and undergo a rapid and highly stereotyped cleavage program have established them a highly studied preparation, particularly in the "golden era" of experimental embryology (circa the end of the nineteenth century).

Ctenophore embryos were one of the first species to have been experimentally investigated at the Stazione Zoologica marine laboratory in Naples, Italy. In 1872 Charles Chun noticed that "damaged" ctenophore embryos that he collected after a large storm formed two "half-embryos" contained within the same vitelline envelope. He surmised that the first two blastomeres had been separated by mechanical means and each had grown up as it normally would have if they had remained together, a result that he and others have experimentally reproduced (Chun 1892; Driesch and Morgan 1895; Martindale 1986). Ctenophore embryos remain as the prime example for the so-called "mosaic" form of development, and thus, ctenophore embryos have played an important historical role in establishing ways of thinking about developmental biology very early in the field's own history.

Virtually all extant ctenophores are holopelagic, lacking a benthic phase of their life cycle and spending their entire life swimming in the water column. Many are deep water, and adults obtain large sizes of over a meter in length, while others range in size from a couple of centimeters. All ctenophores that have been studied to date undergo direct development to rapidly give rise to a miniature functional juvenile (called a cydippid "larva") in less than one day. Ctenophore embryogenesis generates a body plan that is essentially the same as in their adult stages, so they are basically direct developers. All ctenophores are carnivorous and most eat small zooplankton including copepods, rotifers, and even larval fish. Most ctenophores, except for the beroids, possess tentacles with sticky cells called colloblasts that they use to capture prey. The beroids have lost their tentacles and are specialized predators on other ctenophores.

One group, the Platyctenida, loses their swimming structures (comb plates) soon after development is complete and assumes a benthic existence, using their highly characteristic tentacles, bearing a main tentacle with branching tentilla, to capture prey.

Most adult ctenophores have a high capacity to regenerate, and most are self-fertile hermaphrodites with rapid gametogenesis (approximately two days; Greve 1970) and high fecundity (Reeve and Walter 1978), making them ideal "invasive species." They have been known to have spread across the globe in ballast water of ocean-going ships (Reusch et al. 2010), and some species also have been described to undergo a biphasic reproductive pattern called "dissogeny" ("larval" reproduction), where functional gametes are generated within days after development is complete (see below). In fact, there is a species of ctenophore, *Mertensia*, in the Baltic Sea that is thought to consist entirely of subadult sizes, and the population is replaced entirely by this precocious reproductive pattern (Jaspers et al. 2012).

Ctenophores have found numerous ways to use modified cilia for both locomotion and sensory structures (Tamm 2014). All ctenophores move through the water column by the coordinated beating of eight longitudinal rows of comb plates that run along the oral-aboral axis (Fig. 8.1). In fact, their locomotory organs define the group (Ctenophora = "comb bearers"). Each comb plate is composed of thousands of $9+2$ cilia, each with their own membrane, that are attached laterally to one another to form stiff plates that push the animal through the water by a series of metachronal waves that run along each comb row (or ctene row). The power stroke is oriented aborally, making the animals swim mouth first, although ciliary reversal also occurs in all eight rows, providing for a fine motor control of swimming behavior. Comb plate beating is under nervous and mechanical control, integrated primarily through the apical organ, an inertial, gravity sensor located on the aboral pole, that is probably not homologous to an organ bearing the same name in lophotrochozoan larvae (see Vol. 2, Chapters 6, 7, 8, 9, 10, 11, and 12). The apical organ contains a statocyst, which consists of calcium carbonate-containing cells perched upon four groups of balancing cilia. In addition to a relatively sophisticated nervous

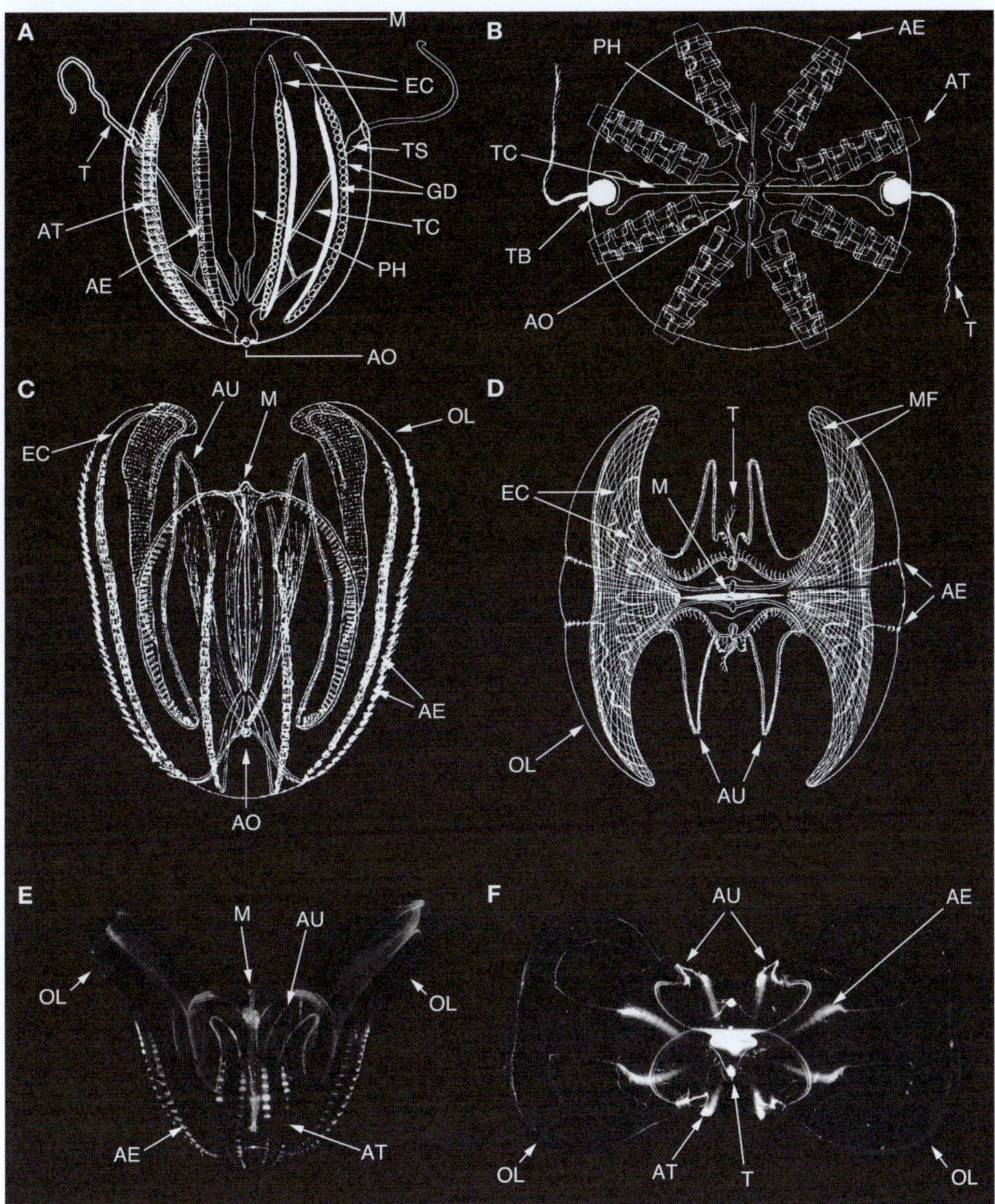

Fig. 8.1 Lobate ctenophore body plan. (**A**) Diagram showing a lateral view of a cydippid-stage ctenophore, oral end up. The apical sense organ is located on the side opposite the mouth. Two feeding tentacles grow out from the tentacle bulbs that are attached to the aboral end of the pharynx by the endodermal tentacular canals. The two pairs of ctene rows adjacent to the tentacular plane are called the adtentacular rows and the ones closest to the pharyngeal plane the adesophageal rows. (**B**) Diagram of an aboral view of a cydippid-stage ctenophore showing the two major body axes, the tentacular plane and the orthogonal esophageal plane. (**C**) Diagram of a lateral view of a cydippid stage making the transition to a lobate stage. The muscular oral lobes and the ciliated auricles extend from the oral pole. (**D**) Diagram of the oral pole of a lobate stage showing the reduced tentacles and elaborated auricles and oral lobes. (**E**) Photograph of a lateral view of an adult lobate stage, oral pole up. This is the same orientation as seen in **C**. (**F**) Photograph of an adult lobate stage seen from the oral pole. This is the same orientation as seen in **D**. Abbreviations: *AE* adesophageal ctene row, *AO* apical organ, *AT* adtentacular ctene row, *AU* auricles, *EC* endodermal canals, *GD* gonads, *M* mouth, *MF* muscle fibers, *OL* oral lobes, *PH* pharynx, *T* tentacle, *TC* tentacular canals, *TS* tentacle sheath

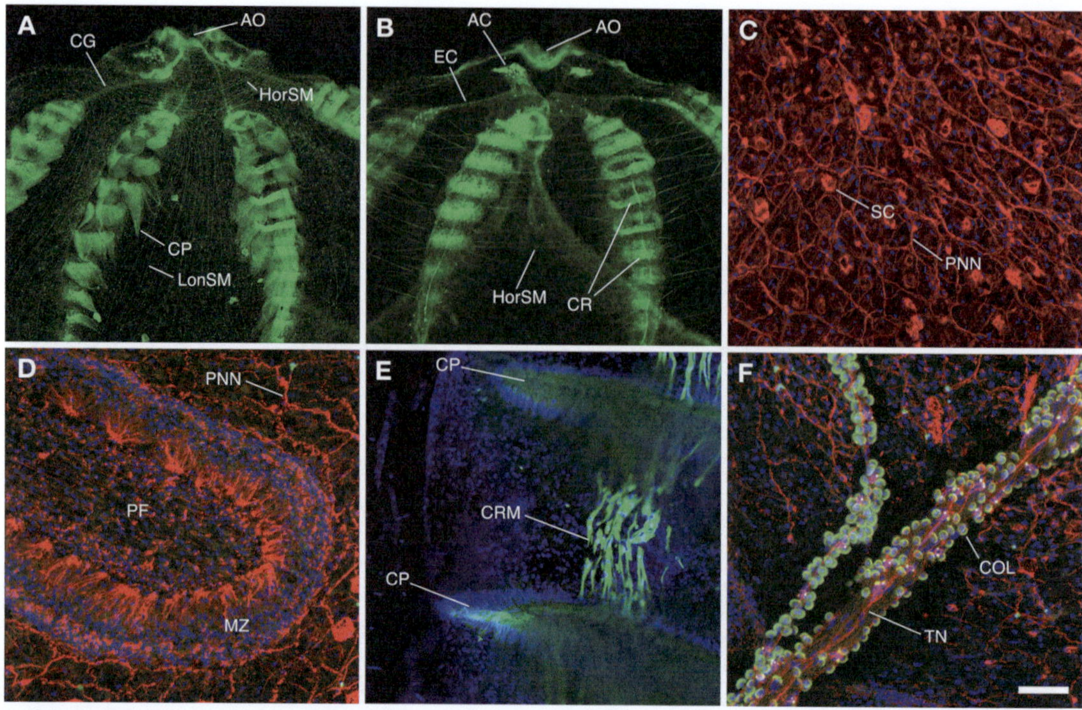

Fig. 8.2 Confocal images of key features of the ctenophore body plan. (**A**) A lateral (apical organ toward top of page) confocal micrograph of a juvenile *Beroe ovata* immunostained with phalloidin (*green*) which binds to filamentous actin. Longitudinal (*LonSM*) and horizontal (*HorSM*) smooth muscle fibers are visible in the mesogloeal layer. The comb plate cilia (*CP*) and the ciliated grooves (*CG*) that originate at the edge of the apical organ (*AO*) are also stained by phalloidin. (**B**) A slightly deeper optical section of the specimen seen in (**A**). The endodermal canal (*EC*) system, including one of the two anal canals, can be seen. Note that a distinct horizontal intercomb plate muscular system is present. (**C**) A confocal surface view of the epidermal nerve net of a juvenile *Pleurobrachia* stained with anti-tyrosylated-tubulin (*red*) and Hoechst nuclear stain (*blue*). The epidermal polygonal nerve net (*PNN*) and associated sensory cells (*SC*) are labeled. (**D**) High magnification confocal view of the polar field region of the apical organ of *Pleurobrachia* sample seen in (**C**). Note the high density of cells in the marginal zone (*MZ*) and the proximity of the polygonal nerve net. (**E**) A high magnification confocal micrograph between two comb plates within the middle of the comb row structure of juvenile *Pleurobrachia* stained with phalloidin (*green*) and Hoechst nuclear stain (*blue*). Each comb row (*CR*) is made up of a large number of support cells at its base. In between each row are muscle cells which contain a large amount of filamentous actin. (**F**) A confocal image of a portion of the tentacles of a juvenile *Pleurobrachia* immunostained with Hoechst nuclear stain (*blue*) and phalloidin (*green*). A rim of cytoplasm around each colloblast (*Col*) cell and the tentacle musculature are labeled with phalloidin. Tentacular nerves (*TN*) are visible running along the length of the tentacle and the polygonal nerve net can be seen underlying the body wall epithelium below. Abbreviations: *AC* anal canal, *AO* apical organ, *CG* ciliated groove, *CP* comb plate, *CR* comb row, *CRM* comb row muscle, *Col* colloblast, *EC* endodermal canal, *HorSM* horizontal smooth muscle, *LonSM* longitudinal smooth muscle, *MZ* marginal zone, *PF* polar field, *PNN* polygonal nerve net, *SC* sensory cell, *Tent* tentacle. The scale bar seen in panel F equals 200 μm for **A** and **B**, 30 μm for **C**, 33 μm for **D** and **F**, and 18 μm for **E** (Images taken by David Kainoa Simmons at the Whitney Lab for Marine Bioscience)

system that is composed of a peripheral polygonal nerve "net," apical organ, and tentacular nerves, all ctenophores have a complex set of muscle and mesenchymal cells that reside in distinct regions of the mesogloea or under the epidermal basal laminae (Fig. 8.2). Several excellent reviews of ctenophore anatomy have been previously published (Hyman 1940; Hernandez-Nicaise 1973, 1991; Horridge 1974; Tamm 1982, 2014; Brusca and Brusca 1990; Martindale and Henry 1997a; Pang and Martindale 2008a; Martindale 2001).

Ctenophores consist of four nearly identical quadrants (Fig. 8.1) separated by two planes, the tentacular plane and the esophageal plane (sometimes called the sagittal plane). Each quadrant contains two comb rows, a half of a tentacle and a quarter of the apical organ, which is located at the aboral pole. However, adjacent quadrants are not identical to one another morphologically, because only two diagonally opposed quadrants have anal canals (regions of the endodermal gut that make open connections to the external world via the anal pores) (Martindale and Henry 1995). Thus, these animals do not have a single plane of mirror symmetry, rather, they have an infinite number of planes of rotational symmetry, a feature that is difficult to compare to a hypothetically simple radially symmetrical ancestor. In fact, the phylogenetic position of ctenophores is still hotly debated, with a growing body of phylogenomic data arguing that they are the earliest branching group of metazoans (Dunn et al. 2008; Hejnol et al. 2009; Ryan et al. 2013). In fact, there are no groups of animals that are truly radially symmetrical (see also Chapter 6; Finnerty et al. 2004).

Ctenophores have at one time or another been placed in a variety of different phylogenetic positions in the Tree of Life, from deuterostomes (Nielsen 1995) to flatworms (Lang 1884; Mortensen 1912a, b). The most consistent historical position is for ctenophores to be sister to cnidarians in a taxon termed Coelenterata (Hyman 1940; Philippe et al. 2009), sister to all bilaterally symmetrical animals (Morris and Simonetta 1991; Nielsen et al. 1996), or sister to Placozoa, Cnidaria, and Bilateria (Collins 2002; Wallberg et al. 2004; Pick et al. 2010). Total genome sequencing of two ctenophores has shed some additional light on the situation and made a compelling case that either sponges or ctenophores are at the base of the metazoan tree (Ryan et al. 2013). The gene content of both sponges and ctenophores is much more similar to one another than either is to other metazoans, generating many questions about the relationship of genomic complexity to morphological complexity (Putnum et al. 2007; Srivastava et al. 2008, 2010). The development of ctenophores and sponges appears so radically different it is difficult to even compare the two, suggesting that one or both groups have become extremely specialized over deep evolutionary time.

EARLY DEVELOPMENT

Most, but not all (Harbison and Miller 1986), ctenophores are self-fertile hermaphrodites (Carré et al. 1990). Male and female gonads are associated with the endodermal canal system, one on each side of each subctene row canal (see Martindale and Henry 1997a, b). Spawning is generally triggered by changes in photoperiod. Motile flagellated sperm are spawned first with oocytes that follow. In most cases spawned oocytes have already undergone first and second polar body formation and have a surrounding acellular vitelline membrane made by the oocyte. In some species the polar bodies are still attached to the egg and mark the animal pole (Freeman 1977). Jelly/mucus is secreted along with the eggs (Dunlap-Pinaka 1974) that make them remain in the water column but that usually washes off after a few minutes in sea water, allowing eggs to settle to the bottom. The delicate vitelline membrane can be easily removed with forceps if experimental manipulations (injections, blastomere deletions) are required (although naked embryos will stick to polystyrene plastic). Ctenophore eggs and embryos are relatively large, varying between 100 µm and over 1 mm in diameter, and are among the most optically transparent embryos described. They are centrolecithal, the yolk being located in the middle of the cell surrounded by a thin layer of ectoplasm containing all of the cellular organelles (Dunlap-Pinaka 1974). The ectoplasm undergoes dynamic rearrangements during the cleavage program to be segregated mainly to ectodermal precursors.

Fertilization in some species such as *Mnemiopsis leidyi* occurs at the time of spawning, although the eggs of some representatives such as *Beroe* tend to outcross with sperm of a different individual, at least early after spawning. This specificity operates at the level of the vitelline membrane (Carré et al. 1991). There does not appear to be a "fast block" to polyspermy. Multiple sperm asters (up to 20) can be seen in ctenophore zygotes with other cytoplasmic mechanisms, insuring that the female pronucleus undergoes syngamy with a single male pronucleus (Carré et al. 1991).

Ctenophore embryos have a unique cleavage program not readily comparable to any other metazoan. It starts with a unipolar cleavage mechanism in which the cleavage furrow starts at the animal pole and completely splits the cytoplasm into two cells (Fig. 8.3). The first two cleavages pass parallel to the animal-vegetal axis

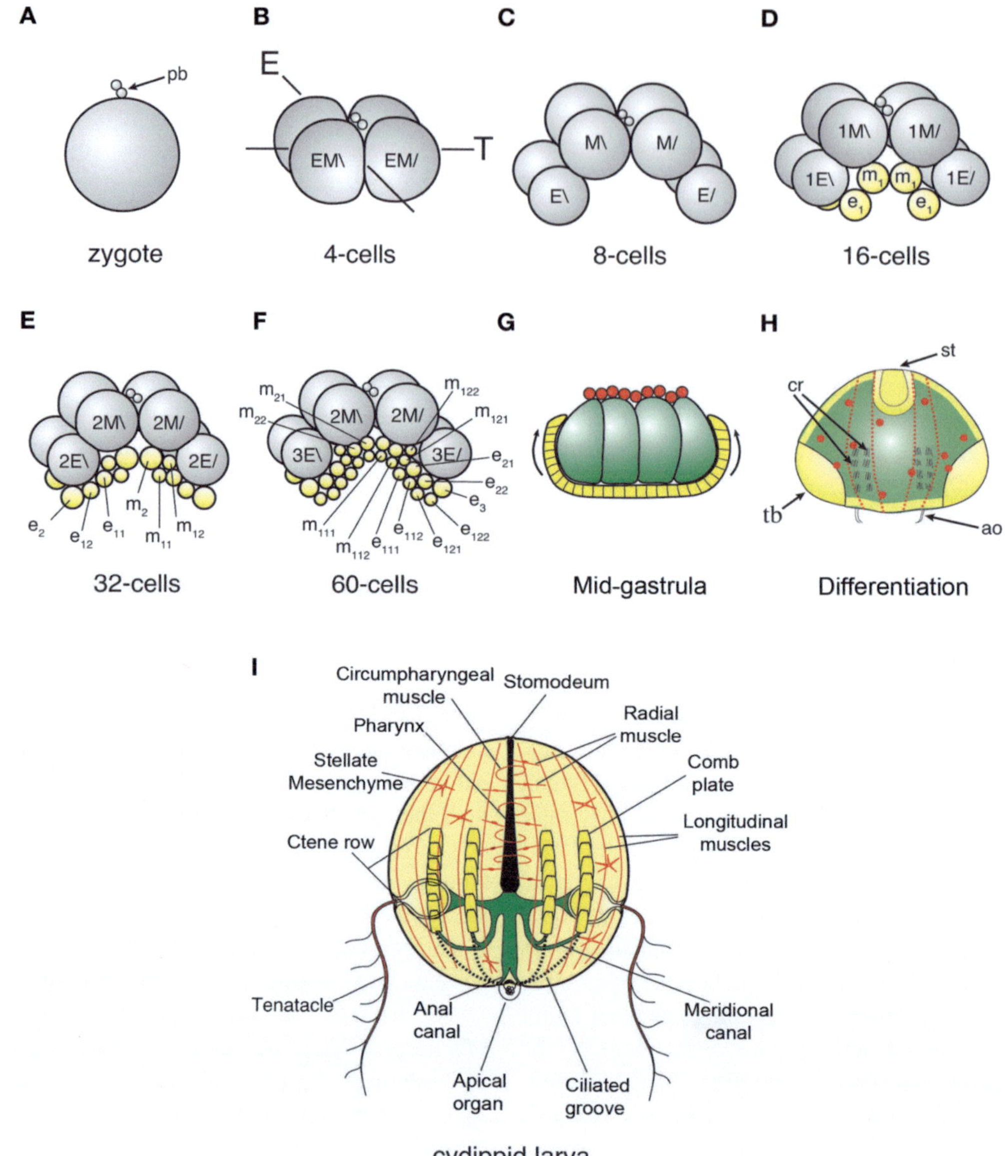

Fig. 8.3 Schematic diagram and fate map of ctenophore development. (**A**) Lateral view of the fertilized egg. The polar bodies (*pb*) mark the animal pole which represents the future oral end of the animal in all panels. (**B**) At second cleavage each blastomere divides asymmetrically to give rise to an EM/and an EM\cell. The EM/and EM\cells lie in opposite, not adjacent locations in the embryo. (**C**) At third cleavage, two distinct cell lineages are born, the E (end) and M (middle) blastomeres. (**D**) At the 16-cell stage, small micromeres arise that are given subscript numerals to indicate the order of their birth (e.g., *e1*). Their sister macromeres are given ordinal numbers for ease of identification. (**E**) At the next round of divisions, the 1E and 1M macromeres give rise to another round of micromeres and each micromere also divides. (**F**) The 60-cell stage arises as all cells divide one additional time except for the 2M macromeres. (**G**) As gastrulation by epiboly begins, a distinct set of micromeres is born at the oral pole. These will give rise to the mesodermal derivatives. (**H**) At the "Mickey mouse" stage, the tentacle bulbs thicken prior to their invagination, the comb row cilia begin to beat, muscle and mesenchymal cells appear, and the apical organ begins to coalesce as components from all four quadrants move toward the aboral pole. (**I**) By approximately 24 h, a fully functional cydippid-stage animal is formed. Cells colored in *yellow* give rise to ectoderm, cells colored *green* give rise to endoderm, and cells in *red* give rise to mesodermal structures. Abbreviations: *ao* apical organ precursors, *cr* ctene row precursors, *pb* polar bodies, *tb* tentacle bulb precursors, *st* stomodeum

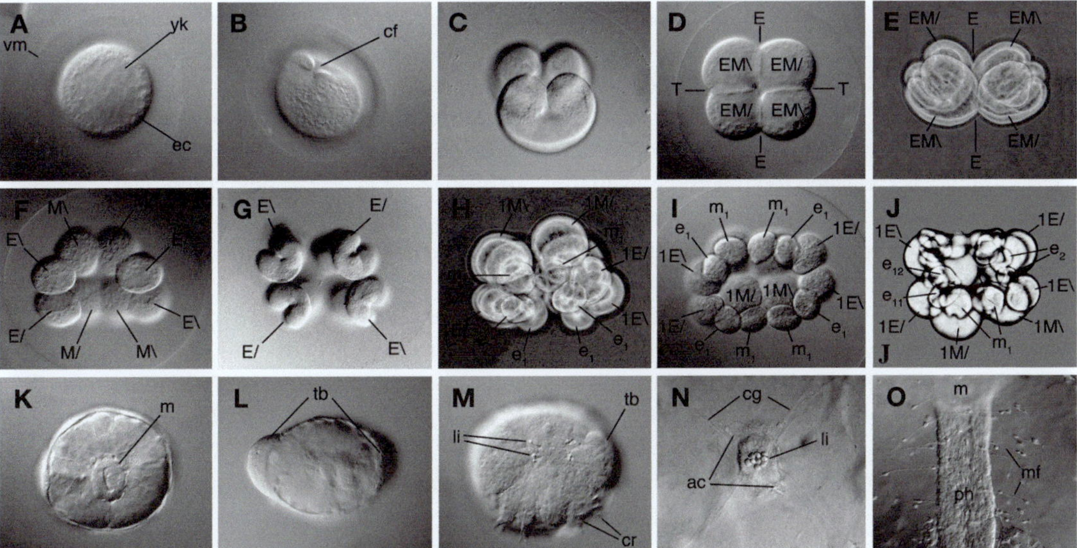

Fig. 8.4 Differential interference contrast light microscope photographs of ctenophore development. The animal pole is situated toward the top of the panel except in panels **I**, **J**, **M**, and **N** (After Martindale and Henry (1997a, b)). (**A**) Fertilized zygote prior to first cleavage inside its acellular vitelline membrane. The ectoplasm surrounds a yolk mass. (**B**) The onset of first cleavage. Note the highly characteristic unipolar cleavage furrow. (**C**) Two cells dividing to four cells via unipolar cleavage furrows. (**D**) The 4-cell stage showing the distribution of the two determined quadrant types, EM/and the EM\. (**E**) An oblique cleavage gives rise to the E and M blastomeres of each quadrant. The E marks the future esophageal plane. (**F**) The 8-cell stage. (**G**) The unipolar cleavage of the E blastomeres gives rise to the 1E macromere and the e_1 micromere. (**H**) The 16-cell stage consisting of four e_1 and four m_1 micromeres and their sister macromeres. (**I**) Aboral (vegetal) view of the 16-cell stage. (**J**) Aboral view of the 16-cell stage. Each macromere has given rise to a new round of micromeres and each existing micromere has divided equally. (**K**) Oral view of the mouth following gastrulation. The large yolky internal cells will give rise to the endodermal portions of the gut. (**L**) Optical section of the tentacle bulbs at the "Mickey mouse" stage of development seen from the oral pole. (**M**) View of the developing apical organ with its mineralized lithocytes, nascent ctene rows, and invaginating tentacle bulbs. (**N**) Higher magnification view of developing apical organ showing the lithocytes, the ciliated grooves that lead to each of the eight ctene rows, and the anal canals which open in opposite quadrants at the anal pores. (**O**) Lateral view of the ectodermal pharynx leading to the mouth. Note the numerous contractile muscle cells surrounding the pharynx and leading out into the mesogloea. Abbreviations: *ac* anal canals, *cf* cleavage furrow, *cg* ciliated grooves, *cr* ctene rows, *ec* ectoplasm, *li* lithocytes, *m* mouth, *mf* muscle fibers, *ph* pharynx, *vm* vitelline membrane, *yk* yolk, *tb* tentacle bulb

and give rise to four equal-sized and visibly identical cells (the fates of these cells are not identical, however, as described below). The third division is oriented obliquely to the animal-vegetal axis and generates four slightly larger cells that all meet at the vegetal pole and are called the "M" cells (for "middle" cells) and four slightly smaller cells that form two pairs of cells that do not touch each other. These cells are called "E" cells (for "end" cells). Both E and M cells then generate three and two rounds (respectively) of highly asymmetric divisions giving rise to micromeres and macromeres at the aboral pole of the embryo.

The micromeres inherit the majority of the ectoplasm, while the larger macromeres retain the yolk and will give rise to the endoderm. A standard nomenclature exists (Martindale and Henry 1997a, b), such that the first round of micromeres from each lineage is given a subscript (e_1, m_1) and their sister macromeres the title 1E and 1M. The macromeres then divide again asymmetrically to generate (e_2, m_2) and 2E and 2M cells. Micromeres divide further to give rise to e_{11} and e_{12}, and m_{11} and m_{12} blastomeres (Figs. 8.3 and 8.4).

As the aboral micromeres continue, they divide to give rise to a micromere cap consisting of hundreds of small cells. As gastrulation takes place, these cells move by epiboly over the larger macromeres toward the animal (oral) pole, and a second set of small micromeres are given off by the macromeres at the oral pole. These oral micromeres migrate up between the macromeres to the aboral pole and give rise to the mesodermal derivatives. Time lapse movies show a buckling of the embryo to give rise to a concave disk with the opening being at the oral pole, but it is not clear if the force required for this shape change is derived from the macromeres or the overlying micromeres. Gastrulation continues as the aboral micromeres converge onto the future oral opening. A sheet of ectodermal cells moves in through the future mouth to give rise to the pharynx and esophagus. The future tentacle buds can be distinguished as ectodermal thickenings on the aboral pole (the so-called "Mickey mouse" stage). The future tentacle buds then invert to give rise to the tentacular canals and meet with mesodermal and endodermal precursors to give rise to the precursors of the tentacle bulbs that generate the constantly growing tentacles. The cilia for the comb plates start to grow out, as do the dome and balancing cilia associated with the apical organ. By 14–16 h postfertilization, the ctenophore body plan can be clearly seen and the swelling of the mesogloea gives rise to an almost perfectly clear and spherical cydippid-stage animal. The motile ctene cilia continue to form, orient into individual plates, and begin to beat in a coordinated fashion. The final step before hatching is the growth of the tentacles out past the body wall, approximately 24 h after fertilization.

Ctenophore Fate Mapping

Cell labeling and fate mapping experiments in conjunction with experimental intervention have revealed a great deal about how ctenophores develop. For example, cell labeling experiments showed that the site of first cleavage gives rise to the site of gastrulation (where mesodermal micromeres are generated) and the future oral pole (Freeman 1977). The first attempt to generate an accurate ctenophore fate map utilized chalk particles (Ortolani 1963; Reverberi and Ortolani 1963),

but more recently a more accurate intracellular lineage analysis up to the 60-cell stage in the lobate ctenophore *Mnemiopsis* (Figs. 8.3 and 8.5) has determined the embryological origin of all major cell types (Martindale and Henry 1999). Ectodermal derivatives (e.g., epidermis, comb plates, and nervous system) are derived from the small aboral micromeres born at the vegetal pole, while mesodermal derivatives (muscle and mesenchymal cells) are derived from the small micromeres generated at the oral pole. Endoderm, including the mineral-containing lithocytes generated in the floor of the apical organ, is derived from the large macromeres at the oral (animal) pole following the production of the mesodermal micromeres. Thus, the mesoderm of ctenophores can be regarded as being endomesodermal, rather than ectomesodermal, in origin. This suggests that there may be sets of genes in common with bilaterian organisms that also form endomesoderm (but see below).

Fate mapping studies further demonstrate that the cleavage process is involved with asymmetric distribution of cell fates. Injection of one of the first two cells with lineage tracer reveals that the first plane of cleavage always corresponds to the esophageal (sagittal) plane, while the second plane of cleavage corresponds to the tentacular plane. Fate mapping analyses have also shown that the second division that gives rise to four cells that all look identical to one another is actually asymmetric. Each daughter cell of the first two blastomeres gives rise to one cell that will make a quadrant of the adult that possesses an anal pore and a sister cell that will give rise to an adjacent quadrant that does not (Martindale and Henry 1995). In fact, further lineage analyses show that the cell that does not make an anal pore makes distinct sets of circumpharyngeal muscle cells (Fig. 8.1). Thus, both cells at the two-cell stage divide asymmetrically to give rise to the progenitors of two diametrically opposed (nonadjacent) quadrants (Fig. 8.6). Experimental analysis reveals that this "diagonal determination" accounts for specific differences in positional information around the oral-aboral axis and has profound effects on regenerative potential (Henry and Martindale 2000). This finding has implications for the naming of ctenophore cells because we now have to identify both classes of opposite

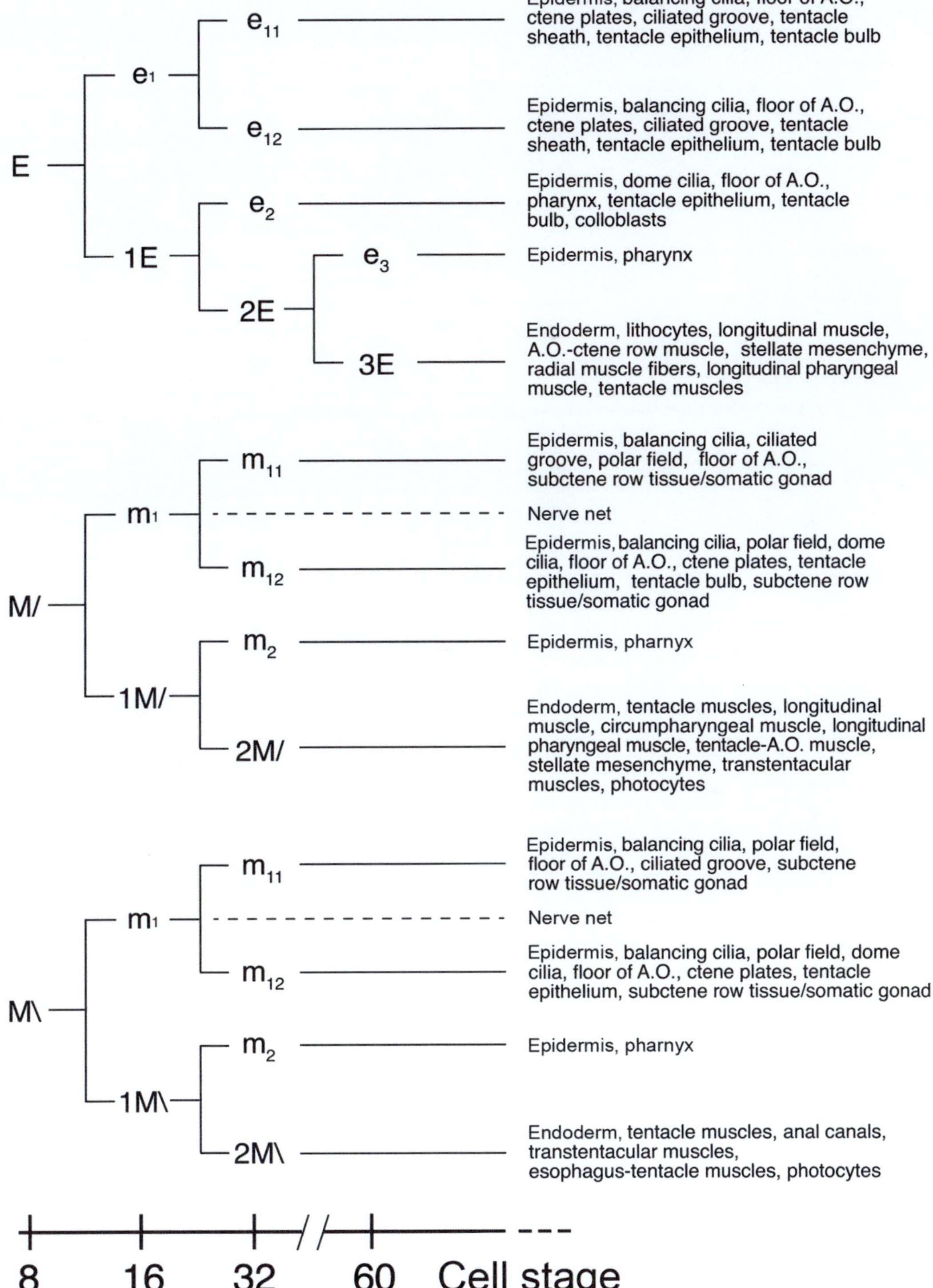

Fig. 8.5 Fate map generated by the intracellular injection of identified blastomeres up through the 60-cell stage (Martindale and Henry 1999). Note that there are differences in the fates of cells derived from the M macromeres (*M/* and *M*) but no differences between E quadrants have thus been identified

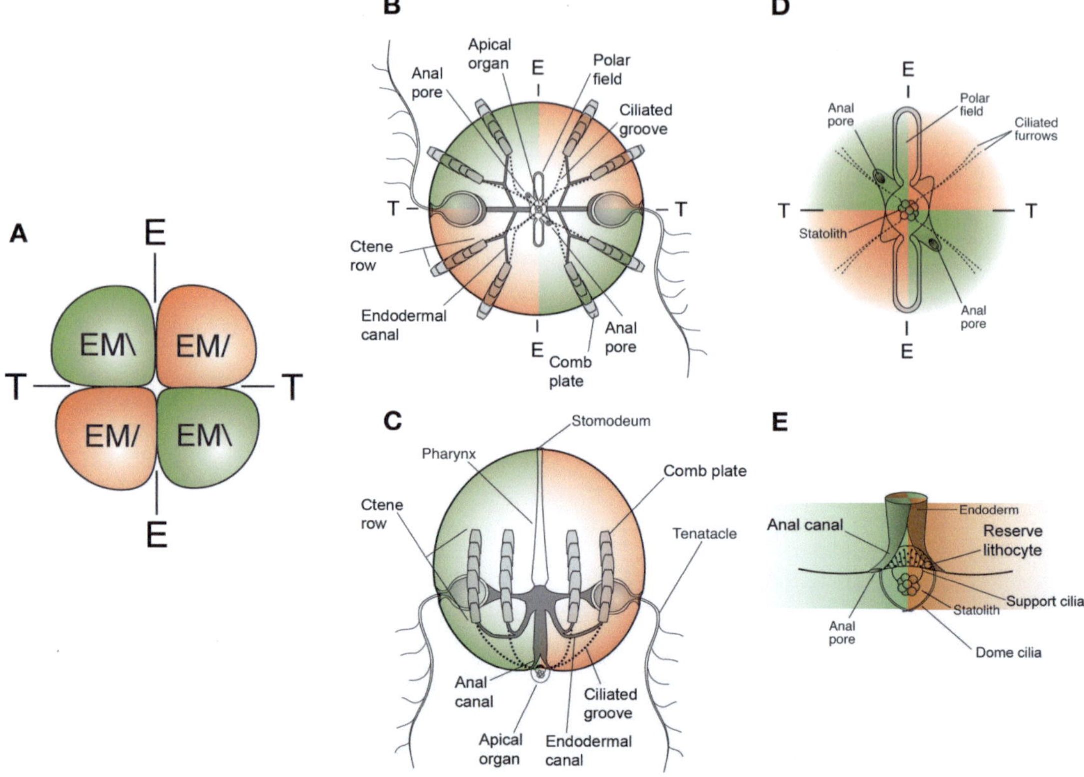

Fig. 8.6 Diagrammatic illustration of the ctenophore anal axis. The EM\(*green*) and EM/(*orange*) blastomeres of the 4-cell stage (**A**) seen from the vegetal pole give rise to distinct quadrants (**B-E**) of the juvenile cydippid. The contributions of the EM\ and EM/blastomeres to the adult body plan are seen from the aboral pole (**B**) and lateral view (**C**). Only the EM\blastomeres give rise to the anal canals seen in aboral (**D**) and lateral (**E**) views. The EM/ quadrants give rise to circumpharyngeal muscle cells that are not generated by the EM\blastomeres (After Martindale and Henry (1995, 1999))

quadrants. A nomenclature has been proposed that uses the symbol "/" (slash) to refer to the two quadrants that do not make the anal pores and "\" (backslash) to refer to the two quadrants that do make anal pores when viewed from the aboral pole (Martindale and Henry 1995, 1999).

The third division also distributes cell fates asymmetrically, giving rise to the E and M lineages (Figs. 8.3 and 8.5). Classical cell fate mapping experiments have shown that the E lineage gives rise to comb plates, ciliated grooves and components of the apical organ (including the mineral-containing lithocytes), and tentacle sheaths. The M lineage also generates components of the tentacles and apical organ but also gives rise to the majority of the peripheral nervous system and the light-producing photocytes (Freeman and Reynolds 1973; Freeman 1976a). The fourth division is asymmetric in both cell fate and in cell size. During cleavage, the ecto-plasm streams into the smaller yolk-free macromeres born at the aboral (vegetal) pole. Comb plate potential gets segregated to the e_1 micromeres; none of the other E lineage blastomeres have the capacity to generate comb plates. Likewise, the ability to make light-producing photocytes is restricted to the oral micromeres of the M lineage. Thus, cell lineage analysis suggests that each of the early cleavage divisions results in the asymmetric distribution of developmental fate.

Experimental Embryology

The cell cycle is relatively short during early ctenophore development (12–15 min), so precise regulation of developmental decision-making appears to be critical for these animals that hatch as functional juveniles in around 24 h. Historically, it has often been assumed that animals that undergo highly stereotyped cleavage programs, such as soil nematodes and spiral-cleaving organisms, also showed precocious specification of cell fates by the segregation of morphogenetic determinants. In fact, Charles Chun's experiments at the end of the nineteenth century demonstrated that ctenophores have a limited ability to regulate during the early cleavage program (Chun 1892). When ctenophore blastomeres are separated at the two-cell stage and grown to cydippid larvae, each one will possess four ctene rows, one tentacle, and a "half" of an apical organ (bearing two balancing cilia rather than all four). When the first four blastomeres are isolated, each gives rise to a pair of ctene rows and a small portion of a tentacle (Chun 1892; Driesch and Morgan 1895; Martindale 1986). These isolated cells do not compensate (regulate) for missing regions to give rise to the normal adult body plan.

The results of blastomere isolation/deletion experiments are easier to interpret if a detailed fate map has been generated. Farfaglio (1963) used chalk particles to show that each of the e_1 micromeres contributed to a pair of comb rows. When he killed an e_1 micromere, a pair of comb rows failed to form. When all four of the e_1 micromeres were removed, no comb plate cilia appeared during the developmental period. He thus concluded that the determinants for comb row formation were segregated to the e_1 micromeres. With a more precise fate map available (Martindale and Henry 1999), the interpretation of these results has changed. Intracellular lineage tracing shows that m_1 micromeres also make a small contribution to the comb plates, yet m_1-derived comb plates do not form after removal of e_1 (Martindale and Henry 1997a, b). This result indicates that e_1 micromeres are somehow required for the normal formation of comb plate cilia from m_1 descendants. Additional blastomere recombination experiments have shown that the influence of e_1 micromeres on comb plate production by m_1 blastomeres, while required, is not sufficient, because endomesodermal precursors from either the 3E or 2M lineages are also required (Henry and Martindale 2001). These data show that, although many early lineage-specific developmental decisions in ctenophores are generated by asymmetric divisions, cell-cell induction is also utilized, particularly later in development.

The Role of the Cleavage Program in the Establishment of Spatial Organization

Several lines of evidence point to the notion that this asymmetric segregation of developmental potential is a causal result of the cleavage program itself, and not a passive sequestration of prelocalized determinants by a stereotyped cleavage program. First, it has been described over a century ago that "ectoplasm" that is normally uniformly distributed around the circumference of the fertilized egg is actively segregated to the aboral micromeres during their formation starting at the eight-cell stage (Chun 1880; Driesch and Morgan 1895; Spek 1926).

If an unfertilized ctenophore egg is experimentally cut into two fragments and then fertilized, each half develops into a normal cydippid with eight comb rows and two tentacles (Driesch and Morgan 1895; Yatsu 1911, 1912; Freeman 1977). Normal cydippid larvae, complete with apical organs, develop from fertilized eggs that had been bisected equatorially at the start of the first cleavage and the aboral portion of the embryo removed (Yatsu 1912; Houliston et al. 1993; Freeman 1977). These results indicate that factors responsible for giving rise to the apical organ and ctene rows are not localized to their presumptive aboral locations until sometime after the onset of the first cleavage. Freeman (1976a, b) extended these experiments by performing a series of complex cleavage arrest experiments, showing that E and M lineage determinants are not prelocalized to the regions that they will eventually arise from at the four-cell stage and that these factors are gradually segregated through the cell cycle to the time that cytokinesis is completed.

A more recent set of cleavage arrest experiments have been performed that extend these earlier findings (Fischer et al. 2014). When cytokinesis in the ctenophore *Mnemiopsis* is permanently arrested with the drug cytochalasin-B, after, but not before the eight-cell stage, E and M lineage-specific markers are expressed in the appropriate cells. For example, in the arrested eight-cell stage, E cells, but not M cells, make miniature motile comb plates, and M cells, but not E cells, produce light (Fischer et al. 2014). If cytokinesis is blocked one cell division later, only the appropriate sister cell expresses the appropriate lineage-specific marker (e.g., the e_1 micromere but not the 1E macromere makes comb plate cilia and the 1M macromere, but not the m_1 micromere, makes light), indicating that something related to the cell's cleavage process is asymmetrically segregating developmental potential to the appropriate cell at each cell division. Interestingly, cytochalasin-B does not prevent nuclear divisions, and the embryo appears to "count" the number of nuclear divisions to express differentiated markers with similar timing and at the same division cycle as untreated controls (Fischer et al. 2014). The nature of this "cleavage clock" is not known but is speculated to be related to the titration of some factor that allows the transcription of marker genes associated with terminal differentiation (Fischer et al. 2014).

Perhaps the most compelling result to indicate that the cleavage process itself is causally involved with highly coordinated segregation of developmental potential comes from centrifugation experiments in which the site of first cleavage is changed from its normal position (as marked with vital dyes) to a site distant from the original position. If the new site of the unipolar first cleavage is marked and the embryo allowed to continue to develop, the mouth now forms at the new site of first cleavage, not the old site where one would expect the mouth to form (corresponding to the original, intrinsic animal-vegetal polarity of the fertilized egg; Freeman 1977). Although most developmental biologists assume that the animal-vegetal axis is maternally established, there are other examples in which the initiation of the site of first cleavage establishes the embryonic, and thus organismal, axial properties, e.g., in some cnidarians and mollusks (see Chapter 6; Vol. 2, Chapter 7; Freeman 1977, 2006). This indicates that there is no fixed maternal organization and that all subsequent developmental patterning events are initiated as a consequence of the cleavage program.

Unfortunately, the nature of the determinants of spatial organization or the precise cell biological mechanisms responsible for their asymmetric distribution are currently unknown. Microtubules or microtubule-associated proteins are one likely place to start looking, as each of the early cleavages result in the asymmetric localization of developmental potential, but it does not appear that the determinants are maternally loaded mRNAs that direct lineage-specific differentiation (Fischer et al. 2014).

LATE DEVELOPMENT

Adult Regeneration

Considering the lack of ability of ctenophore embryos to regulate it is perhaps surprising that most adult ctenophores have an outstanding capacity to regenerate. Coonfield (1936, 1937) demonstrated that *Mnemiopsis* can regenerate all major structures, including comb rows, tentacles, and apical organ, and that animals cut in one-half or one-fourth animals can reconstitute a complete animal. Since then, the dynamics of comb plate formation have been studied in different species (Tamm 2012a, b). Regenerative capacity varies tremendously in different ctenophore taxa. The "creeping" platyctene ctenophores (e.g., *Coeloplana*, *Vallicula*, and *Ctenoplana*), who lose their comb plates as juveniles and assume a benthic existence, routinely divide asexually by a process of fission throughout their life (Tanaka 1931; Dawydoff 1938; Freeman 1967). Interestingly, the beroids, which have lost their tentacles (Podar et al. 2001), have a very limited capacity to regenerate and cannot replace aboral structures, including their apical organs, following surgical removal. It is unclear if the lack of regenerative ability of beroids is related to their loss of tentacles and the stem cells that reside in tentacles bulbs. Regeneration in some species can also be incomplete. For example, when adult *Mnemiopsis* are bisected along the oral-aboral axis, wound healing can occur but regeneration aborts, and the animals remain as stable "half-animals" (Coonfield 1936; Freeman 1967; Martindale 1986). These results jibe well with the stability of half-animals generated during the embryonic period and support the notion that the half-animal phenotype is a metastable state that can be maintained throughout the animal's life under certain conditions.

Bisecting ctenophore embryos has also been used to learn more about the timing of the adult regenerative response (Martindale 1986). If adjacent quadrants are isolated during early development, half-animals are produced in 100 % of the cases (Chun 1892; Driesch and Morgan 1895; Martindale 1986), but if adult animals are bisected, normal adults are made in the majority of cases (Coonfield 1936; Martindale 1986). There must be a stage in development in which this transition in regenerative ability occurs. A time course of surgical bisections revealed that this time period occurs over a relatively short interval, on the order of 30 min, well after gastrulation is complete but before comb plate beating becomes coordinated in metachronal waves. It is not known whether this represents a fundamental change in the neural organization of the embryo or if it is associated with some other event, e.g., related to the stem cell precursors (Martindale 1986). Further investigations are required to understand this dramatic change in the response to the same experimental operation.

Evidence for Quadrant-Specific Positional Information

Intracellular fate mapping experiments indicated that the second division is asymmetric, with one daughter giving rise to circumpharyngeal muscle cells and the other daughter to an anal pore (Martindale and Henry 1999). Surgical manipulations of early embryos have revealed that this division from two to four cells is even more asymmetric than previously appreciated. As has already been mentioned, if ctenophore embryos are cut in half at the four-cell stage, such that any two adjacent quadrants remain together, the animal will grow up to remain as a "half-animal" possessing four ctene rows, one tentacle, and a half of an apical organ with two rather than four groups of balancing cilia (Chun 1892; Driesch and Morgan 1895; Martindale 1986). If two opposite (diagonal) quadrants are isolated, however, the embryos will grow up to be complete "whole" animals possessing eight comb plates, two tentacles, and a whole apical organ (Henry and Martindale 2000). These data reveal that opposite and adjacent quadrants have completely differently "identities" (one pair initiates a regulative response, while the other does not) when put in

juxtaposition. The results of these experiments are compatible with a polar coordinate-type model (French et al. 1976) of positional information (Fig. 8.7; Martindale and Henry 1997c, 2000); however, the molecular basis for these differences in cellular identity is currently unknown.

Post-generation

Deletions of identified blastomeres during ctenophore development generally result in the absence of the structure derived from the deleted cell; however, if these animals are kept alive, many of these structures can be "reformed" during the adult period through a process called "post-generation" (Martindale 1986; Martindale and Henry 1996; Henry and Martindale 2000). The stereotyped development in ctenophores allows one to identify what lineage of cells is responsible for the replacement of adult cell types in the absence of those deleted cells. For example, when all four e_1 micromeres are deleted at the 16-cell stage, new ctene rows will eventually develop from m_1 micromere derivatives, but this requires the presence of e_2 micromere derivatives (Henry and Martindale 2000). If one assumes that the m_1 cells are merely responding to signals from the e_1 and e_2 lineages and asks what structures are uniquely generated by both of these lineages, this appears to be the tentacles and the apical organ (Martindale and Henry 1999). The apical organ is a complex neural structure that has been shown to have a role in patterning the outcome of regenerative events (Freeman 1967; Martindale 1986). The tentacles are also interesting structures that are known to contain stem cells that operate throughout the life of the animal to generate new tentacle tissue, and this expresses genes that regulate stem cells in bilaterian animals (see below). Furthermore, the e_1 and e_2 lineages are known to be important for organizing the formation of tentacles because when they are both deleted, no tentacles form (Martindale and Henry 1997a, b). One interesting observation is that the Beroidae ctenophores do not have tentacles, and they have a reduced capacity to regenerate. It thus would be interesting to compare the differences in the development of e_1 and e_2 lineages between tentaculate and atentaculate ctenophore embryos.

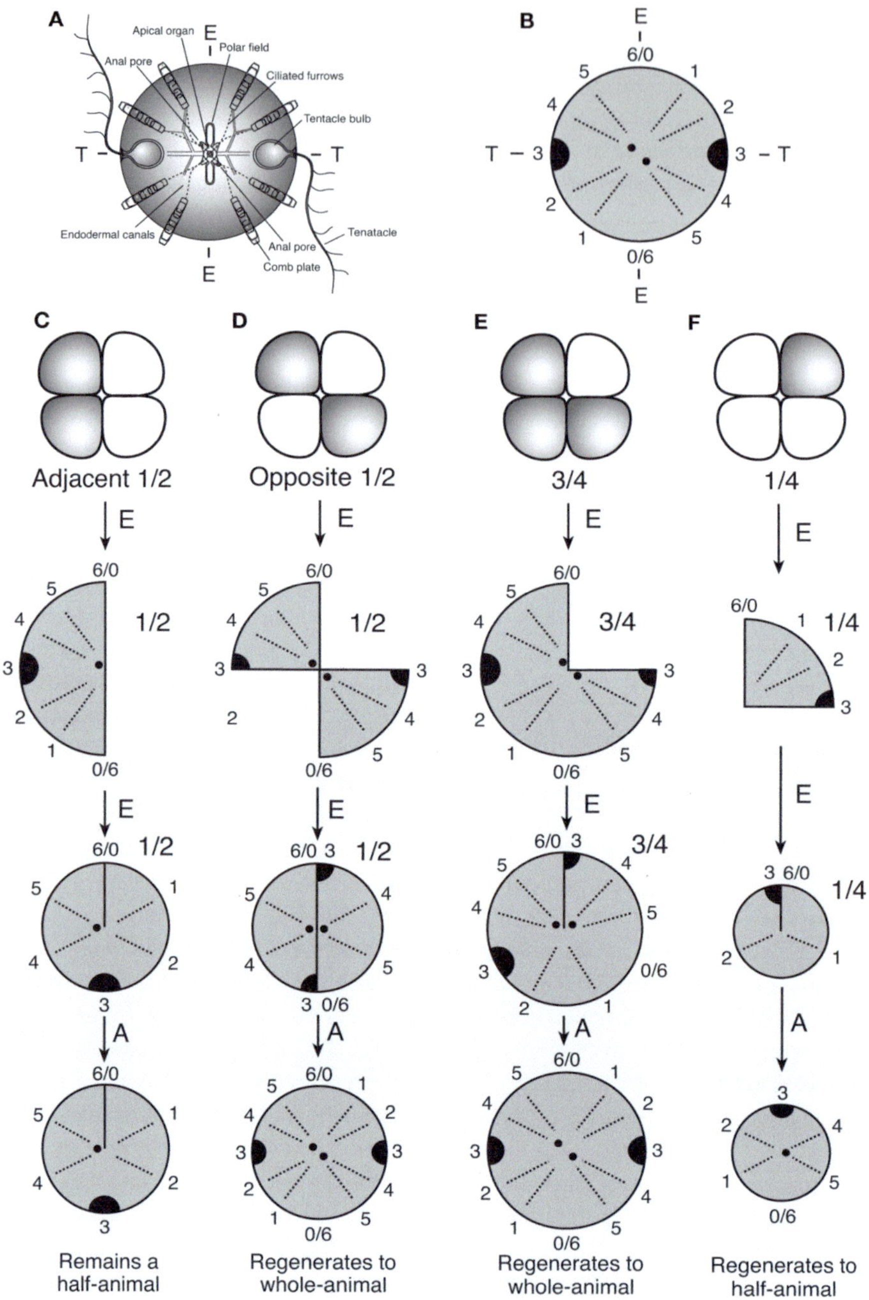

Fig. 8.7 Illustration of the existence of quadrant-specific identities and the stability of the ctenophore body plan. (**A**) Aboral view of the ctenophore body plan showing the\(anal pore-containing) and/(lacking anal pores) quadrants. Each quadrant is generated from one cell at the 4-cell stage with the first cleavage defining the esophageal axis (*E*) and the second cleavage defining the tentacular axis (*T*). (**B**) A hypothetical scheme of "positional information" along the circumference of the ctenophore body based on the polar coordinate model (French et al. 1976). The rules of the model predict that when parts of the body with different noncontinuous positional values are put in juxtaposition, a regenerative response is initiated to "fill in the differences." The ctenophore model predicts that "/quadrants" (*0–3*) and "\quadrants" (*4–6*) have different positional values. (**C**) The first test of the ctenophore polar coordinate model predicts the observation that in half-animals generated by isolating any two adjacent blastomeres will fail to regenerate. (**D**) In contrast, isolating any two opposite quadrants results in the formation of an entire normal whole animal. (**E**) Removal of one quadrant regardless if it is a/or\quadrant always results in the formation of a whole animal. (**F**) Isolation of a single quadrant always results in the formation of a stable half-animal

Dissogeny

One interesting feature of postembryonic life of some ctenophore species is their ability to generate progeny at very early stages of "larval" life. This process was termed "dissogeny" by Chun (1880, 1892) and has been reported for several different ctenophore species (Garbe 1901; Hirota 1972; Martindale 1987). The reason that ctenophores are able to generate rather limited numbers of progeny at very young ages appears to be an adaptation to ephemeral ecological conditions and/or high predation rates (Reeve and Walter 1978; Stanlaw et al. 1981); however, it is not even known if it is a response to changing environmental conditions or a final attempt to propagate the species. Not all individuals of a single spawn become precociously reproductive, and those which are reproductive at an early age will also become reproductive at later adult stages (Martindale 1987). It is also of interest to note that only four of the eight possible gonads, the four associated with the ctene rows adjacent to the esophagus (the adesophageal gonads), become precociously reproductive (Fig. 8.8), a finding that is true in experimentally generated "half-animals" (Chun 1880, 1892; Martindale 1987). The significance of dissogeny in ctenophores has been recently highlighted by two events: (1) the finding that an entire population of cydippid ctenophore of the arctic species *Mertensia ovum* found in the Baltic Sea (Jaspers et al. 2012) might sustain itself by "larval reproduction" (large adult forms have never been

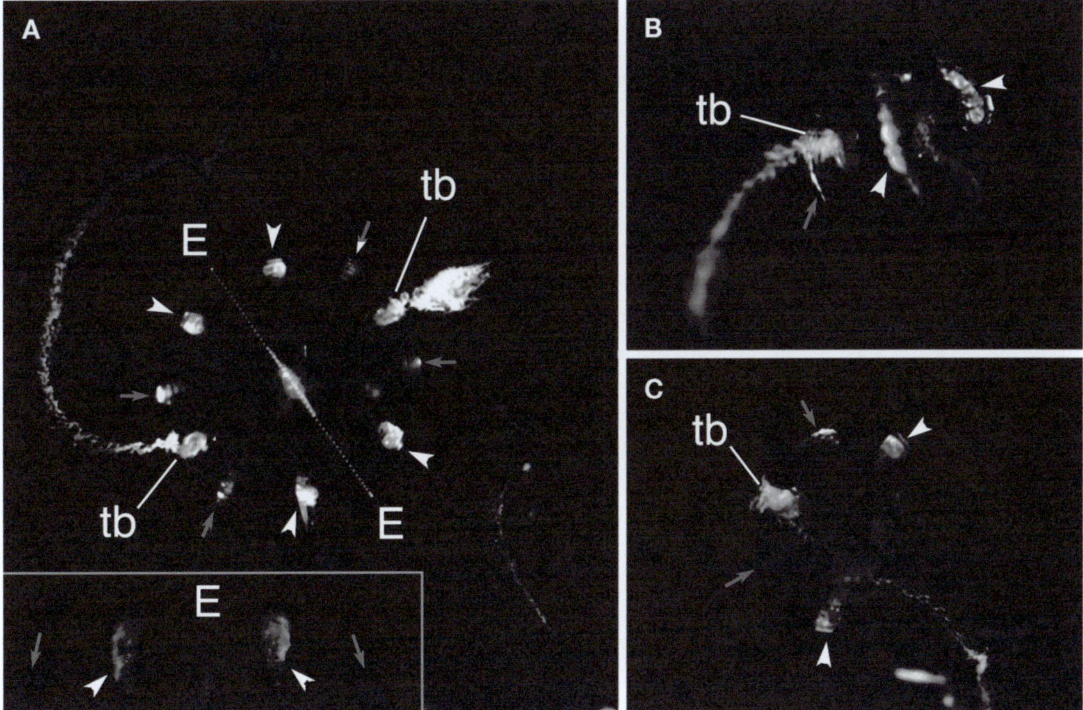

Fig. 8.8 Dissogony or larval reproduction in ctenophores. In some cases, thought to be related to early nutritional status, small cydippid-stage animals become reproductive and generate viable embryos. (**A**) *Mnemiopsis* cydippid-stage animal approximately 6 mm in diameter becomes sexually mature. In all cases described, only the adesophageal (*large arrow heads*), but not the adtentacular (*gray arrows*), gonads that reside under the ctene rows enlarge and form gametes. The inset shows a high magnification view of the adesophageal gonads. Each adesophageal ctene row has both a male and female gonad associated with it. The ovaries are close to the esophageal axis (*E*), while the testes (*arrow heads*) are situated on the opposite side. This is the same configuration seen during adult lobate stages. (**B**) A lateral view of a cydippid-stage half-animal produced by separating adjacent blastomeres at the 4-cell stage. The same asymmetrical maturation of gonads is seen in half-animals. Only the adesophageal (*large arrowheads*) gonads become reproductive. (**C**) Oral view of a half-animal showing the difference in size between adtentacular (*gray arrows*) and adesophageal gonads (*large arrow heads*). *Tb* tentacle bulb

recovered) and (2) that dissogony might be playing a role in the ballast water-driven invasion of the lobate ctenophore *Mnemiopsis* into the Black and Caspian Seas (Jaspers et al. 2012).

GENE EXPRESSION

Although there are a limited number of genes whose expression has been examined in adult ctenophores (Jager et al. 2008; Alié et al. 2011), gene expression patterns in developing ctenophores have only been described for a single species, *Mnemiopsis leidyi*. In situ hybridization experiments for over 50 genes (Table 8.1) have been published for *Mnemiopsis* including a number of important gene families and pathways such as homeodomain genes (Pang and Martindale 2008b, 2009; Ryan et al. 2010; Simmons et al. 2012), forkhead genes (Yamada and Martindale 2002), T-box genes (Yamada et al. 2007), nuclear receptors (Reitzel et al. 2011), and sox genes (Schnitzler et al. 2014). In addition, several signaling pathways have been investigated including the Wnt (Pang et al. 2010) and TGF-B pathways (Pang et al. 2011). An example of the spatially complex patterns of expression from just one gene family is shown in Fig. 8.9.

Much of the recent interest in the expression of these genes is due to the sequencing of the *Mnemiopsis leidyi* genome (Ryan et al. 2013) that has allowed a systematic analysis of entire gene families and signaling and metabolic pathways. One of the surprising findings of this, and a subsequent ctenophore genome paper (Moroz et al. 2014), is the absences of many key transcription factors, signaling pathways, and regulators of these pathways. For example, ctenophores do not have Hox genes, *hedgehog* (or its ligand smoothened), FGF, *notch*, members of the Wnt PCP (e.g., Flamingo and Strabismus), and JAK/STAT pathways or antagonists of the TGF-B and Wnt pathways (Pang et al. 2010, 2011; Ryan et al. 2013). Moreover, many genes in "highly conserved" bilaterian pathways such as endomesoderm formation, including *snail, twist, nodal, GATA, MyoD, Lbx, NK4, NK3, NK2, Myf5, Noggin, vMrf4, Myogenin, Eomesoderm*, and *troponin*, are not present in ctenophores (Ryan et al. 2013), raising the question of whether muscle

Table 8.1 Published expression patterns for genes in three regions of the *Mnemiopsis leidyi* embryo

Apical organ	Tentacle bulb	Pharynx
MLBmp5–8a	*MlTgf1aa*	*MLBmp5–8a*
MlTgf1aa	*MlTgf2a*	*MlTgf1aa*
MlTgfRIIa	*MlTgfbBa*	*MlTolloida*
MlTgfRIaa	*MlTolloida*	*MlTgfRIIa*
MlTgfRIba	*MlTgfRIIa*	*MlTgfRIaa*
MlSmad6a	*MlTgfRIaa*	*MlTgfRIba*
MlSmad1aa	*MlTgfRIba*	*MlTgfRIca*
MlSmad2a	*MlTgfRIca*	*MlSmad6a*
MlWnt6b	*MlSmad6a*	*MlSmad4a*
MlWntXb	*MlSmad1aa*	*MlSmad2a*
MlFzdAb	*MlSmad2a*	*MlFzdAb*
MlFzdBb	*MlWnt9b*	*MlFzdBb*
MlDshb	*MlWntAb*	*MlSfrpb*
MlBcatb	*MlWnt6b*	*MlDshb*
MlTcfb	*MlFzdAb*	*MlBcatb*
MlIsletc	*MlFzdBb*	*MlTcfb*
MlLhx1/5c	*MlSfrpb*	*MlLhx1/5c*
MlLhx3/4c	*MlDshb*	*MleSox1d*
MlLmxc	*MlBcatb*	*MleSox2d*
MleSox1d	*MlTcfb*	*MleSox3d*
MleSox2d	*MlLhx1/5c*	*MleSox4d*
MleSox3d	*MlLmxc*	*MleSox6d*
MleSox4d	*MleSox1d*	*MlNR2e*
MleSox6d	*MleSox2d*	*ctenoBF-1f*
MleOpsin1d	*MleSox3d*	*MlBrag*
MleOpsin2d	*MleSox4d*	*MlTbx1g*
MlNR1e	*MleSox6d*	*MlTbxEg*
MlBrag	*MlNR1e*	*MlNKL1h*
MlTbx2/3g	*MlNR2e*	*MlBarh*
MlBarh	*ctenoBF-1f*	*MlTlx-likeh*
MlTlx-likeh	*MlBrag*	*MlPrd1h*
MlPrd1h	*MlTbx1g*	*MlPrd3h*
MlPrd3h	*MlTbxDg*	
	MlTbxEg	
	MlNKL1h	
	MlBshh	
	MlBarh	
	MlTlx-likhh	
	MlPrd2h	
	MlPrd3h	

Superscript letters indicate the references of the works as follows: [a]Pang et al. (2011), [b]Pang et al. (2010), [c]Simmons et al. (2012), [d]Schnitzler et al. (2014), [e]Reitzel et al. (2011), [f]Yamada and Martindale (2002), [g]Yamada et al. (2007), and [h]Pang and Martindale (2008a)

and mesoderm in general are homologous with bilaterian mesoderm. This is rather surprising

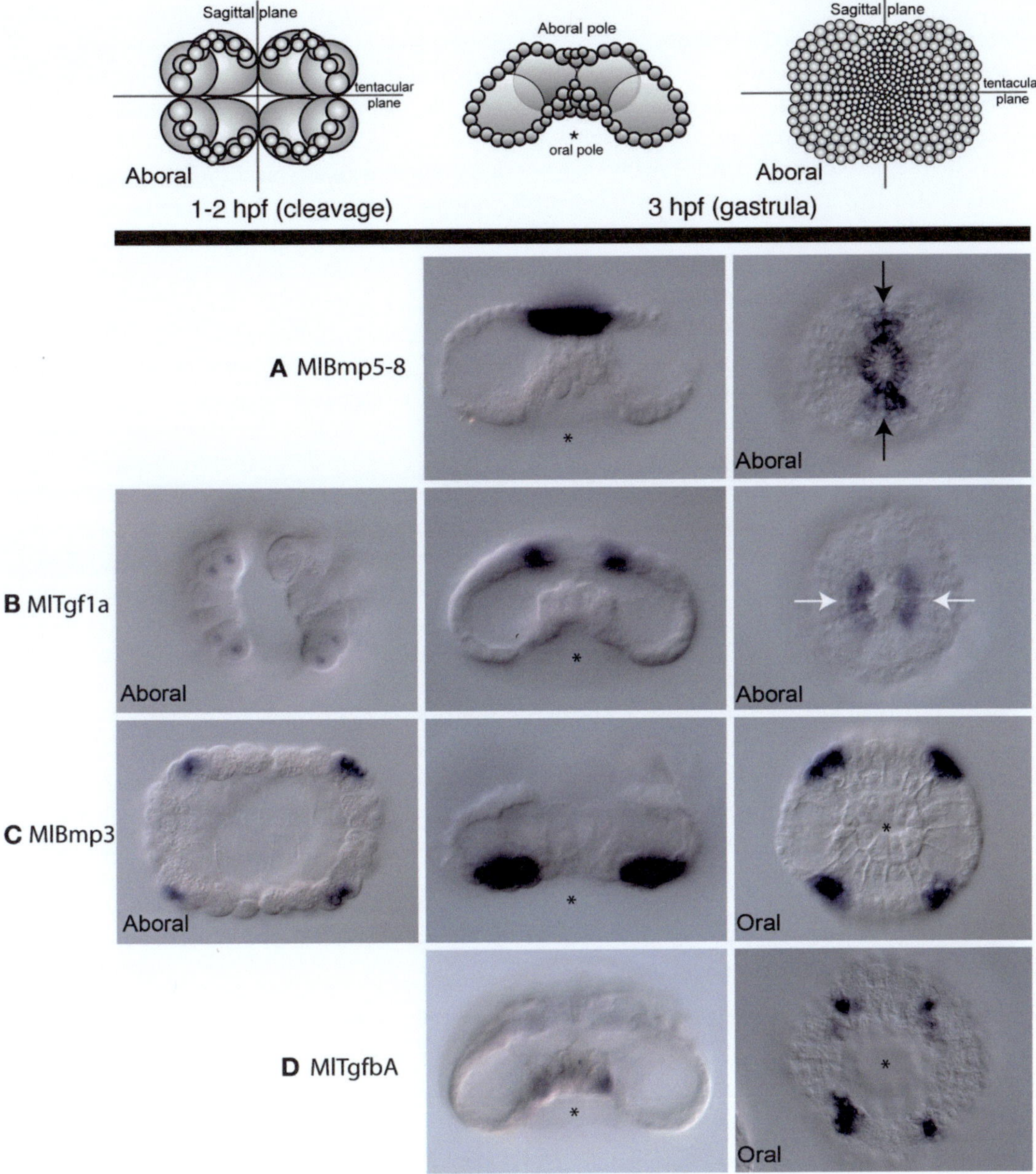

Fig. 8.9 In situ hybridizations of early TGF-ß family member expression in *Mnemiopsis* embryos. Early TGF-ß mRNA expression showing distinct domains of expression along the oral-aboral axis. Four of the *Mnemiopsis* TGF-ß genes are detected early in development, prior to and during gastrulation. The schematic at the top depicts the stages of embryos during cleavage and gastrulation, at 1–2 and 3 h postfertilization (*hpf*), respectively. Embryos are lateral views, otherwise oral-aboral as stated. The *asterisk* marks the position of the blastopore. Embryos are 220 μm in diameter (Taken from Pang et al. (2011), doi:10.1371/journal.pone.0024152.g006). (**A**) *MlBmp5–8* expression in the aboral ectoderm, with more expression detected in the sagittal plane (*black arrows*). (**B**) *MlTgf1a* expression is detected in late cleavage stages around the nuclei of aboral micromeres. By gastrulation, the aboral expression remains; however their expression is primarily along the tentacular plane (*white arrows*). (**C**) *MlBmp3* is detected in four groups of ectodermal cells from early to mid-gastrulation. (**D**) *MlTGFbA* is detected in four groups of ectodermal cells just adjacent to the blastopore at gastrulation

Mnemiopsis Leidyi as a Ctenophore Model for EvoDevo Research

Mnemiopsis is a lobate ctenophore that has been a primary developmental model for ctenophore development. Unlike many ctenophores that live at depth and are difficult to collect, *Mnemiopsis* is abundant in near-shore waters including bays and boat harbors along the east coast of North America, the Gulf of Mexico, and the Caribbean Sea. It occurs year-round in Florida and the Gulf of Mexico, as does the atentaculate *Beroe ovata*, but tends to be common in northern waters up to Cape Cod in the summer months (mid-June–August). *Mnemiopsis* are delicate animals, however, and adults should be scooped from the water in beakers or buckets and not handled directly or with nets if needed for reproductive experiments.

Adults with oral feeding lobes (see below right) are sexually mature and if nutrition is high will spawn thousands of viable embryos. The large size and optical clarity of these embryos make them ideal for studying ctenophore development, and if handled gently, the vast majority of embryos will develop normally in filtered seawater. Spawning can be initiated by manipulating the light cycle. Development is rapid and cydippid-stage (see below left) animals with feeding tentacles hatch out of their vitelline membranes in approximately 24 h, depending on temperature. *Mnemiopsis* continues to grow with a typical cydippid shape (the cydippid body plan is thought to be ancestral for Ctenophora) until the oral feeding lobes begin to develop. Regenerative ability in *Mnemiopsis* commences shortly before hatching and continues throughout life. The early onset of regeneration in this species makes it possible to attempt to manipulate gene expression by injections into embryonic stages. Lobate stages are difficult to work with due to their large size and "floppy" tissue composition, and molecularly one needs to be careful as there are many epibionts that normally raft on, and in, adult ctenophores that can contaminate DNA preparations. *Mnemiopsis* is the only ctenophore to date in which in situ hybridization protocols have been published on embryonic stages and the only one in which microinjection techniques have proven successful. The *Mnemiopsis* genome was made publicly available in 2011 (http://research.nhgri.nih.gov/mnemiopsis) and an analysis was published in 2013. Since then, a second ctenophore genome has been released the following year.

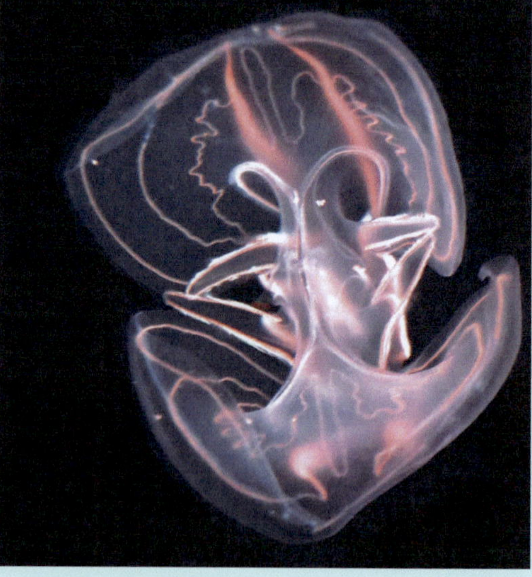

because lineage tracing showed that ctenophore mesoderm comes from endodermal precursors just as it does in most bilaterians.

In fact, perusal of two ctenophore genomes reveals that there is good reason to argue that the ctenophore neurons (and nervous system) have evolved independently from those found in bilaterians. Although many of the genes found in postsynaptic membranes are found in all animals (regardless of whether they have neurons or not), only a single bilaterian neurotransmitter receptor, glutamate (which also occurs in plants), has been found in ctenophores (Ryan et al. 2013; Moroz et al. 2014). Instead, a large number of putative neuropeptides have been identified (Moroz et al. 2014) that have been postulated to be independently evolved neuromodulators in ctenophores.

Despite the surprising findings regarding mesodermal and neural gene regulatory networks, there is some evidence that somatic and potentially germ line stem cells might utilize similar genes as bilaterians. Expression of *vasa*, *nanos*, and sox genes in the stem cell precursors of adult (Jager et al. 2008; Alié et al. 2011) and embryonic (Reitzel et al. 2015.) ctenophores is consistent with maintenance of the pluripotent state, although these ideas have not been functionally tested.

One interesting analysis (Schnitzler et al. 2012) involved the structure of a family of calcium-activated photoproteins that *Mnemiopsis* generates in a defined cell type called "photocytes." Genome analyses have shown that *Mnemiopsis* has a total of ten of these photoproteins on two distinct genomic scaffolds. These photoprotein families evolved at the base of the Metazoa and have been lost in various lineages, including placozoans, some hydrozoans, and bilaterians. The *Mnemiopsis* genome also revealed the presence of opsins, proteins involved in the conversion of light into chemical energy (Schnitzler et al. 2012). In situ hybridization analysis showed that they are expressed in the same cells (as well as some cells in the floor of the apical organ) as the photoproteins. These observations suggest that there may be some level of autoregulation of light emission from these cells.

Considering the fact that each cell division during early development in ctenophores results in the asymmetric segregation of developmental potential, it is potentially surprising that no lineage-specific localization of transcripts has been detected. In situs generally show either uniform (or absent) expression during early cleavage stages to be followed by localized expression in different structural domains. This could, of course, be due to the fact that none of the 50 or so genes studied to date are involved in early cell fate decisions or that in situ hybridization is not sensitive enough to detect these asymmetries. Perhaps newer transcriptomic approaches will reveal some differences, or perhaps the molecules that are influencing cell fate are not transcripts at all but maternally loaded proteins. Cleavages are quite rapid in most ctenophores and injected endogenous mRNAs labeled with fluorescent proteins do not accumulate to visible levels until gastrulation stages, when the cell cycle slows down. These observations suggest that maternal "determinants" might not be transcripts and that there are different mechanisms for segregating developmental potential in these embryos. It is also unclear how patterning in these embryos relates to the dramatic regenerative ability of most ctenophores. Hopefully additional work on these fascinating embryos will yield some insight into the evolution of cell fate specification in metazoan embryos.

Over all, the results of all of these gene expression studies have shown that in ctenophores, like in other taxa such as anthozoan cnidarians (Chapter 6), there is much more cell-type "complexity" found at the molecular level than is apparent at the morphological level. For example, different domains of the apical organ, the pharynx, and the tentacle bulbs express different genes in distinct and overlapping domains, even though these regions look identical morphologically. Unfortunately, these early expression studies do not have true cellular resolution to identify co-expression of multiple genes needed to develop hypotheses about gene regulatory networks; however, these techniques currently exist, and it is only a matter of time before people begin to investigate these ideas experimentally. Morpholino-based antisense oligonucleotide gene knockdown techniques have been published that demonstrate that the ctenophore gene *brachyury* is involved in pharyngeal morphogenesis (Yamada et al. 2010) and thus appear promising for unraveling gene function. Unfortunately, RNAi interference approaches appear unlikely to become established in ctenophores due to the absence of key members of the micro RNA biogenesis pathway (Maxwell et al. 2012). Thus, the stage is set for more detailed investigations of axial organization and cell-type evolution in this important group of animals.

OPEN QUESTIONS

- Does ctenophore development represent some deep history that is somehow shared by other metazoan taxa? Or is their development due to the fact that ctenophores are holopelagic, direct-developing organisms that put a premium on getting to a free-living feeding stage as quickly as possible?
- Does the rapid pace (e.g., short cell cycles) of ctenophore development provide constraints on mechanisms establishing cell fate during early embryogenesis?
- Is the microtubule component of the cytoskeleton, which has been adapted for so many other functions (e.g., sensory and locomotory structures), involved in the asymmetric segregation of developmental fate at each of the early cleavages?
- Can the physiological polyspermy observed in ctenophores teach us anything about how cells evolved responses to excess DNA in the absence of canonical RNAi pathways?
- Can a better understanding of the development of individual cell types in ctenophores tell us anything about the origins of complex traits (such as mesodermal cell types and nervous systems)?
- Is the regenerative potential seen in most living ctenophores characteristic of ancient metazoans or a feature related to their delicate body plan?
- What is the molecular basis for "quadrant-specific positional information" in these animals?
- Are there interesting differences in fate maps and mechanisms of cell fate specification in different ctenophore species, e.g., related to the presence of tentacles or adult regenerative potential?

References

Alié A, Leclère L, Jager M, Dayraud C, Chang P, Le Guyader H, Quéinnec E, Manuel M (2011) Somatic stem cells express *Piwi* and *Vasa* genes in an adult ctenophore: ancient association of "germline genes" with stemness. Dev Biol 350(1):183–197

Brusca RC, Brusca GJ (1990) The invertebrates. Sinauer Ass, Sunderland

Carré D, Sardet C, Rouvière C (1990) Fertilization in ctenophores. In: Dale B (ed) Mechanism of fertilization, vol 45, NATO ASI series, Series H. Springer, Berlin, pp 626–636

Carré D, Rouvière C, Sardet C (1991) *In Vitro* fertilization in ctenophores: sperm entry, mitosis, and the establishment of bilateral symmetry in *Beroe ovata*. Dev Biol 147:381–391

Chun C (1880) Die Ctenophoren des Golfes von Neapel. Fauna Flora Golfes von Neapel 1:1–311

Chun C (1892) Die Dissogonie, eine neue Form der geschlechtlichen Zeugung. Festsch. Zum siebzigsten Geburtstage Rudorf Leuckarts 1:77–108

Collins AG (2002) Phylogeny of Medusozoa and the evolution of cnidarian life cycles. J Evol Biol 15(3):418–432

Coonfield B (1936) Regeneration in *Mnemiopsis leidyi* Agassiz. Biol Bull 71:421–428

Coonfield B (1937) The regeneration of plate rows in *Mnemiopsis leidyi* Agassiz. Proc Natl Acad Sci 23:152–158

Dawydoff C (1938) Multiplication asexuëe, par lacération, chez les Ctenoplana. C R Acad Sci Paris 206:127–128

Driesch H, Morgan TH (1895) Zur Analysis der ersten Entwickelungsstadien des Ctenophoreneies. Arch Entwicklungsmech Organ 2:204–224

Dunlap-Pinaka H (1974) Ctenophora. In: Giese AC, Pearse JS (eds) Reproduction of marine invertebrates, vol 1, Acoelomate and pseudocoelomate metazoans. Academic, New York, pp 201–265

Dunn CW, Hejnol A, Matus DQ, Pang K, Browne WE, Smith SA, Seaver E, Rouse GW, Obst M, Edgecombe GD, Sørensen MV, Schmidt-Rhaesa A, Haddock SHDG, Okusu A, Kristensen R, Wheeler WC, Martindale MQ, Giribet G (2008) Broad phylogenomic sampling improves resolution of the animal tree of life. Nature 452:745–749. doi:10.1038/nature06614

Farfaglio G (1963) Experiments on the formation of the ciliated plates in ctenophores. Acta Embryol Morphol Exp 6:191–203

Finnerty JR, Pang K, Burton P, Paulson D, Martindale MQ (2004) Deep origins for bilateral symmetry: hox and *Dpp* expression in a sea anemone. Science 304:1335–1337

Fischer A, Pang K, Henry JQ, Martindale MQ (2014) A cleavage clock regulates features of lineage-specific differentiation in the development in a basal branching metazoan, the ctenophore *Mnemiopsis leidyi*. Evodevo 5:4. doi:10.1186/2041-9139-5-4

Freeman G (1967) Studies on regeneration in the creeping ctenophore, *Vallicula multiformis*. J Morphol 123:71–84

Freeman G (1976a) The role of cleavage in the localization of developmental potential in the ctenophore *Mnemiopsis leidyi*. Dev Biol 49:143–177

Freeman G (1976b) The effects of altering the position of cleavage planes on the process of localization of developmental potential in ctenophores. Dev Biol 51:332–337

Freeman G (1977) The establishment of the oral-aboral axis in the ctenophore embryo. J Embryol Exp Morphol 42:237–260

Freeman G (2006) Oocyte and egg organization in the patellogastropod *Lottia* and its bearing on axial specification during early embryogenesis. Dev Biol 295:141–155

Freeman G, Reynolds GT (1973) The development of bioluminescence in the ctenophore *Mnemiopsis leidyi*. Dev Biol 31:61–100

French V, Bryant PJ, Bryant SV (1976) Pattern regulation in epimorphic fields. Science 193:969–981

Garbe A (1901) Untersuchungen über die Entstehung der Geschlechtsorgane bei den Ctenophoren. Z Wiss Zool 69:472–491

Greve W (1970) Cultivation experiments on North Sea ctenophores. Helgoländer Meeresun 20:304–317

Harbison GR, Miller RL (1986) Not all ctenophores are hermaphrodites. Studies on the systematics, distribution, sexuality and development of two species of *Ocyropsis*. Mar Biol 90:413–424

Hejnol A, Obst M, Stamatakis A, Ott M, Rouse G, Edgecombe G, Martínez P, Baguñà J, Bailly X, Jondelius U, Wiens M, Muller WEG, Seaver E, Martindale MQ, Giribet G, Dunn CW (2009) Rooting the bilaterian tree with scalable phylogenomic and supercomputing tools. Proc R Soc B 276:4261–4270. doi:10.1098/rspb.2009.0896

Henry JQ, Martindale MQ (2000) Regulation and regeneration in the ctenophore *Mnemiopsis*. Dev Biol 227:720–733

Henry JQ, Martindale MQ (2001) Multiple inductive signals are involved in the development of the ctenophore *Mnemiopsis leidyi*. Dev Biol 238:40–46

Hernandez-Nicaise M-L (1973) The nervous system of ctenophores. III. Ultrastructure of synapses. J Neurocytol 2:249–263

Hernandez-Nicaise M-L (1991) Ctenophora. In: microscopic anatomy of invertebrates, vol 2, Placozoa, Porifera, Cnidaria, and Ctenophora. Wiley-Liss, New York, pp 359–418

Hirota J (1972) Laboratory culture and metabolism of the planktonic ctenophore, *Pleurobrachia bachei* A. Agassiz. In: Takenouti AY (ed) Biological oceanography of the northern North Pacific Island. Idemitsu Shoten, Tokyo, pp 465–484

Horridge GA (1974) Recent studies on the Ctenophora. In: Muscatine L, Lenhoff HM (eds) Coelenterate biology. Academic, New York, pp 439–468

Houliston E, Carré D, Johnston JA, Sardet C (1993) Axis establishment and microtubule-mediated waves prior to first cleavage in *Beroe ovata*. Development 117:75–87

Hyman LH (1940) The invertebrates. Protozoa through Ctenophora. McGraw Hill, New York, pp 662–695

Jager M, Quéinnec E, Chiori R, Le Guyader H, Manuel M (2008) Insights into the early evolution of SOX genes from expression analyses in a ctenophore. J Exp Zool B Mol Dev Evol 310(8):650–667

Jaspers C, Haraldsson M, Bolte S, Reusch TB, Thygesen UH, Kiørboe T (2012) Ctenophore population recruits entirely through larval reproduction in the central Baltic Sea. Biol Lett 8:809. doi:10.1098/rsbl.2012.0163

Lang A (1884) Die Polycladen (Seeplanarien) des Golfes von Neapel und der angrenzenden Meerabschnitte. Fauna Flora Golf Neapel 11:1–688

Martindale MQ (1986) The expression and maintenance of adult symmetry properties in the ctenophore, *Mnemiopsis mccradyi*. Dev Biol 118:556–576

Martindale MQ (1987) Larval reproduction in the ctenophore, *Mnemiopsis mccradyi* (order Lobata). Mar Biol 94:409–414

Martindale MQ (2001) Chapter 4. Phylum Ctenophora. In: Young CM, Sewell MA, Rice ME (eds) The atlas of marine invertebrate larvae. Academic, San Diego, pp 109–122

Martindale MQ, Henry JQ (1995) Diagonal development: establishment of the anal axis in the ctenophore *Mnemiopsis leidyi*. Biol Bull 189:190–192

Martindale MQ, Henry JQ (1995) Modifications of cell fate specification in equal-cleaving nemertean embryos: alternate patterns of spiralian development. Dev 121:3175–3185

Martindale MQ, Henry JQ (1996) Development and regeneration of comb plates in the ctenophore *Mnemiopsis leidyi*. Biol Bull 191:290–292

Martindale MQ, Henry JQ (1997a) The Ctenophora. In: Gilbert S, Raunio A (eds) Embryology, the construction of life. Sinauer Press, Sunderland, pp 87–111

Martindale MQ, Henry JQ (1997b) Experimental analysis of tentacle formation in the ctenophore *Mnemiopsis leidyi*. Biol Bull 193:245–247

Martindale MQ, Henry JQ (1997c) Reassessing embryogenesis in the Ctenophora: the inductive role of e1 micromeres in organizing ctene row formation in the "mosaic" embryo, Mnemiopsis leidyi. Dev 124:1999–2006

Martindale MQ, Henry JQ (1999) Intracellular fate mapping in a basal metazoan, the ctenophore *Mnemiopsis leidyi*, reveals the origins of mesoderm and the existence of indeterminate cell lineages. Dev Biol 214:243–257

Maxwell EK, Ryan JF, Schnitzler CE, Browne WE, Baxevanis AD (2012) MicroRNAs and essential components of the microRNA processing machinery are not encoded in the genome of the ctenophore *Mnemiopsis leidyi*. BMC Genomics 13:714

Moroz LL, Kocot KM, Citarella MR, Dosung S, Norekian TP, Povolotskaya IS et al (2014) The ctenophore genome and the evolutionary origins of neural systems. Nature 510:109–114

Morris SC, Simonetta AM (eds) (1991) The early evolution of metazoa and the significance of problematic taxa. Cambridge University Press, New York, p 308

Mortensen T (1912a) Ctenophora. Danish Ingolf-Exped, vol 5(2), H. Hagerup, Copenhagen

Mortensen T (1912b) On regeneration in ctenophores. Dan Naturhist For Kobenhavn Vidensk Medd 66:45–51

Nielsen C (1995) Animal evolution: interrelationships of the living phyla. Oxford University Press, Oxford

Nielsen C, Scharff N, Eibye-Jacobsen D (1996) Cladistic analyses of the animal kingdom. Biol J Linn Soc 57(4):385

Ortolani G (1963) Origine dell'organo apicale e dei derivati mesodermici nello sviluppo embrionale di Ctenofori. Acta Embryol Morphol Exp 7:191–200

Pang K, Martindale MQ (2008a) Ctenophores. Curr Biol 18(24):R1119–R1120

Pang K, Martindale MQ (2008b) Developmental expression of homeobox genes in the ctenophore Mnemiopsis leidyi. Dev Genes Evol 218:307–319

Pang K, Martindale MQ (2009) Comb jellies (Ctenophora): a model for basal metazoan evolution and development. In: Crotty DA, Gann A (eds) Emerging model organisms, vol 1. Cold Spring Harbor Laboratory Press, Cold Spring Harbor, pp 167–195

Pang K, Ryan JF, Baxevanis AD, Martindale MQ (2010) Genomic insights into Wnt signaling in an early diverging metazoan, the ctenophore *Mnemiopsis leidyi*. Evodevo 1:10

Pang K, Ryan JF, Baxevanis AD, Martindale MQ (2011) Evolution of the TGF-β signaling pathway and its potential role in the ctenophore, *Mnemiopsis leidyi*. PLoS One 6(9):e24152. doi:10.1371/journal.pone.0024152

Philippe H, Derelle R, Lopez P, Pick K, Borchiellini C, Boury-Esnault N, Manuel M (2009) Phylogenomics revives traditional views on deep animal relationships. Curr Biol 19(8):706–712

Pick KS, Philippe H, Schreiber F, Erpenbeck D, Jackson DJ, Wrede P, Wiens M, Alie A, Morgenstern B, Manuel M, Worheide G (2010) Improved phylogenomic taxon sampling noticeably affects non-bilaterian relationships. Mol Biol Evol 27(9):1983–1987

Podar M, Haddock SHD, Sogin ML, Harbison GR (2001) A molecular phylogenetic framework for the phylum Ctenophora using 18S rRNA genes. Mol Phylogenet Evol 21:218–230

Putnum N, Srivastava M, Hellsten U, Dirks B, Chapman J, Salamov A, Terry A, Shapiro H, Lindquist E, Kapitonov VV, Jurka J, Genikhovich G, Grigoriev I, Sequencing Team JGI, Steele R, Finnerty JR, Technau U, Martindale MQ, Rokhsar D (2007) Sea anemone genome reveals the gene repertoire and genomic organization of the eumetazoan ancestor. Science 317:86–94. doi:10.1126/science.1139158

Reeve MR, Walter MA (1978) Nutritional ecology of ctenophores-a review of recent research. New York. Adv Mar Biol 15:249–287

Reitzel AM, Pang K, Ryan JF, Mullikin JC, Martindale MQ, Baxevanis AD, Tarrant A (2011) Nuclear receptors from the ctenophore *Mnemiopsis leidyi* lack a zinc-finger DNA-binding domain: lineage-specific loss or ancestral condition in the emergence of the nuclear receptor superfamily? Evodevo 2:3

Reitzel A, Pang K, Martindale MQ (2015) The expression of *vasa, nanos,* and *piwi* in the ctenophore, *Mnmiopsis leidyi.* In Prep

Reusch TBH, Bolte S, Saprwel M, Moss A, Javidpour J (2010) Microsatellites reveal origin and genetic diversity of Eurasian invasions by one of the world's most notorious invader, *Mnemiopsis leidyi* (Ctenophora). Mol Ecol 19:2690–2699

Reverberi G, Ortolani G (1963) On the origin of the ciliated plates and mesoderm in the ctenophore. Acta Embryol Morphol Exp 6:175–199

Ryan JF, Pang K, NIH Intramural Sequencing Center, Mullikin JC, Martindale MQ, Baxevanis AD (2010) The homeodomain complement of the ctenophore *Mnemiopsis leidyi* suggests that Ctenophora and Porifera diverged prior to the ParaHoxozoa. Evodevo 1:9

Ryan JF, Pang K, Schnitzler CE, Nguyen A, Moreland RT, Simmons DK, Koch BJ, Havlak P, NISC Comparative Sequencing Program, Smith SA, Putnam N, Dunn CW, Wolfsberg TG, JE, Mullikin JC, Martindale MQ, Baxevanis AD (2013) Total genome sequencing of the genome of the ctenophore *Mnemiopsis leidyi* using new generation approaches. Science 342, doi:10.1126/science.1242592

Schnitzler C, Pang K, Powers M, Reitzel AM, Ryan JF, Simmons D, Park M, Gupta J, Brooks SY, Blakesley RW, Haddock SH, Mullikin JC, Martindale MQ, Baxevanis AD (2012) Bioluminescence and the evolution of photoproteins: a ctenophore genome lights the way. BMC Biol 10:107. doi:10.1186/1741-7007-10-107

Schnitzler CE, Simmons DK, Pang K, Martindale MQ, Baxevanis AD (2014) Expression of multiple Sox genes through embryonic development in the ctenophore *Mnemiopsis leidyi* is spatially restricted to zones of cell proliferation. Evodevo 5:15. doi:10.1186/2041-9139-5-15

Simmons DK, Pang K, Martindale MQ (2012) Lim homeobox genes in the ctenophore *Mnemiopsis leidyi*: the evolution of neural cell type specification. Evodevo 3:2

Spek J (1926) Über gesetzmässige Substanzverteilung bei der Furchung des Ctenophoreneis und ihre Beziehung du dem Determinationsproblem. Arch Entwicklungsmech Organismen 107:54–73

Srivastava M, Begovic E, Chapman J, Putnam NH, Hellsten U, Kawashima T, Kuo A, Mitros T, Salamov A, Carpenter ML, Signorovitch AY, Moreno MA, Kamm K, Grimwood J, Schmutz J, Shapiro H, Grigoriev IV, Buss LW, Schierwater B, Dellaporta SL, Rokhsar DS (2008) The *Trichoplax* genome and the nature of placozoans. Nature 454:955–960. doi:10.1038/nature07191

Srivastava M, Simakov O, Chapman J, Fahey B, Gauthier ME, Mitros T, Richards GS, Conaco C, Dacre M, Hellsten U, Larroux C, Putnam NH, Stanke M, Adamska M, Darling A, Degnan SM, Oakley TH, Plachetzki DC, Zhai Y, Adamski M, Calcino A, Cummins SF, Goodstein DM, Harris C, Jackson DJ, Leys SP, Shu S, Woodcroft BJ, Vervoort M, Kosik KS, Manning G, Degnan BM, Rokhsar DS (2010) The

Amphimedon queenslandica genome and the evolution of animal complexity. Nature 466:720–726

Stanlaw KA, Reeve MR, Walter MA (1981) Growth, food, and vulnerability to damage of the ctenophore *Mnemiopsis mccradyi* in its early life history stages. Limnol Oceanogr 26:224–234

Tamm SL (1982) Ctenophora. In: Shelton GG (ed) Electrical conduction and behavior in 'simple' invertebrates. Clarendon Press, Oxford, pp 266–358

Tamm SL (2012a) Patterns of comb row development in young and adult stages of the ctenophores *Mnemiopsis leidyi* and *Pleurobrachia pileus*. J Morphol 273(9): 1050–1063

Tamm SL (2012b) Regeneration of ciliary comb plates in the ctenophore *Mnemiopsis leidyi*. J Morphol 273(1): 109–120

Tamm SL (2014) Cilia and the life of ctenophores. Invertebr Biol 133:1–46

Tanaka H (1931) Reorganization in regenerating pieces of Coeloplana. Kyoto Imp Univ Coll Sci Ser B 7:223–246

Wallberg A, Thollesson M, Farris JS, Jondelius U (2004) The phylogenetic position of the comb jellies (Ctenophora) and the importance of taxonomic sampling. Cladistics 20(6):558

Yamada A, Martindale MQ (2002) The ctenophore *Brain Factor-1* forkhead gene ortholog (*ctenoBF-1*) is expressed in the presumptive oral region and feeding apparatus: implications for axial organization in the Metazoa. Dev Genes Evol 212:338–348

Yamada A, Pang K, Martindale MQ, Tochinai S (2007) Surprisingly complex T-box gene complement in diploblastic metazoans. Evol Dev 9:220–230

Yamada A, Martindale MQ, Fukui A, Tochinai S (2010) Highly conserved functions of the *Brachyury* gene on morphogenetic movements: insight from the early-diverging phylum Ctenophora. Dev Biol 339: 212–222

Yatsu N (1911) Observation and experiments on the ctenophore egg. II. Notes on the early cleavage stages and experiments on cleavage. Ann Zool Jpn 7:333–346

Yatsu N (1912) Observation and experiments on the ctenophore egg. III. Experiments on germinal localization of the egg *Beroe ovata*. Ann Zool Jpn 8:5–13

Andreas Hejnol

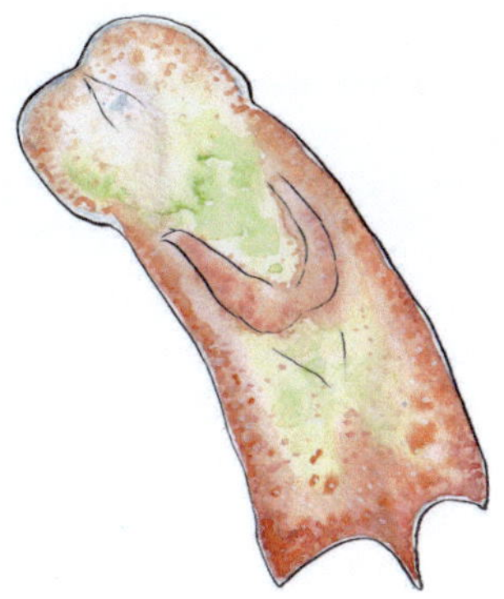

Chapter vignette artwork by Brigitte Baldrian.
© Brigitte Baldrian and Andreas Wanninger.

A. Hejnol
Sars International Centre for Marine Molecular
Biology, University of Bergen,
Thormøhlensgate 55, Bergen 5008, Norway
e-mail: andreas.hejnol@uib.no

A. Wanninger (ed.), *Evolutionary Developmental Biology of Invertebrates 1:*
Introduction, Non-Bilateria, Acoelomorpha, Xenoturbellida, Chaetognatha
DOI 10.1007/978-3-7091-1862-7_9, © Springer-Verlag Wien 2015

INTRODUCTION

Acoelomorpha, comprising Acoela and Nemertodermatida (Ehlers 1985), and Xenoturbellida (with one single hitherto described species, *Xenoturbella bocki*) are simple, aquatic, acoelomate worms that measure between 100 μm and 1 cm. Acoelomorpha and *Xenoturbella* are found to cluster together as the monophyletic Xenacoelomorpha in some recent molecular phylogenetic analyses. With only few exceptions all species are marine, with most of them living in the interstitial environment. Xenoturbellids and acoelomorphs possess a simple nervous system that generally is a basiepidermal nerve net; however, in some cases this net is condensed into basiepidermal neurite bundles at different parts of the body or is submerged under the epidermis where condensed brains and submuscular cords are formed (Achatz and Martinez 2012; Hejnol 2015). Some Acoela possess eye spots, while most nemertodermatids, *Xenoturbella*, and Acoela lack eyes (Rieger et al. 1991). Recent internal phylogenetic analyses suggest that eyes were absent from the ground pattern of Acoelomorpha (Jondelius et al. 2011). A prominent gravitational sensory organ, the statocyst, is present in all xenacoelomorph taxa, albeit with differing ultrastructure (Ferrero 1973; Ehlers 1991).

The digestive tract is epithelial in xenoturbellids and nemertodermatids, while in acoels, it forms via a syncytium without a cavity (Smith and Tyler 1985). The digestive system has a single opening that corresponds to the mouth opening of other Bilateria (Hejnol and Martindale 2008a). The mesoderm of acoels consists of a limited number of cell types – which are myocytes, cells associated with gonads, and neoblasts (Chiodin et al. 2013) – whereas nephridia and a blood vascular system are absent. The musculature forms an orthogon composed of ring and longitudinal muscles that surround the whole body. Specialized musculature is present in the copulatory structures of acoels (Ladurner and Rieger 2000; Hooge 2001). Altogether, acoelomorphs and xenoturbellids are rather simply organized animals, and this is also the reason why their phylogenetic position has been of great interest and is still under debate (Fig. 9.1). After the exclusion of Acoelomorpha

from the Platyhelminthes (Carranza et al. 1997; Ruiz-Trillo et al. 1999, 2002; Jondelius et al. 2002; Wallberg et al. 2007; Egger et al. 2009; Paps et al. 2009), molecular phylogenies have produced ambiguous results concerning their definite placement within the tree of life. Phylogenomic studies that include *Xenoturbella* place Acoelomorpha together with *Xenoturbella* (Xenacoelomorpha) either as sister to all remaining Bilateria (Hejnol et al. 2009; Srivastava et al. 2014) or as sister to Ambulacraria – i.e., within the deuterostomes (Philippe et al. 2011). It is important to mention that in these studies *Xenoturbella* was an unstable taxon, while Acoelomorpha could be placed as sister group to all remaining Bilateria with higher confidence (Hejnol et al. 2009; Srivastava et al. 2014). The placement as sister to all remaining Bilateria (Nephrozoa) would indicate a rather simple organized last common ancestor of Bilateria (Baguñá and Riutort 2004; Hejnol and Martindale 2008b), while in case of a placement inside deuterostomes, acoelomorphs would have lost many bilaterian characteristics in the stem lineage. Although the most recent studies with increased taxon sampling and more complete matrices confirm the sister group relationship to the Nephrozoa (Srivastava et al. 2014), more studies are necessary and should lead to a phylogenetic placement with higher confidence (Dunn et al. 2014).

DEVELOPMENT OF XENOTURBELLIDA

Embryological studies of this enigmatic group have so far not been conducted. A recent publication on the sole species described, *Xenoturbella bocki*, indicates direct development because the hatchling is a completely ciliated juvenile worm (Nakano et al. 2013). The hatchling has no digestive tract or mouth opening and has basiepidermal nerve cells and a musculature that is not yet organized into longitudinal and ring musculature as in the adult (Nakano et al. 2013). The lack of a mouth opening in the early stage is comparable to the lack of the mouth in early stages of nemertodermatids (Meyer-Wachsmuth et al. 2013; Børve and Hejnol 2014).

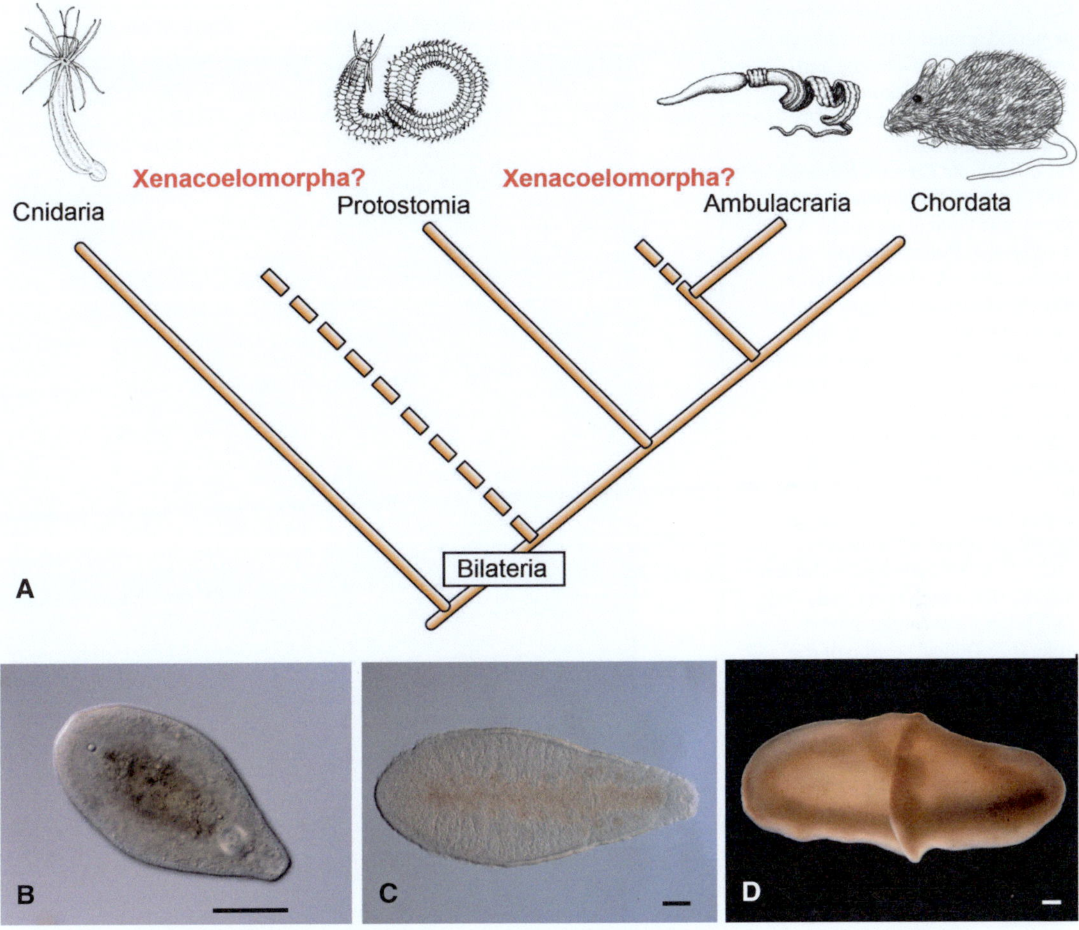

Fig. 9.1 Acoelomorpha and Xenoturbellida, here united as Xenacoelomorpha, and their phylogenetic position. (**A**) Debated phylogenetic positions of Xenacoelomorpha. Xenacoelomorpha is placed by most molecular and morphological studies as sister group to all remaining Bilateria. One study suggests a position of the Xenacoelomorpha as sister to Ambulacraria (Philippe et al. 2011). (**B**) Adult of the acoel *Isodiametra pulchra*. (**C**) Adult of the nemertodermatid *Meara stichopi*. (**A** and **B** from Chiodin et al. (2013)); (**C** from Børve and Hejnol (2014)). (**D**) Adult of *Xenoturbella bocki* (Photograph by Greg Rouse). Scale bars equal 100 μm

DEVELOPMENT OF ACOELOMORPHA

Most studies of the development of acoelomorphs have been conducted in acoels and only few on nemertodermatids. All acoel species investigated show a consistent pattern regarding their cleavage program and gastrulation that has been termed "duet cleavage." The investigated nemertodermatid species deviate from this duet pattern by having four micromeres instead of two and being less stereotypic after the 16-cell stage (Fig. 9.2; Børve and Hejnol 2014). Despite their stereotypic cleavage pattern, acoels seem to be able to regulate blastomere deletions in the early embryo (Boyer 1971). Experimental approaches have not yet been carried out in nemertodermatid embryos. The development in acoels generally lasts between 3 and 5 days from egg deposition to hatching, while nemertodermatids undergo a longer development, with the duration ranging from 10 days in *Nemertoderma westbladi* (12 °C; Jondelius et al. 2004) to 9 weeks in *Meara stichopi* (6 °C; Børve and Hejnol 2014) (Table 9.1).

Fig. 9.2 Early cleavage pattern of nemertodermatid *Meara stichopi* embryos. Nuclear labeling and spindles with propidium iodide (magenta), cell cortices, and spindle with BODIPY FL-phallacidin (*green*). *Left row*: Maximum intensity projections. *Right row*: Optical sections. (**A**) 2-cell stage. Polar body (*pb*) at animal pole. (**A′**) Optical section through the same embryo. Nucleus (*nc*) and centrosomes (*ct*). Both blastomeres are equal in size. (**B**) After 4.5 days, the 4-cell stage has large blastomeres at the vegetal pole, and two smaller daughter blastomeres at the animal pole. (**B′**) Section of the embryo in (**B**). The spindles are already arranged for the future direction of cell division. (**C**) 8-cell stage with four larger cells at the vegetal pole and four blastomeres at the animal pole. (**C′**) Optical section of the embryo in (**C**), with spindles arranged to the future plane of division. (**D**) 16-cell stage at 7 days after fertilization. The size differences between the blastomeres are less prominent and the arrangement is variable. (**D′**) BODIPY FL-phallacidin-labeled cell borders and centrosomes, while the chromatin is labeled by propidium iodide. (**E**) 24-cell stage after 8.5 days. (**E′**) Median section of the embryo shown in (**E**). The blastocoel (*bc*) is bordered with the phallacidin-labeled cell cortex of the outer blastomeres. (**F**) 64-cell stage 10.5 days after fertilization. (**F′**) Cells that have been internalized (gastrulation, blastomeres labeled with *arrowheads*) during the transition from the 24- to the 64-cell stage. Sister blastomeres are connected by *white bars*; animal pole is indicated with an *asterisk*. Scale bars equal 30 µm (Figure from Børve and Hejnol (2014))

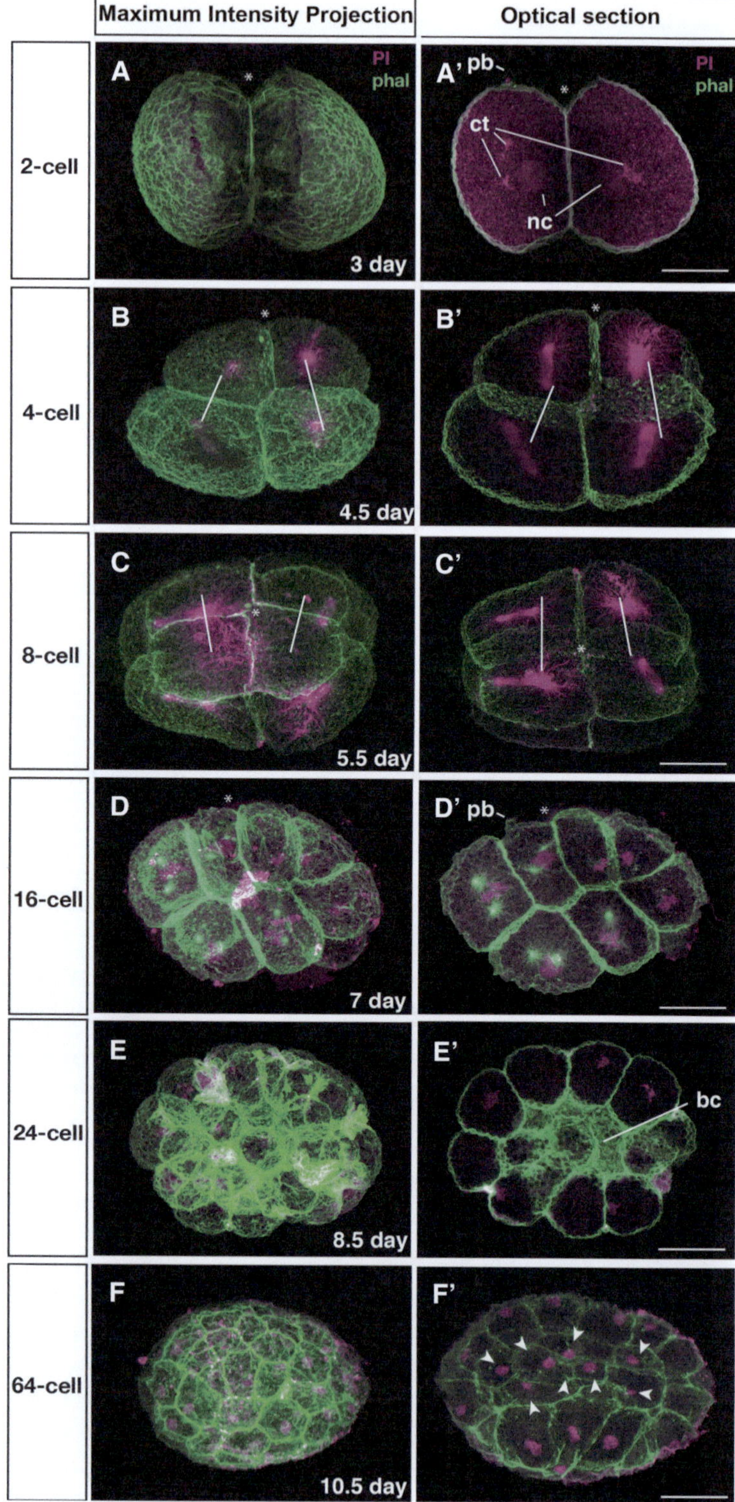

Table 9.1 Morphological studies of acoelomorph development

Reference	Species	Stages and methods
Gardiner (1895)	*Polychoerus caudatus* (Acoela)	Cleavage, gastrulation, light microscopy, and histology
Gardiner (1898)	*Polychoerus caudatus* (Acoela)	Oogenesis, polar body formation, fertilization, histology
Georgévitch (1899)	*Convoluta (Symsagittifera) roscoffensis* (Acoela)	Cleavage, gastrulation, light microscopy, and histology
Bresslau (1909)	*Convoluta (Symsagittifera) roscoffensis* (Acoela)	Cleavage, gastrulation, light microscopy, and histology
Apelt (1969)	*Archaphanostoma agile, Archocelis macrorhabditis, Diopisthoporus brachypharyngeus* (Acoela)	Cleavage, light microscopy
Boyer (1971)	*Childia groenlandica* (Acoela)	Cleavage and gastrulation, cell ablations with needle
Henry et al. (2000)	*Neochildia fusca* (Acoela)	Fate map with intracellular dye lineage tracing
Ladurner and Rieger (2000)	*Convoluta (Isodiametra) pulchra* (Acoela)	Muscle development with fluorescent labels
Ramachandra et al. (2002)	*Neochildia fusca* (Acoela)	Neurogenesis, histochemistry, and histology
Jondelius et al. (2004)	*Nemertoderma westbladi* (Nemertodermatida)	Cleavage, light microscopy
Semmler et al. (2008)	*Symsagittifera roscoffensis* (Acoela)	Muscle development with fluorescent labels and confocal microscopy
Børve and Hejnol (2014)	*Meara stichopi* (Nemertodermatida)	Cleavage, organogenesis, light microscopy, and fluorescent dyes

Acoela

Cleavage

Cleavage is total and the fertilized zygote divides equally and meridionally into the two blastomeres A and B (Fig. 9.3A). All researchers use the nomenclature developed for naming blastomeres of the spiralian embryo (Conklin 1897) because the acoel cleavage pattern was thought to be derived from spiral cleavage (Ax 1984). Both blastomeres of the 2-cell stage give off small blastomeres at the animal pole – the mitotic spindles have a slightly oblique angle to the animal-vegetal axis of the embryo. The micromeres are positioned between the vegetal macromeres in a typical pattern (Fig. 9.3B). During the next cleavage round, the vegetal macromeres divide equatorially into two equal-sized blastomeres (the animal micromere 2a and the vegetal macromere 2A). The micromeres at the animal pole divide equally with an oblique spindle to the animal blastomere $1a^1$ and the vegetal blastomere $1a^2$ (and $1b^1$ and $1b^2$, respectively; Fig. 9.3C). The next cell division starts with the vegetal blastomeres dividing equally and meridionally, thus giving rise to the endomesodermal precursors 3A and 3B. The micromeres of the second duet divide also meridionally and give rise to the micromeres $2x^1$ and $2x^2$. The animal micromeres undergo another cell division round in which the micromere $1a^2$ divides into a vegetal blastomere $1a^{22}$ which is in contact with the vegetal-most macromeres 3A and 3B. The animal descendant $1a^{21}$ is in contact with the animal-most micromeres. The 16-cell stage is the stage in which the endomesodermal precursors 3A and 3B are specified.

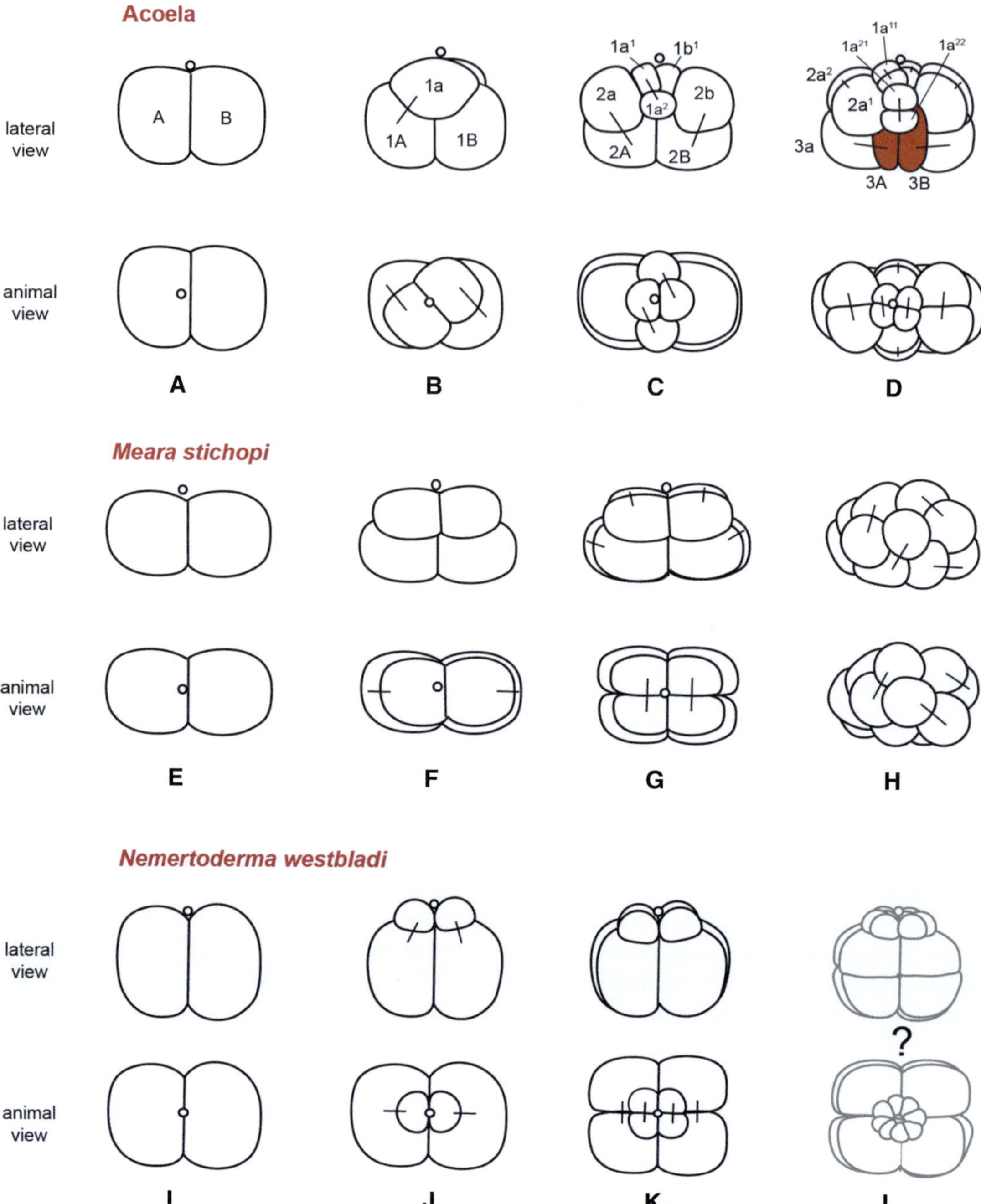

Fig. 9.3 Cleavage in Acoelomorpha. (**A–D**) "Duet"-cleavage program in Acoela. The pattern is consistent in all species investigated so far. *Brown-colored* blastomeres are the endomesodermal precursors that gastrulate. (**E–H**) Cleavage pattern in the nemertodermatid *Meara stichopi*. First three cell divisions follow a stereotypic pattern. In later stages, cells cannot be identified. (**I–L**) Cleavage pattern in the nemertodermatid *Nemertoderma westbladi*. The pattern is similar to *Meara stichopi* but differs in blastomere sizes. The cleavage stage in L remains unclear and needs further investigations (Figure from Børve and Hejnol (2014))

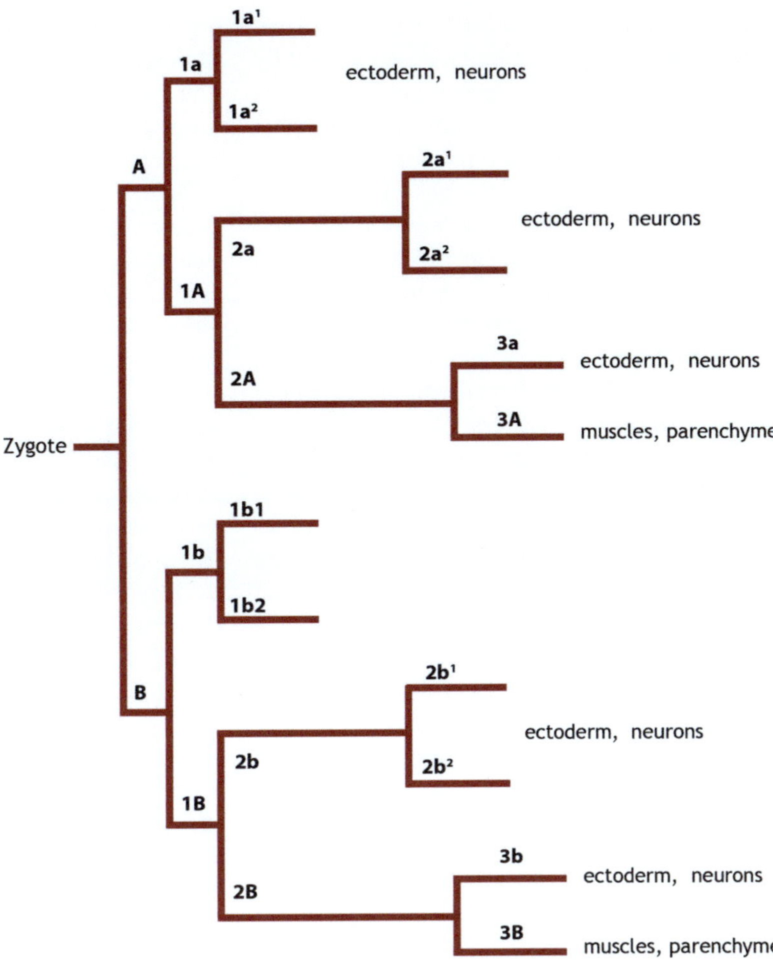

Fig. 9.4 Fate map of the acoel embryo. Early blastomere contributions to major tissues in the acoel *Neochildia fusca* after Henry et al. (2000)

Gastrulation

Gastrulation begins with the internalization of the endomesodermal precursors 3A and 3B (Fig. 9.3D). The cells undergo a shape change to form a flask shape, with the larger internalized portion filling the entire internal space of the embryo. The narrow tip of the cells is still connected to the outside. The ectodermal micromeres continue to divide and proliferate slowly over the tip of the internalized macromeres. The fate map of *Neochildia fusca* indicates that all mesoderm and endoderm are derived from the gastrulating cells (Fig. 9.4; Henry et al. 2000). The micromeres of the ectoderm give rise to the epidermal cells and the nervous system (Henry et al. 2000). Descendants of all original micromeres of the three duets (1×, 2×, 3×) give rise to elements of the nervous system. The neural precursors do segregate from the ectoderm later during development, which is also indicated by the internalization of the expression of neural markers and Hox genes (Hejnol and Martindale 2009). The endomesodermal precursors continue to divide. The specification and separation of the mesodermal precursors remain unclear.

Organogenesis

After gastrulation and proliferation of all cells, the embryo consists of a ball of cells that is surrounded by an outer epithelial layer. The embryo is first round and later extends to an oval shape, and no blastomeres can yet be characterized by light microscopy. The investigation of myogenesis in the acoel species *Isodiametra pulchra* using fluorescently labeled phallotoxins reveals the pattern of the establishment of the ring and longitudinal musculature (Ladurner and Rieger 2000). First, the ring musculature develops from individual elements that are distributed around the embryo. The first filaments of the longitudinal musculature develop later. It seems that in another acoel species, *Symsagittifera roscoffensis*, such sequential pattern of circular versus longitudinal muscle formation might be absent (Semmler et al. 2008). Before any muscle is formed, the whole epidermis is ciliated and no nervous system has been identified at this stage (Ladurner and Rieger 2000). Ladurner and Rieger (2000) report that the first musculature is observed at the equator of the embryo, followed by muscles at the future anterior pole. After the orthogonal muscle grid is established, more specialized muscles such as the musculature surrounding the future mouth opening and the diagonal muscles are formed (Ladurner and Rieger 2000). So far, no study of neurogenesis with fluorescent markers has been conducted on acoel embryos – however, neural structures seem to be formed after the musculature has been established (Ladurner and Rieger 2000). Studies that investigated the expression of neural markers in acoel embryos all indicate that in the future anterior region, a subepidermal cluster of neural cells is located in a bilateral arrangement (Ramachandra et al. 2002; Hejnol and Martindale 2008a, b, 2009; Chiodin et al. 2013). These precursors are likely descendants of ectodermal cells that are internalized after gastrulation (Hejnol and Martindale 2009). From all previous investigations of acoel embryogenesis, it seems that the syncytial endoderm and the mouth opening is established relatively late during development.

Postembryonic Development

Acoels hatch as juvenile worms with all major organ systems present except the reproductive organs (Ladurner and Rieger 2000; Hejnol and Martindale 2008b; Semmler et al. 2008, 2010). The musculature of the reproductive organs is formed after hatching, and the growth of the juvenile is connected with an elaboration of the musculature and nervous system. All growth in acoels is accomplished by specialized, self-renewing stem cells, the so-called neoblasts (De Mulder et al. 2009; Srivastava et al. 2014). This means that as soon as a cell has differentiated to its final fate, it does not undergo further mitotic divisions.

Molecular Approaches

Acoels gained more attention after molecular phylogenies had placed these worms as sister group to all remaining bilaterians (Nephrozoa) (Ruiz-Trillo et al. 1999). Due to this intermediate phylogenetic position in the stem lineage of the Bilateria, acoels can provide insights into the evolution of morphological, developmental, and genomic characters (Hejnol and Martindale 2008b). In recent years, molecular studies in acoels have been conducted in several species of a rather derived acoel taxon, the Convolutidae (*Symsagittifera roscoffensis* and *Convolutriloba* species). Other species such as *Isodiametra pulchra* (Isodiametridae) and *Hofstenia miamia* (Hofsteniidae) are easy to culture from egg to adult and are suitable for RNA inference (RNAi) methods such as soaking and injections of double-stranded RNA. They belong to taxa outside the Convolutidae (Jondelius et al. 2011) and show more plesiomorphic characters in nervous system architecture (Achatz and Martinez 2012; Hejnol 2015). A major focus on the molecular studies in acoels has been their stem cell system and regeneration (De Mulder et al. 2009; Moreno et al. 2010; Srivastava et al. 2014), and here, RNAi by soaking and adult injections have been useful for investigating the role of developmental genes during homeostasis and regeneration. The focus of molecular embryological studies has been on the role of Hox genes, digestive tract development, and development of mesodermal tissues.

Acoels are interesting regarding the role of Hox genes since they possess only a single Hox

Bilateria

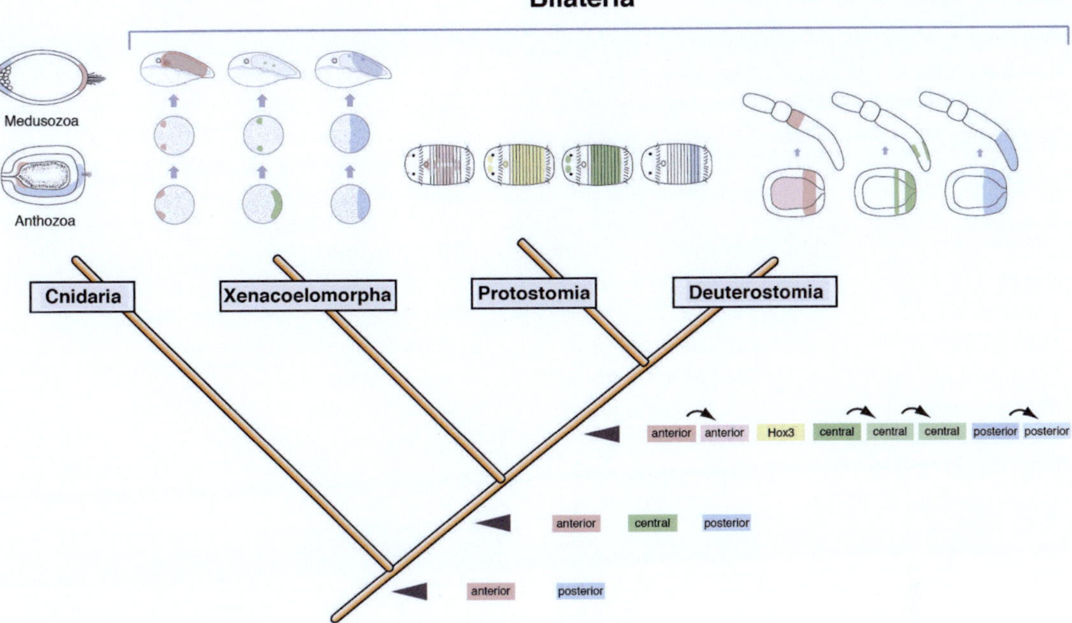

Fig. 9.5 Hox gene expression in (Xen-)Acoelomorpha compared to Cnidaria and Bilateria. Hox gene expression of anterior, central, and posterior class genes in the acoel *Convolutriloba longifissura*. Differences to and variations within Cnidaria are indicated by the planula stage expression in the Medusozoa and the basal Anthozoa. The Protostomia expression is represented by the annelid *Capitella teleta* (Fröbius et al. 2008) and the expression in the Deuterostomia by the hemichordate *Saccoglossus kowalevskii* (Aronowicz and Lowe 2006). *Arrows* indicate extension of the Hox cluster in Bilateria by gene duplication (Figure from Hejnol and Martindale (2009))

gene of each class: one anterior class, one central class, and one posterior class Hox gene (Fig. 9.5; Cook et al. 2004; Hejnol and Martindale 2009; Moreno et al. 2009; Sikes and Bely 2010). It has been hypothesized that this is a plesiomorphic condition for acoelomorphs and Bilateria, since cnidarians possess only anterior and possibly a central/posterior Hox gene (Chourrout et al. 2006; Ryan et al. 2007; DuBuc et al. 2012).

Hox genes in acoels seem to have an early role during gastrulation and a later role during nervous system development (Hejnol and Martindale 2009). The posterior Hox ortholog also plays a role in specifying the posterior fate of cells in the juvenile (Moreno et al. 2010). So far, the only ParaHox gene that has been identified is *cdx*, which is broadly expressed in the nervous system of the juvenile and later in the maturing male gonopore (Hejnol and Martindale 2009; Moreno et al. 2011). Other studies have investigated the

question of the homology of the single digestive tract opening (mouth) in acoels to the bilaterian mouth or anus and found that the acoel mouth is likely homologous to the mouth of the remaining bilaterians. In the same study, hindgut genes have been found to be expressed in the male gonopore of acoels (Hejnol and Martindale 2008a). However, further studies are needed to test this hypothesis about their common origin.

Since Acoelomorpha – if sister group to all remaining Bilateria – would be one of the two taxa that show true mesodermal tissues, several studies have investigated the role of bilaterian mesoderm candidate genes (Chiodin et al. 2011, 2013). These studies confirmed that mesodermal markers are expressed in the musculature, gonads, and neoblasts. The orthogonal structure of the acoel nervous system has been subject of investigations, but studies are so far preliminary (Ramachandra et al. 2002) (Table 9.2).

Table 9.2 Molecular studies in Acoela

Reference	Species	*Genes;* stages
Ramachandra et al. (2002)	*Neochildia fusca*	*brn-1, brn-3*; organogenesis, adults
Hejnol and Martindale (2008a)	*Convolutriloba longifissura*	*cdx, vax, emx, evx, pax6*; juvenile
Hejnol and Martindale (2008b)	*Convolutriloba longifissura*	*bra, gsc, cdx, otp, foxA, nk2.1, six3/6, dlx, pitx, bmp2/4*; embryo, juvenile, adult
De Mulder et al. (2009)	*Isodiametra pulchra*	*piwi-1*; juvenile, adult, regeneration, RNAi soaking
Hejnol and Martindale (2009)	*Convolutriloba longifissura*	*antHox, centHox, postHox, cdx, six3/6, soxB1*; embryo, juvenile
Moreno et al. (2009)	*Symsagittifera roscoffensis*	*antHox, centHox, postHox*; juvenile
Sikes and Bely (2010)	*Convolutriloba retrogemma*	*antHox, centHox, postHox, otx*; budding
Moreno et al. (2010)	*Isodiametra pulchra*	*postHox*; regeneration, RNAi soaking
Semmler et al. (2010)	*Symsagittifera roscoffensis*	*soxB1*; juvenile
Moreno et al. (2011)	*Symsagittifera roscoffensis*	*cdx*; juvenile, adult
Chiodin et al. (2011)	*Symsagittifera roscoffensis*	*actin, troponin I, tropomyosin,*
Chiodin et al. (2013)	*Isodiametra pulchra*	*mLIM, pitx, foxA1, foxA2, foxC, GATA456, Mef2, six1/2, twist1, twist2, tbr, tropomyosin*; embryo, juvenile, adult
Srivastava et al. (2014)	*Hofstenia miamia*	*piwi-1, RNR, wnt-1, wnt-3, wnt-4, wnt-5, sFRP-1, sFRP-2, sFRP-3, fz-1, fz-2, fz-3, fz-4, fz-5, fz-6, fz-7, fz-8, fz-9, fz-10, fz-11, notum, netrin-1, netrin-2, admp, bmp, synapsin, PC2, TrpC-1*; juvenile, regeneration, RNAi injection

Nemertodermatida

Nemertodermatids comprise only nine described species, and their overall anatomy is similar to that of acoels (Sterrer 1998). However, they show important differences in nervous system and gut morphology that have been interpreted as plesiomorphies for the Acoelomorpha (Ehlers 1985). Nemertodermatids possess an epithelial gut with cavity, similar to that of Cnidaria, *Xenoturbella*, and all remaining Bilateria (Smith and Tyler 1985; Rieger et al. 1991; Tyler 2001). The nemertodermatid nervous system is completely basiepidermal – similar to that of *Xenoturbella* – and only in some cases condensed to intraepithelial neurite bundles (Westblad 1937, 1949; Rieger et al. 1991; Raikova et al. 2000, 2004; Børve and Hejnol 2014; Hejnol 2015). These morphological plesiomorphies make nemertodermatids a more suitable group to reconstruct ancestral states of acoelomorphs and their molecular patterning. The cleavage pattern in nemertodermatids differs from that of acoels and seems to be less stereotypic (Fig. 9.2E–L; Jondelius et al. 2004; Børve and Hejnol 2014). The fate of early blastomeres has not yet been studied. No expression data exist for any developmental genes in Nemertodermatida.

OPEN QUESTIONS

- Development and life cycle of Xenoturbellida
- Detailed studies of fate map and molecular mechanisms including expression of key developmental genes in Acoela, Nemertodermatida, and Xenoturbellida
- Fate map of Nemertodermatida and Xenoturbellida
- Virtually all aspects of organogenesis in Acoelomorpha and Xenoturbellida

References

Achatz JG, Martinez P (2012) The nervous system of *Isodiametra pulchra* (Acoela) with a discussion on the neuroanatomy of the Xenacoelomorpha and its evolutionary implications. Front Zool 9:27. doi:10.1186/1742-9994-9-27

Apelt G (1969) Fortpflanzungsbiologie, Entwicklungszyklen und vergleichende Frühentwicklung acoeler Turbellarien. Mar Biol 4:267–325

Aronowicz J, Lowe CJ (2006) Hox gene expression in the hemichordate *Saccoglossus kowalevskii* and the evolution of deuterostome nervous systems. Integr Comp Biol 46:890–901. doi:10.1093/Icb/Icl045

Ax P (1984) Das phylogenetische system. Gustav Fischer Verlag, Stuttgart

Baguñá J, Riutort M (2004) The dawn of bilaterian animals: the case of acoelomorph flatworms. Bioessays 26:1046–1057

Børve A, Hejnol A (2014) Development and juvenile anatomy of the nemertodermatid *Meara stichopi* (Bock) Westblad 1949 (Acoelomorpha). Front Zool 11:50. doi:10.1186/1742-9994-11-50

Boyer BC (1971) Regulative development in a spiralian embryo as shown by cell deletion experiments on the acoel, *Childia*. J Exp Zool 176:97–105. doi:10.1002/jez.1401760110

Bresslau E (1909) Die Entwicklung der Acoelen. Verhandlungen der Deutschen Zoologischen Gesellschaft 314–323

Carranza S, Baguñá J, Riutort M (1997) Are the Platyhelminthes a monophyletic primitive group? An assessment using 18S rDNA sequences. Mol Biol Evol 14:485–497

Chiodin M, Achatz JG, Wanninger A, Martinez P (2011) Molecular architecture of muscles in an acoel and its evolutionary implications. J Exp Zool B Mol Dev Evol 316:427–439. doi:10.1002/jez.b.21416

Chiodin M, Børve A, Berezikov E, Ladurner P, Martinez P, Hejnol A (2013) Mesodermal gene expression in the acoel *Isodiametra pulchra* indicates a low number of mesodermal cell types and the endomesodermal origin of the gonads. PLoS One 8:e55499. doi:10.1371/journal.pone.0055499

Chourrout D, Delsuc F, Chourrout P, Edvardsen RB, Rentzsch F, Renfer E, Jensen MF, Zhu B, de Jong P, Steele RE, Technau U (2006) Minimal ProtoHox cluster inferred from bilaterian and cnidarian Hox complements. Nature 442:684–687. doi:10.1038/nature04863

Conklin EG (1897) The embryology of *Crepidula*. J Morphol 13:1–230

Cook CE, Jiménez E, Akam M, Saló E (2004) The Hox gene complement of acoel flatworms, a basal bilaterian clade. Evol Dev 6:154–163. doi:10.1111/j.1525-142X.2004.04020.x

De Mulder K, Kuales G, Pfister D, Willems M, Egger B, Salvenmoser W, Thaler M, Gorny A-K, Hrouda M, Borgonie G, Ladurner P (2009) Characterization of the stem cell system of the acoel *Isodiametra pulchra*. BMC Dev Biol 9:69. doi:10.1186/1471-213X-9-69

DuBuc TQ, Ryan JF, Shinzato C, Satoh N, Martindale MQ (2012) Coral comparative genomics reveal expanded Hox cluster in the cnidarian-bilaterian ancestor. Integr Comp Biol 52:835–841. doi:10.1093/icb/ics098

Dunn CW, Giribet G, Edgecombe GD, Hejnol A (2014) Animal phylogeny and its evolutionary implications. Ann Rev Ecol Evol Syst 45:371–395. doi:10.1146/annurev-ecolsys-120213-091627

Egger B, Steinke D, Tarui H, De Mulder K, Arendt D, Borgonie G, Funayama N, Gschwentner R, Hartenstein V, Hobmayer B, Hooge M, Hrouda M, Ishida S, Kobayashi C, Kuales G, Nishimura O, Pfister D, Rieger R, Salvenmoser W, Smith J III, Technau U, Tyler S, Agata K, Salzburger W, Ladurner P (2009) To be or not to be a flatworm: the acoel controversy. PLoS One 4:e5502. doi:10.1371/journal.pone.0005502

Ehlers U (1985) Das phylogenetische System der Plathelminthes. Gustav Fischer Verlag, Stuttgart

Ehlers U (1991) Comparative morphology of statocysts in the Plathelminthes and the Xenoturbellida. Hydrobiologia 227:263–271

Ferrero E (1973) A fine structural analysis of the statocyst in Turbellaria Acoela. Zool Scr 2:5–16

Fröbius AC, Matus DQ, Seaver EC (2008) Genomic organization and expression demonstrate spatial and temporal Hox gene colinearity in the lophotrochozoan *Capitella* sp. I. PLoS One 3:e4004. doi:10.1371/journal.pone.0004004

Gardiner EG (1895) Early development of *Polychoerus caudatus*, Mark. J Morph 11:155–176

Gardiner EG (1898) The growth of the ovum, formation of the polar bodies, and the fertilization in *Polychoerus caudatus*. J Morph 15:73–104

Georgévitch J (1899) Etude sur le développement de la *Convoluta roscoffensis* Graff. Arch Zool Expérim 3:343–361

Hejnol A (2015) Acoelomorpha. In: Schmidt-Rhaesa A, Harzsch S, Purschke G (eds) Structure and evolution of invertebrate nervous systems. Oxford University Press, Oxford

Hejnol A, Martindale MQ (2008a) Acoel development indicates the independent evolution of the bilaterian mouth and anus. Nature 456:382–386. doi:10.1038/nature07309

Hejnol A, Martindale MQ (2008b) Acoel development supports a simple planula-like urbilaterian. Philos Trans R Soc Lond B Biol Sci 363:1493–1501. doi:10.1098/rstb.2007.2239

Hejnol A, Martindale MQ (2009) Coordinated spatial and temporal expression of Hox genes during embryogenesis in the acoel *Convolutriloba longifissura*. BMC Biol 7:65. doi:10.1186/1741-7007-7-65

Hejnol A, Obst M, Stamatakis A, Ott M, Rouse GW, Edgecombe GD, Martinez P, Baguna J, Bailly X, Jondelius U, Wiens M, Müller WEG, Seaver E, Wheeler WC, Martindale MQ, Giribet G, Dunn CW (2009) Assessing the root of bilaterian animals with scalable phylogenomic methods. Proc Royal Soc Series B 276:4261–4270. doi:10.1098/rspb.2009.0896

Henry JQ, Martindale MQ, Boyer BC (2000) The unique developmental program of the acoel flatworm, *Neochildia fusca*. Dev Biol 220:285–295. doi:10.1006/dbio.2000.9628

Hooge M (2001) Evolution of body-wall musculature in the Platyhelminthes (Acoelomorpha, Catenulida, Rhabditophora). J Morphol 249:171–194

Jondelius U, Ruiz-Trillo I, Baguñà J, Riutort M (2002) The Nemertodermatida are basal bilaterians and not members of the Platyhelminthes. Zool Scr 31:201–215

Jondelius U, Larsson K, Raikova OI (2004) Cleavage in *Nemertoderma westbladi* (Nemertodermatida) and its phylogenetic significance. Zoomorphology 123:221–225

Jondelius U, Wallberg A, Hooge M, Raikova OI (2011) How the worm got its pharynx: phylogeny, classification, and Bayesian assessment of character evolution in acoela. Syst Biol 60:845–871

Ladurner P, Rieger R (2000) Embryonic muscle development of *Convoluta pulchra* (Turbellaria-Acoelomorpha, Platyhelminthes). Dev Biol 222: 359–375. doi:10.1006/dbio.2000.9715

Meyer-Wachsmuth I, Raikova OI, Jondelius U (2013) The muscular system of *Nemertoderma westbladi* and *Meara stichopi* (Nemertoderma, Acoelomorpha). Zoomorphology 132:239–252

Moreno E, Nadal M, Baguñà J, Martínez P (2009) Tracking the origins of the bilaterian Hox patterning system: insights from the acoel flatworm *Symsagittifera roscoffensis*. Evol Dev 11:574–581. doi:10.1111/j.1525-142X.2009.00363.x

Moreno E, De Mulder K, Salvenmoser W, Ladurner P, Martinez P (2010) Inferring the ancestral function of the posterior Hox gene within the bilateria: controlling the maintenance of reproductive structures, the musculature and the nervous system in the acoel flatworm *Isodiametra pulchra*. Evol Dev 12:258–266. doi:10.1111/j.1525-142X.2010.00411.x

Moreno E, Permanyer J, Martinez P (2011) The origin of patterning systems in bilateria-insights from the Hox and ParaHox genes in Acoelomorpha. Genomics Proteomics Bioinforma 9:65–76. doi:10.1016/S1672-0229(11)60010-7

Nakano H, Lundin K, Bourlat SJ, Telford MJ, Funch P, Nyengaard JR, Obst M, Thorndyke MC (2013) *Xenoturbella bocki* exhibits direct development with similarities to Acoelomorpha. Nat Commun 4:1537. doi:10.1038/ncomms2556

Paps J, Baguña J, Riutort M (2009) Bilaterian phylogeny: a broad sampling of 13 nuclear genes provides a new Lophotrochozoa phylogeny and supports a paraphyletic basal Acoelomorpha. Mol Biol Evol 26: 2397–2406. doi:10.1093/molbev/msp150

Philippe H, Brinkmann H, Copley RR, Moroz LL, Nakano H, Poustka AJ, Wallberg A, Peterson KJ, Telford MJ (2011) Acoelomorph flatworms are deuterostomes related to *Xenoturbella*. Nature 470: 255–258. doi:10.1038/nature09676

Raikova OI, Reuter M, Jondelius U, Gustafsson MKS (2000) The brain of the Nemertodermatida (Platyhelminthes) as revealed by anti-5HT and anti-FMRFamide immunostainings. Tissue Cell 32:358–365

Raikova OI, Reuter M, Gustafsson MKS, Maule AG, Halton DW, Jondelius U (2004) Basiepidermal nervous system in *Nemertoderma westbladi* (Nemertodermatida): GYIRFamide immunoreactivity. Zoology (Jena) 107:75–86. doi:10.1016/j.zool.2003.12.002

Ramachandra NB, Gates RD, Ladurner P, Jacobs DK, Hartenstein V (2002) Embryonic development in the primitive bilaterian *Neochildia fusca*: normal morphogenesis and isolation of POU genes *Brn-1* and *Brn-3*. Dev Genes Evol 212:55–69. doi:10.1007/s00427-001-0207-y

Rieger R, Tyler S, Smith JPS, Rieger GE (1991) Platyhelminthes: turbellaria. In: Harrison FW, Bogitsch BJ (eds) Microscopic anatomy of invertebrates. Wiley, New York, pp 7–140

Ruiz-Trillo I, Riutort M, Littlewood DTJ, Herniou EA, Baguña J (1999) Acoel flatworms: earliest extant bilaterian Metazoans, not members of Platyhelminthes. Science 283:1919–1923

Ruiz-Trillo I, Paps J, Loukota M, Ribera C, Jondelius U, Baguñá J, Riutort M (2002) A phylogenetic analysis of myosin heavy chain type II sequences corroborates that Acoela and Nemertodermatida are basal bilaterians. Proc Natl Acad Sci U S A 99:11246–11251. doi:10.1073/pnas.172390199

Ryan JF, Mazza ME, Pang K, Matus DQ, Baxevanis AD, Martindale MQ, Finnerty JR (2007) Pre-bilaterian origins of the Hox cluster and the Hox code: evidence from the sea anemone, *Nematostella vectensis*. PLoS One 2:e153. doi:10.1371/journal.pone.0000153

Semmler H, Bailly X, Wanninger A (2008) Myogenesis in the basal bilaterian *Symsagittifera roscoffensis* (Acoela). Front Zool 5:14. doi:10.1186/1742-9994-5-14

Semmler H, Chiodin M, Bailly X, Martinez P, Wanninger A (2010) Steps towards a centralized nervous system in basal bilaterians: insights from neurogenesis of the acoel *Symsagittifera roscoffensis*. Dev Growth Differ 52:701–713. doi:10.1111/j.1440-169X.2010.01207.x

Sikes JM, Bely AE (2010) Making heads from tails: development of a reversed anterior-posterior axis during budding in an acoel. Dev Biol 338:86–97. doi:10.1016/j.ydbio.2009.10.033

Smith J, Tyler S (1985) The acoel turbellarians: kingpins of metazoan evolution or a specialized offshoot? In: Conway Morris S, George JD, Gibson R, Platt HM (eds) The origins and relationships of lower invertebrates. Calderon Press, Oxford, pp 123–142

Srivastava M, Mazza-Curll KL, van Wolfswinkel JC, Reddien PW (2014) Whole-body acoel regeneration is controlled by Wnt and Bmp-Admp signaling. Curr Biol 24:1107–1113. doi:10.1016/j.cub.2014.03.042

Sterrer W (1998) New and known nemertodermatida (Platyhelminthes-Acoelomorpha) – a revision. Belg J Zool 128:55–92

Tyler S (2001) The early worm: origins and relationships of the lower flatworms. In: Littlewood DTJ, Bray RA (eds) Interrelationships of the platyhelminthes. Taylor & Francis Ltd., London, pp 3–12

Wallberg A, Curini-Galletti M, Ahmadzadeh A, Jondelius U (2007) Dismissal of acoelomorpha: acoela and nemertodermatida are separate early bilaterian clades. Zool Scr 36:509–523

Westblad E (1937) Die Turbellarien-Gattung Nemertoderma Steinböck. Acta Soc pro Fauna et Flora Fenn 60:45–89

Westblad E (1949) On *Meara stichopi* (Bock) Westblad, a new representative of Turbellaria archoophora. Arkiv Zoologi 1:43–57

Chaetognatha

Steffen Harzsch, Carsten H.G. Müller, and Yvan Perez

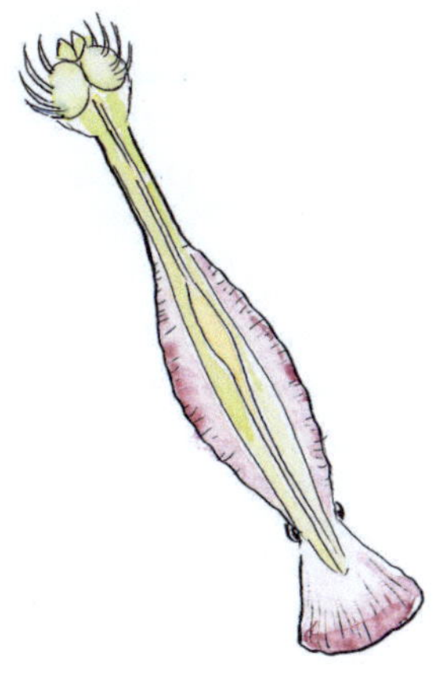

Chapter vignette artwork by Brigitte Baldrian.
© Brigitte Baldrian and Andreas Wanninger.

S. Harzsch (✉)
Department of Cytology and Evolutionary Biology,
Zoological Institute and Museum,
Ernst-Moritz-Arndt-University Greifswald,
Soldmannstr. 23, Greifswald D-17487, Germany
e-mail: steffen.harzsch@uni-greifswald.de

C.H.G. Müller
Department of General and Systematic Zoology,
Zoological Institute and Museum,
Ernst-Moritz-Arndt-University Greifswald,
Anklamer Str. 20, Greifswald D-17487, Germany

Y. Perez
Aix-Marseille Université, CNRS, IRD,
Avignon Université, IMBE UMR 7263,
Marseille 13331, France

A. Wanninger (ed.), *Evolutionary Developmental Biology of Invertebrates 1:*
Introduction, Non-Bilateria, Acoelomorpha, Xenoturbellida, Chaetognatha
DOI 10.1007/978-3-7091-1862-7_10, © Springer-Verlag Wien 2015

INTRODUCTION

According to palaeontological evidence, the Chaetognatha (arrow worms), a group of small marine predators that are major components of the zooplankton throughout our world oceans, were present already in the Early Cambrian (approx. 540–520 Myr years ago), namely, as Chengjiang biota (Vannier et al. 2007), but have also been documented in the Middle Cambrian Burgess Shale (Szaniawski 2005). The so-called protoconodonts, spine-like, small, shelly microfossil elements that were abundant in the Cambrian, are today convincingly interpreted as parts of the chaetognath grasping apparatus or, at least, as belonging to protoconodont animals most closely related to Chaetognatha (e.g., Szaniawski 1982, 2002, 2005; Vannier et al. 2007; but see Conway Morris 2009; Szaniawski 2009 for a controversial discussion). The presence of protoconodonts in the lowermost Cambrian and the complexity of their feeding apparatus points to a Precambrian origin of these animals (Vannier et al. 2007). These authors also suggested placing them among the earliest active predator metazoans and argued that the ancestral chaetognaths were planktonic with possible ecological preferences for hyperbenthic niches close to the sea bottom. Today, the taxon Chaetognatha comprises more than 150 described species from all geographical and vertical ranges of the ocean. They are characterised by an elongated, streamlined body; the presence of horizontally projecting fins; and, at the anterior end, two groups of moveable, cuticularised grasping spines used in capturing prey such as copepods (Fig. 10.1). With a body length between just a few millimetres up to 120 mm, these glassily transparent carnivores are among the most abundant planktonic organisms, but several epibenthic species are also known (Bieri 1991; Shinn 1997; Nielsen 2001; Kapp 2007).

There is a long-standing interest into the body organisation of Chaetognatha. General information on their morphology and anatomy has been summarised in the contributions by Hertwig (1880), John (1933), Hyman (1959), Ghirardelli (1968), Goto and Yoshida (1987), Bone and Goto (1991), Kapp (1991), Nielsen (2001, 2012), Ax (2001), and Perez et al. (2014); the most detailed review of their microscopic anatomy is that of Shinn (1997). Their embryonic development is known from the fundamental studies by Hertwig (1880), Doncaster (1903), Elpatiewsky (1909), Burfield (1927), and John (1933) on *Sagitta* and *Spadella* (Fig. 10.2). The phylogenetic position of the chaetognaths within the Bilateria is heavily debated, and in the last 170 years, since (Darwin 1844) described them as remarkable for "the obscurity of their affinities", various phylogenetic affiliations have been proposed (summarised by, e.g., Ghirardelli 1968, 1995; Bone et al. 1991; Papillon et al. 2004; Harzsch and Müller 2007; Harzsch and Wanninger 2009; Edgecombe et al. 2011; Nielsen 2012). Despite the ever-increasing number of molecular phylogenetic studies and an emerging consensus for protostome affinities based on broad phylogenomic datasets, today the relationships of the Chaetognatha are still among the most enigmatic issues of metazoan phylogeny (reviewed in Perez et al. 2014). The chaetognath genome is likely the product of a unique evolutionary history and stands for the long isolation of this group. Furthermore, morphological characters indicate a long evolutionary distance that separates the Chaetognatha from its closest (unknown) metazoan relative and suggest that this taxon in many aspects seems to have explored its own evolutionary pathways in generating tissue and organ diversity (reviewed in Perez et al. 2014). These authors pointed out the following unusual features of the Chaetognatha which have played an important role in the discussion on their phylogenetic position: (i) The ribosomal cluster of Chaetognatha is duplicated so that two classes of paralogous 18S rRNA and 28S rRNA genes are present, both of which diverge extremely from other Metazoa. (ii) The sequence of the intermediate filament protein (nuclear lamins) gene of *Sagitta elegans* is very unusual compared to a field of approx. 20 protostome and deuterostome sequences, suggesting a particularly high evolutionary drift of the chaetognath sequence. (iii) The unique mosaic organisation of a posterior/median Hox gene shows a long, isolated evolution of the chaetognath Hox complex. (iv) Chaetognatha display a very unusual mitochondrial genome

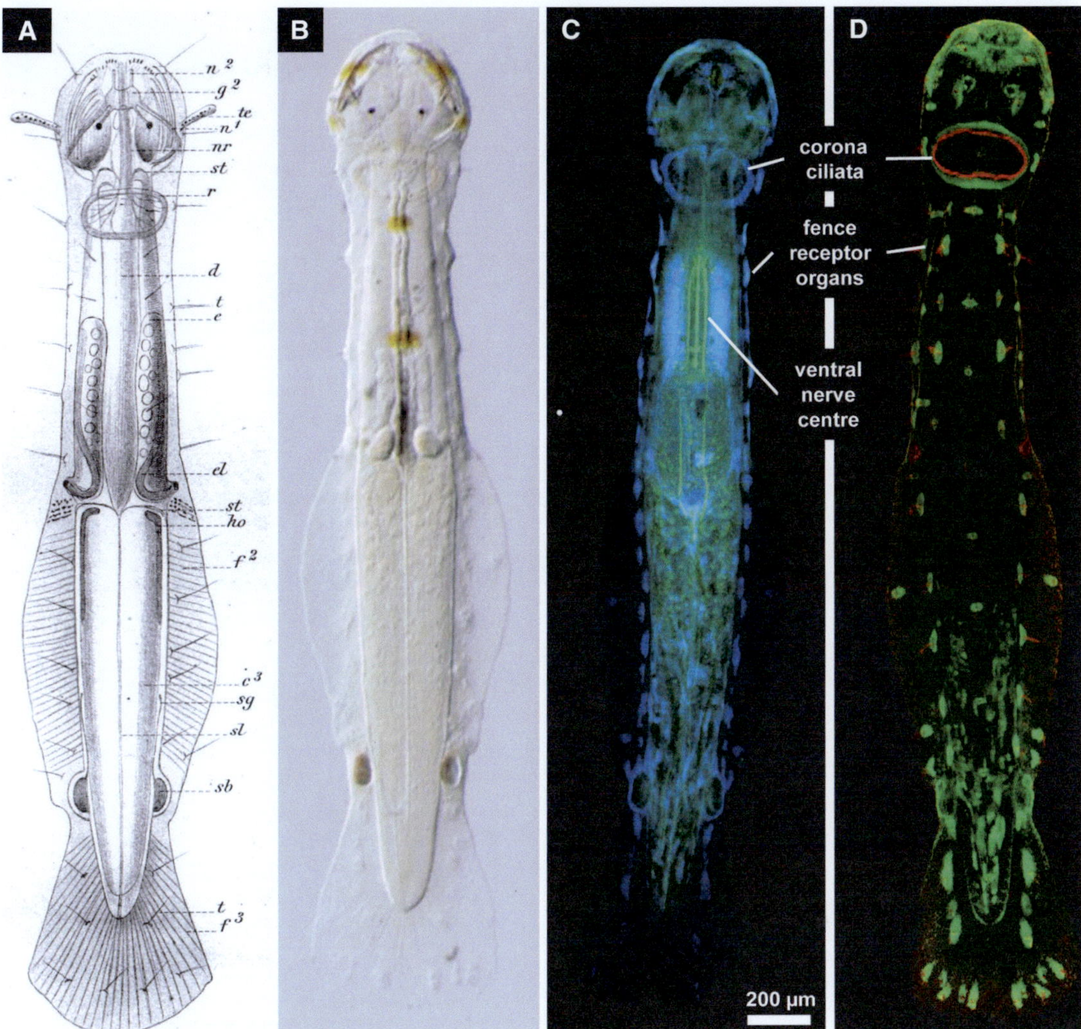

Fig. 10.1 (**A**) *Spadella cephaloptera*; drawing from Hertwig (1880) with the author's original nomenclature (*from top to bottom*; note that the current identification of the various organs may be different; consider, e.g., the "Riechorgan"). Abbreviations: *n2* "Nerv zwischen Bauchganglion und Schlundganglion" (frontal connective between brain and vestibular ganglion), *g2* "Oberes Schlundganglion" (brain), *te* "Tentakelartiger Fortsatz der Kopfkappe" (tentacular head process), *n1* "Commissur zwischen Bauchganglion und Schlundganglion" (main connective between brain and ventral nerve centre), *nr* "Riechnerv" ("olfactory nerve", nerve that links the corona ciliata to the brain), *st* "Querseptum" (transverse septum), *r* "Riechorgan" ("olfactory organ", the corona ciliata), *d* "Darm" (gut), *t* "Tastorgan" (mechanosensory organ), *e* "Eierstock" (ovary), *el* "Eileiter" (oviduct), *ho* "Hoden" (testes), *c3* "Schwanzhöhle" (coelom), *sg* "Samengang" (seminal duct), *sl* "Längsseptum" (longitudinal septum), *sb* "Samenblase" (seminal vesicle), *f3* "Schwanzflosse" (tail). (**B**) Light micrograph of a live specimen of *S. cephaloptera*. (**C**) *S. cephaloptera*, immunohistochemical localisation of RFamide-related peptides (*green*) and histochemistry to label cell nuclei (*blue*; Reprinted with permission from Harzsch et al. 2009). (**D**) *S. cephaloptera*; immunohistochemical localisation of acetylated alpha-tubulin (*red*) and histochemistry to label nuclei (*green*; Reprinted with permission from Harzsch and Wanninger 2009)

which is one of the smallest known in metazoans and contains only 14 of the 37 usual genes. (v) Chaetognatha display unusual ways of neuromuscular innervations with axonal varicosities that lack specialised junctions so that the presynapses are separated from the underlying muscles by a thick extracellular matrix through which the transmitter must pass. (vi) Whereas obliquely

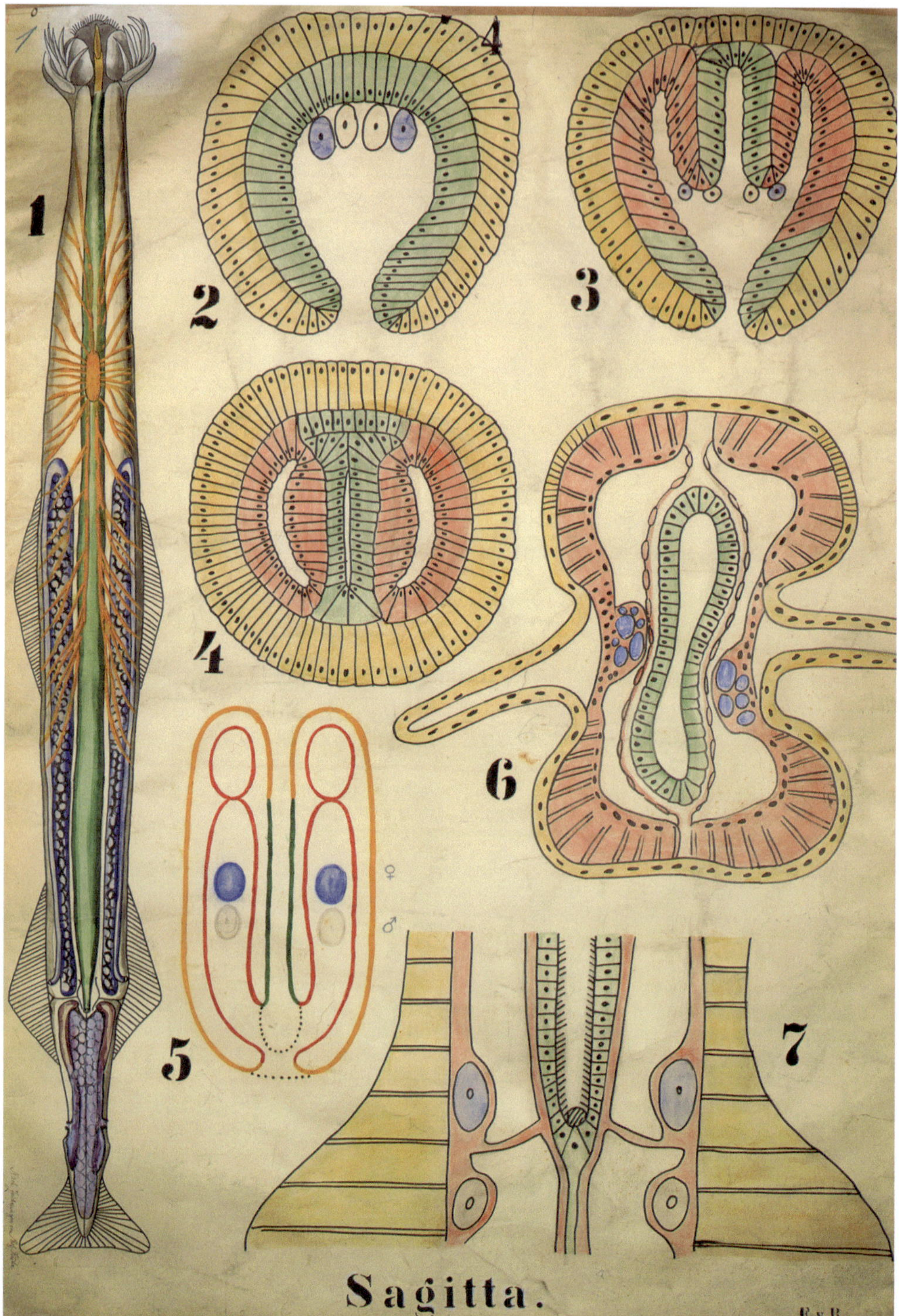

Fig. 10.2 An example from the wall chart collection of the Zoological Institute of the Berlin University which was founded by its first director Franz Eilhard Schulze in 1884. This chart most likely was drawn at the Zoological Institute by Erika von Bruchhausen around 1920–1930 under the directors Karl Heider and Richard Hesse (information is based on the contribution by Stefan Richter "The collection of the Zoological Institute of the Berlin University – History and Importance". http://www2.hu-berlin.de/biologie/zoologie/Lehrsammlung.htm). According to the textbook by Korschelt and Heider (1890) "Lehrbuch der vergleichenden Entwicklungsgeschichte der wirbellosen Thiere" (p. 245), figures 2, 3, and 4 most likely are based on the study by Hertwig (1880)

striated musculature, as is typical for many other invertebrates, is absent in Chaetognatha, this taxon has "experimented" with various types of cross-striated muscles and evolved a unique type of secondary muscles with two different kinds of sarcomeres that alternate within the muscles. (vii) The chaetognath integument mainly consists of a multilayered, non-cuticularised epidermis that is made up by two different cell types (single layer of vacuolated and secretory distal epidermal cells and multilayer of flattened, filamentous proximal epidermal cells). This multilayered, dual-type epidermis is a unique, hence apomorphic, character of Chaetognatha. (viii) The intra- and basiepidermal neuronal plexus is a plesiomorphic character retained from the metazoan or bilaterian ground pattern. Due to the thickness of the epidermis, however, this intraepidermal aspect of the chaetognath plexus is particularly complex. Therefore, the combination of an orthogonal basiepidermal neuronal plexus and a heavily diversified intraepidermal plexus may represent a further derived, apomorphic state characterising the Chaetognatha, however inevitably linked to the possession of a multilayered epidermis. (ix) The structure of the ciliated photoreceptors in the eyes shows a unique architecture with a basal body, a conical body, and an outer segment with parallel stacks of perforated membrane lamellae. (x) The circumesophageal arrangement of the adult cephalic nervous system including, in addition to the brain, vestibular and subesophageal ganglia, is suggestive of protostome affinities. However, the situation in the hatchlings clearly shows that we do not only face one but even two brain components that have a basically circumoral arrangement.

In summary, both the genome and morphological characters represent many autapomorphies of this group in addition to a character mix of protostome and deuterostome features, suggesting that Chaetognatha is likely an early offshoot of the protostome lineage (cf. Nielsen 2012; Perez et al. 2014), but so far it is not possible to unequivocally decide whether they are basal lophotrochozoans (spiralians), basal ecdysozoans, or the sister group to all others protostomes. Perez et al. (2014) suggested a new basal rooting of the Bilateria, which considers the Chaetognatha as the sister group to the Lophotrochozoa (Spiralia) and Ecdysozoa (soft polytomy), an assemblage that may be called Hyponeuria to indicate the ventralisation of the nervous system as one of the most important steps in metazoan evolution.

EARLY DEVELOPMENT

Germ Plasm, Primary Germ Cells, and Fertilisation

Aspects of mating, reproduction, and growth of Chaetognatha were reviewed by Ghirardelli (1968), Pearre (1991), Alvarino (1992), and Shinn (1997). There is a long and controversial debate in the literature on the mode of fertilisation in these animals (see, e.g., Pearre 1991), so that the novice in this field is well advised to begin his studies with the most recent references.

Chaetognaths are considered protandrous hermaphrodites, and the testes mature before the ovaries (Pearre 1991). They develop directly, so that the hatchlings exhibit a body organisation that is in most respects similar to the adult (Fig. 10.3). Spermatogenesis is intracoelomic in the tail with the two testes located adjacent to the dorsal longitudinal muscles. Spermatogonia from the testes pass into the tail coeloms to undergo mitosis and meiosis and to differentiate into spermatozoa (Shinn 1997). Oogenesis occurs in two discrete ovaries in the trunk (Fig. 10.4; Shinn 1997). After an elaborate mating behaviour (e.g., Ghirardelli 1968; Goto and Yoshida 1985), the sperm cluster is deposited onto the mate's body. Several authors have suggested the possibility of self-fertilisation, but this has never been demonstrated, and the reproduction even in meso- and bathypelagic species only depends upon cross-fertilisation (Terasaki and Miller 1982). For internal fertilisation, the sperm migrates into the female gonopores (Ghirardelli 1953, 1968) and is stored in the seminal receptacles. Fertilisation occurs in the ovaries prior to ovulation and is mediated by specialised somatic "accessory fertilisation cells" (e.g., Shinn 1994a, 1997; Goto 1999; Carré et al. 2002). In a unique fertilisation process, sperm passes through these accessory

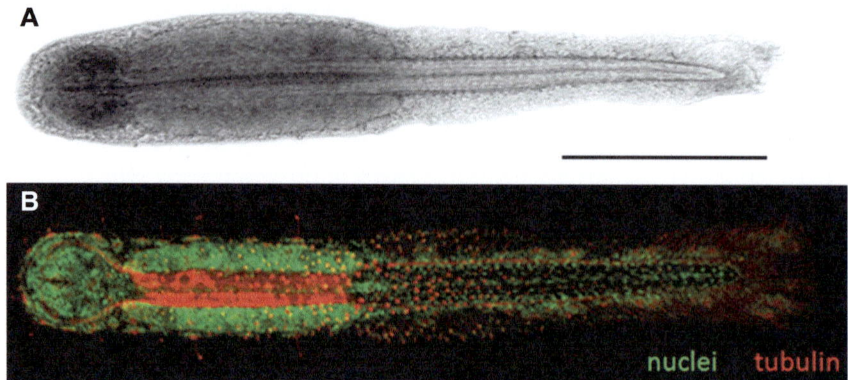

Fig. 10.3 (**A**) Hatchling of *Spadella cephaloptera*. (**B**) 2-day-old juvenile; immunohistochemical labeling of tubulin (*red*) and histochemical labeling of nuclei (*green*; Reprinted from Perez et al. 2014). Scale bar equals 250 μm

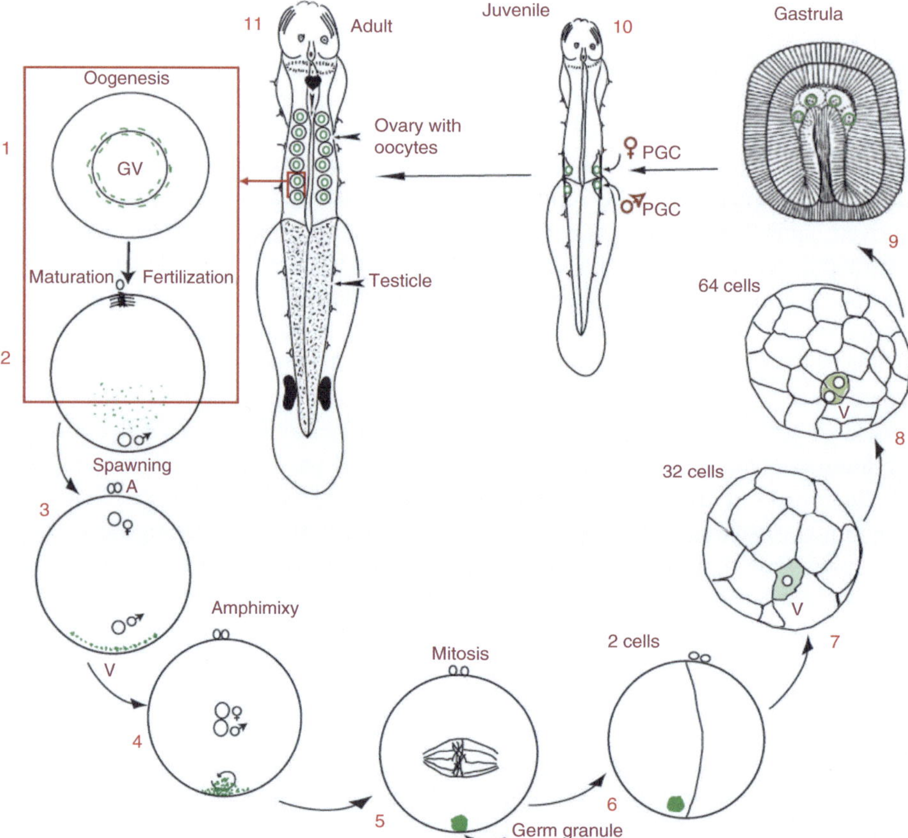

Fig. 10.4 Germ plasm and germ cells during the life cycle of chaetognaths (Reprinted from Carré et al. 2002). During oogenesis (*1*), germ plasm/nuage material (*green*) is within and around the germinal vesicle (*GV*). During maturation and internal fertilisation at the vegetal pole (*2*), germ plasm presumably fragments into minute granules. After spawning (*3*), many small granules line the vegetal cortex (*V*) and then aggregate during amphimixy (*4*). At mitosis (*5*), small germ granules aggregate into a single large granule. This large granule is segregated into one of the first two blastomeres and continues to be inherited by only one vegetal blastomere until the 32-cell stage (*7*). The germ granule then fragments and is distributed into two blastomeres at the 64-cell stage (*8*). The germ plasm is then found in the four presumptive primary germ cells (*PGCs*) at the tip of the archenteron in the gastrula (*9*). The four primary germ cells (*PGC*) become the male (posterior) and female (anterior) germ cells in the hatchling (*10*), which give rise to the spermatocytes and the oocytes in the adult

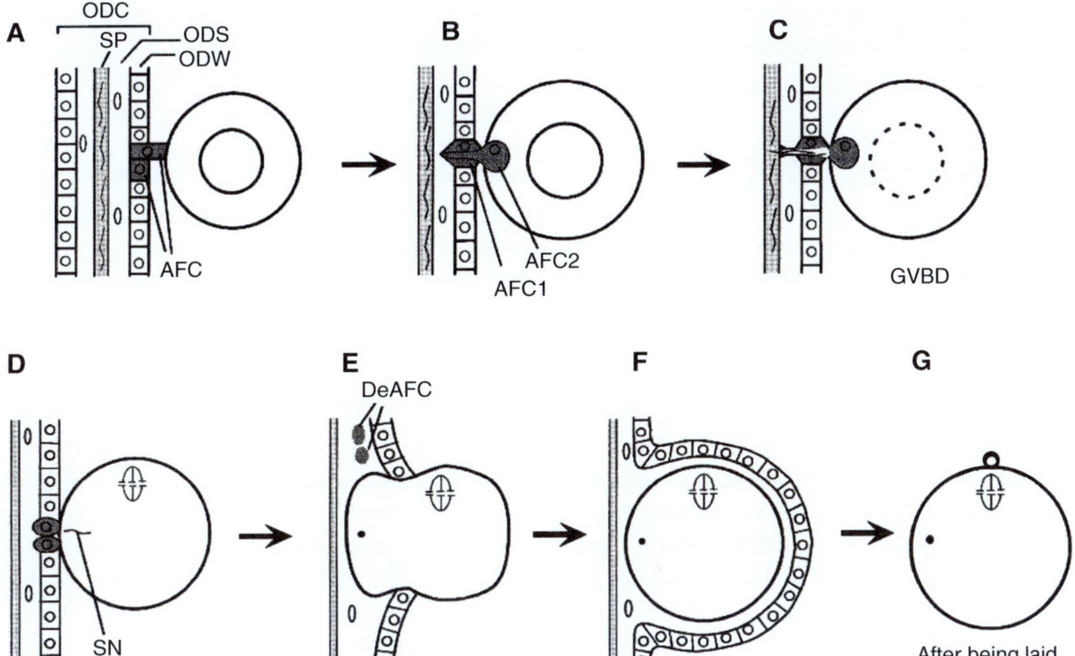

Fig. 10.5 A schematic diagram showing the process of chaetognath fertilisation (from Goto 1999 as based on his own studies and those of Shinn 1994b, 1997). (**A**) Growing oocyte attaches to pairs of specialised oviducal cells that differentiate into accessory fertilisations cells (*AFCs*). (**B**) AFC2 sinks in the oocyte. Cytoplasmic process of AFC2 occludes the fertilisation canal formed in AFC1. (**C**) The cytoplasmic process disappears from the fertilisation canal through which sperm enters into the egg. (**D**) A single sperm cell enters an oocyte after germinal vesicle breakdown (*GVBD*). AFC2 moves outside of the oocyte. (**E**) The fertilised egg moves into the oviducal syncytial complex through a pore that is formed by degeneration of AFCs. Sperm chromatin is condensed into a round shape. (**F**) The fertilised egg is stored in the oviducal complex for the first meiotic metaphase until laid in sea water. (**G**) Resumption of meiosis occurs after the eggs have been deposited in sea water. The sperm chromatin remains a spherical shape until a late stage of meiosis. Abbreviations: *AFC* accessory fertilisation cell, *DeAFC* degenerating AFC, *ODC* oviducal complex, *ODS* oviducal syncytium, *ODW* ovarian wall, *SP* sperm, *SN* sperm

fertilisation cells to reach the oocytes (Figs. 10.4 and 10.5; Goto 1999; Carré et al. 2002). Egg laying occurs immediately after fertilisation in *Sagitta hispida*, or, as in *Spadella cephaloptera*, the zygotes may remain in the oviduct arrested in their first meiotic metaphase for several hours. In general, the zygotes are then either released into the water column in pelagic species (Sagittidae) or sometimes retained in brood pouches (Eukrohniidae) or attached to rocks or plants in the case of benthic species (Spadellidae; for review see Pearre 1991). The establishment of laboratory cultures has been essential for exploring the embryology and aspects of growth in both planktonic and benthic species of Chaetognatha (Kuhl and Kuhl 1965; Reeve 1970; Reeve and Walter 1972; Reeve and Lester 1974; Kotori 1975; Feigenbaum 1976; Goto and Yoshida 1997; Papillon et al. 2005).

It has long been known that the oocytes bear a large granule, presumed to be a germ plasm that is passed on to the four primary germ cells (reviewed in Shimotori and Goto 1999; Carré et al. 2002). A single germ granule forms in the vegetal cortex of the zygote at the time of the first mitosis (Fig. 10.4). This germ granule associates with the cleavage furrow and is segregated into one of the first two blastomeres. It is then translocated from the cell cortex to the mitotic spindle and remains in the single-most vegetal blastomere until the 32-cell stage (Carré et al. 2002). At the 64-cell stage, the germ granule is partitioned as nuage material into two founder primary germ cells (PGCs) and further partitioned into four PGCs located at the tip of the archenteron during its invagination (Fig. 10.4). These PGCs differentiate into the oocytes and spermatocytes of the two ovaries and testes.

Cleavage and Gastrulation

Cleavage in chaetognaths is total and equal (Fig. 10.6). It was by many embryologists considered as resembling the radial cleavage pattern of deuterostomes (reviewed in, e.g., Kapp 2000; Brusca and Brusca 2003). However, a recent analysis of the developmental fate of the first four blastomeres led Shimotori and Goto (2001) to propose that chaetognaths are more similar to protostomes in their developmental programme. Using intracellular dye injections into the blastomeres of the two-cell stage, Shimotori and Goto (1999) had shown that the first cleavage plane parallels the anterior-posterior axis and bears an angular relationship to the sagittal and frontal planes. These authors pointed out that the appearance of the four-cell stage is similar to that of equal cleavage in spiralians because the embryo consists of the animal and vegetal cross-furrow cells in tetrahedral arrangement, reflecting an oblique orientation of the mitotic spindle in relation to the zygote axis (Figs. 10.6 and 10.7). Already Elpatiewsky (1909) had observed that the second cleavage of *Sagitta bipunctata* resulted in a counterclockwise (leiotropic) displacement of the two animal cross-furrow cells with respect to the two vegetal cross-furrow cells and that the next cleavage was in an opposite direction (dexiotropic). Shimotori and Goto (2001) concluded for *Paraspadella gotoi* that the direction of displacement of the second cleavage, the alignment of the future body axes, and the tetrahedral position of the four blastomeres are similar to those of spiralians so that "the developmental fates of the first four blastomeres in chaetognath embryos may have some similarities to spiralians". They designated the blastomeres of the four-cell embryo of *Paraspadella gotoi* as a, b, c, and d cells based on the nomenclature used in spiralians (Fig. 10.7A, B), but used lower case letters to avoid confusion. The cell containing the germ plasm was designated

the d cell. Hence, the two-cell stage consists of the ab cell and the cd cell containing the germ plasm, which generates the a and b (animal cross-furrow) cells and the c and d (vegetal cross-furrow) cells, respectively (Fig. 10.7A, B; Shimotori and Goto 2001). Injecting single blastomeres of *Paraspadella gotoi* embryos with lineage tracing dyes, Shimotori and Goto examined the fate of the first four blastomeres to assess to which organ systems they give rise to (Fig. 10.7C–F).

Gastrulation describes the process of germ layer formation and is intimately linked to the formation of the digestive system that includes mouth opening, gut, and anus. The blastopore is the site of tissue internalisation, and two major phylogenetic concepts are based on the fate of the blastopore. In protostomes, the blastopore later on becomes the adult mouth, while the anus (if present) is formed secondarily at a different site (reviewed in Hejnol and Martindale 2009; Martindale and Heijnol 2009). In deuterostomes, the blastopore becomes the anus while the mouth is formed secondarily at a different site in the animal hemisphere of the embryos. In Chaetognatha, an invagination of the blastopore initiates the formation of the endoderm and subsequently a second opening forms the stomodeum opposite to the blastopore (Fig. 10.6; Hertwig 1880; Doncaster 1903; John 1933; Kuhl and Kuhl 1965; Kapp 2000). For a short period the embryo possesses both a stomodeum and a blastopore in a typical deuterostome fashion, but it is important to note that the blastopore is not the future anus in Chaetognatha. The mouth and blastopore then close, so that young hatchlings for a given period have neither a mouth opening nor an anus (Fig. 10.3; Shimotori and Goto 1999; Shinn and Roberts 1994). Both openings are re-established some time after hatching (approx. 48 h in *Spadella cephaloptera*). Hence, Chaetognatha in principle conform to the developmental pattern of deuterostomes as laid out above, but nevertheless display some variation. The evolutionary meaning of this fact is discussed below.

In addition to features of spiralian cleavage and deuterostome-like gastrulation, chaetognath embryology includes features that were suggested

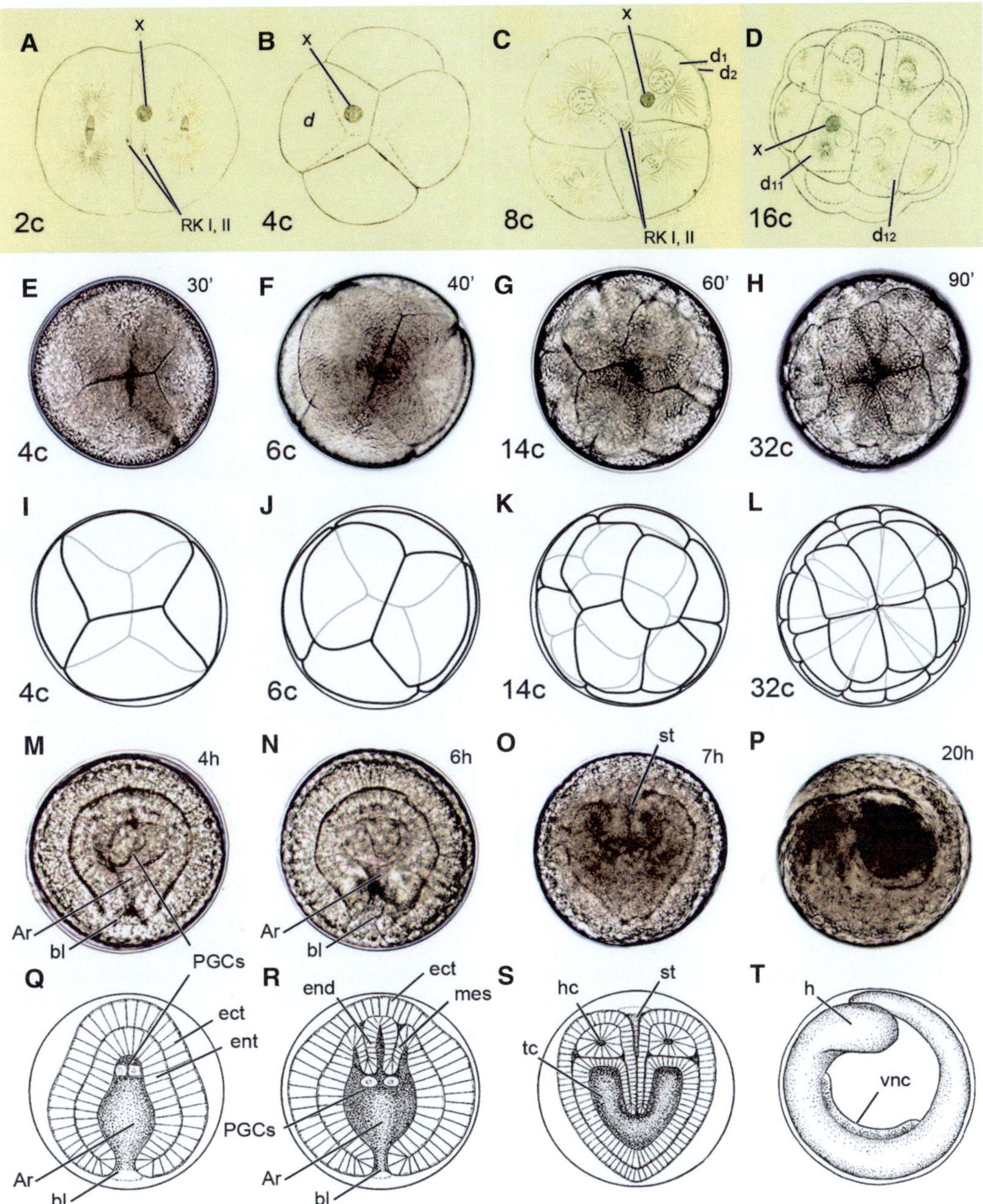

Fig. 10.6 The developmental sequence in *Sagitta bipunctata* and *Spadella cephaloptera*. (**A–D**) Early cleavage in *Sagitta bipunctata* (Drawings from Elpatiewsky 1909). (**E–L**) Early cleavage in *Spadella cephaloptera*. (**M–O**) Gastrulation and mesoderm formation in *Spadella cephaloptera*. (**P**) Specimen of *Spadella cephaloptera* just before hatching. (**Q–S**) Gastrulation and mesoderm formation in *Sagitta bipunctata* from Burfield (1927). A few hours after the beginning of cleavage, the blastopore (*bl*) completely closes. Two mesodermal folds progress backward directly into the archenteron (*Ar*) and mark off the endoderm (*ent*) from a pair of meso-dermal cavities (*hc* head coelom, *tc* trunk coelom). The anterior region of the ectoderm (*ect*) forms the stomo-deum (*st*). (**T**) Specimen of *Sagitta bipunctata* just before hatching (After Doncaster 1903). Early cleavage is total and equal. Labels denote the 2-cell (*2c*), 4-cell (*4c*), 8-cell (*8c*), etc., stages and the developmental time in minutes and hours from fertilisation at room temperature. Abbreviations: *Ar* archenteron, *bl* blastoporus, *d* descen-dants of the d cell, *ect* ectoderm, *ent* endoderm, *h* head, *hc* head coelom, *mes* mesoderm, *PGCs* primordial germ cells, *RK* "Richtungskörper" (polar body), *st* stomodeum, *tc* trunk coelom, *vnc* ventral nerve centre, *X* germ granule

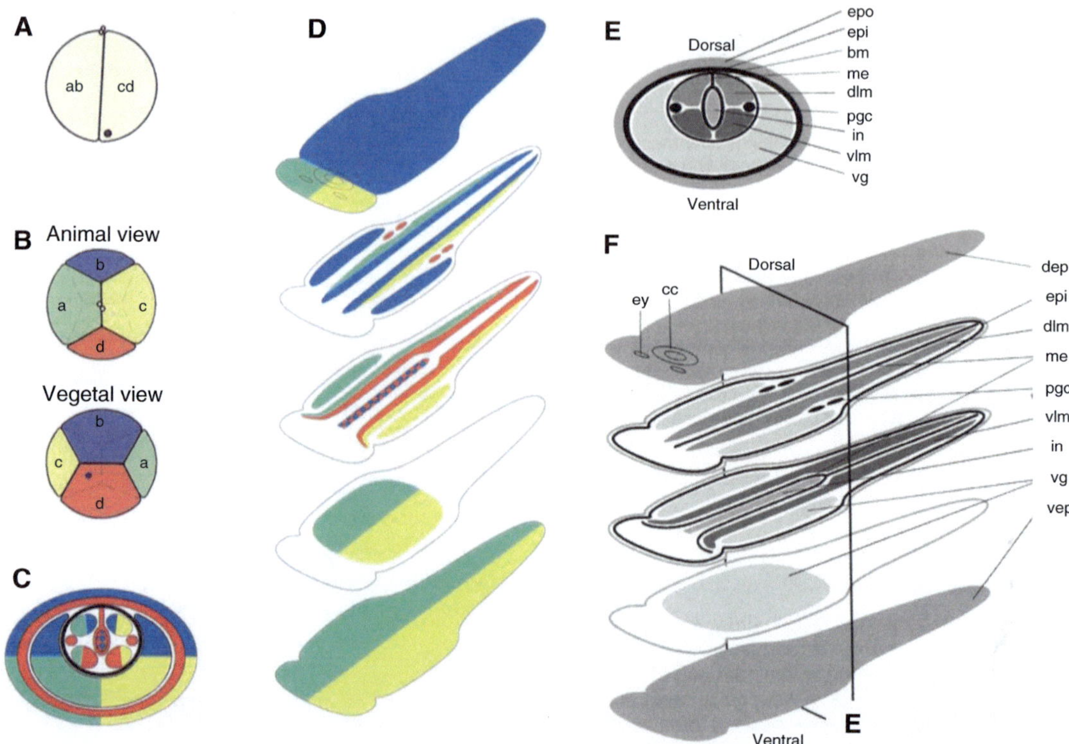

Fig. 10.7 (**A–D**) Diagrams summarising the fate map of *Paraspadella gotoi* at the 4-cell stage (From Shimotori and Goto 2001). The *a*, *b*, *c*, and *d* cells and the regions derived from each blastomere are coloured *green*, *blue*, *yellow*, and *red*, respectively. (**A**) Lateral view of the two-cell embryo. (**B**) Animal and vegetal view of the 4-cell embryo. Two open small circles represent the polar bodies. The *blue dot* indicates the germ plasma. (**C**) Transverse section of the body. (**D**) A series of horizontal sections. Hatched *blue* and *light red* indicates a mixture of clones derived from the *b* and *d* cells. (**E**, **F**) Diagrams of the gross anatomy of a hatchling (From Shimotori and Goto 2001). (**E**) Diagram of a transverse plane at the body region indicated by *lettered box E* in (**F**). (**F**) Diagrams of a series of frontal sections. The chaetognath body is divided indistinctly externally into three segments: head, trunk, and tail. The epidermis is composed of outer and inner layers. In the trunk, the ventral ganglion occupies a large space and surrounds the mesodermal tissues. Longitudinal muscles are arranged in four major bands, two dorsolateral and two ventrolateral. Two dorsal longitudinal muscles extend from the centre of the head to the end of the tail, while two ventral muscles obviously curve with the anterior ends and extend from the posterior laterals of the head to the ends of the tail. Another mesodermal tissue, mesentery, surrounds the intestine and these tissues are in the centre of four muscle bands. The four primordial germ cells (*pgc*) are situated near the posterior end of the intestine. Abbreviations: *bm* basal matrix, *cc* corona ciliata, *dep* dorsal surface of outer epidermis, *dlm* dorsal longitudinal muscles, *epi* inner epidermis, *epo* outer epidermis, *ey* eye, *in* intestine, *me* mesentery, *pgc* primordial germ cells, *vep* ventral surface of outer epidermis, *vg* ventral ganglion, *vlm* ventral longitudinal muscles

to be unusual and difficult to relate to other Bilateria, so that Kapp (2000) coined a new term, heterocoely, to describe their mode of coelom formation. This author, reviewing the classical studies by Hertwig (1880), Doncaster (1903), and John (1933), concluded that the process of mesoderm formation seems to be unique to this taxon. However, beyond taking these classical studies as a foundation, chaetognath embryogenesis and especially germ layer and coelom formation urgently need to be reanalysed by contemporary methods including electron and confocal laser scanning microscopy in order to allow for a phylogenetic comparison of chaetognath development (cf. Shinn and Roberts 1994).

After the invagination of the endoderm at the blastopore as described above, the four PGCs become localised opposite to the blastopore inside of the cavity enclosed by endoderm, the archenteron (Figs. 10.4 and 10.6; note that Kapp 2000 argues against this term as it implies a function as a gut anlagen, which is not the case in chaetognaths because blastopore and the new mouth opening will close again shortly after gastrulation). The primary body cavity lying between the ecto- and endoderm, the blastocoel, is very narrow (Fig. 10.6). The mesoderm forms such that the endoderm folds inwards in two places into the archenteron and the blastocoel subsequently vanishes. Endodermal folds separate part of the archenteron opposite to the blastopore into three hollows. The upper tip of the middle space is the area where the new mouth will form and this second opening subsequently develops, as already described above. The middle hollow shrinks as its walls move towards each other. These cell layers will eventually form the intestine. The other two hollows will be part of the head and the trunk/tail paired coelomic cavities (Fig. 10.6). In the course of development, the mesodermal cells are arranged in two bilateral groups and obliterate both the cephalic and the trunk coelomic cavities. Few morphological and cytological changes occur until hatching with the exception of (i) the gradual elongation of the developing intestine which forms a thin median septum, (ii) the migration of the primordial germ cells until the middle of the body at the level of the future posterior septum, (iii) the beginning of the longitudinal and transverse muscle differentiation, and (iv) the development of the ventral nerve centre which represents the most prominent structure at hatching.

Chaetognath Gastrulation in the Light of Their Presumed Phylogenetic Position

Chaetognath gastrulation displays a mosaic of protostome features and distinct aspects conventionally assigned to the deuterostomes. Nonetheless, in the past a placement within the deuterostomes, as based on developmental aspects (Kuhl 1938; Hyman 1959), was accepted in several textbooks (Kapp 1991; Brusca and Brusca 2003; sum-

marised in Takada et al. 2002). Such a placement was also based on the fact that differentiation of the coelom seemingly resembles enterocoely, but Perez et al. (2014) recently pointed out that other aspects of chaetognath gastrulation, especially the process of mesoderm and coelom formation, seem to be apomorphies of this taxon and thus not suggestive of a close affinity to any other bilaterian phylum. Gastrulation patterns play essential roles in discussions on bilaterian evolution (e.g., Arendt and Nübler-Jung 1997; Nielsen 2001, 2005a, b; Hejnol and Martindale 2009; Martindale and Heijnol 2009; Lacalli 2010; Nielsen 2010; Martín-Duran et al. 2012). Hejnol and Martindale (2009) have emphasised that, on closer examination, a considerable variation in gastrulation patterns is evident between bilaterian subtaxa, especially in protostomes, which led these authors to stress the extreme variability in blastopore fates. In ctenophores and cnidarians, the mouth and the blastopore have a common origin and these animals gastrulate at the animal pole (see Chapters 6 and 8). Nevertheless, bilaterians gastrulate at the vegetal pole (Hejnol and Martindale 2009; Martindale and Heijnol 2009). The authors propose that in Bilateria a separation of the signalling centres that determine the sites of mouth formation versus the site of germ layer specification has taken place and that this ancient separation explains the variation of the spatial relation of blastopore and mouth in Bilateria. Furthermore, they critically review the concept of amphistomy, which refers to the lateral closure of a slit-like, elongate blastopore. The latter, by staying open at both ends, can account in principle for the formation of both the bilaterian mouth and the anus (e.g., Arendt and Nübler-Jung 1997; Nielsen 2001). Amphistomy in a strict sense, however, appears to occur in only a few protostomian taxa.

Perez et al. (2014) suggested that the Chaetognatha display their own variant of deuterostomy and may represent a new evolutionary experiment in which a mode of gastrulation has evolved that is clearly distinct from the other protostomes, but without necessarily meaning that chaetognaths are phylogenetically ingroup deuterostomes. Similarly, Valentine (1997) and

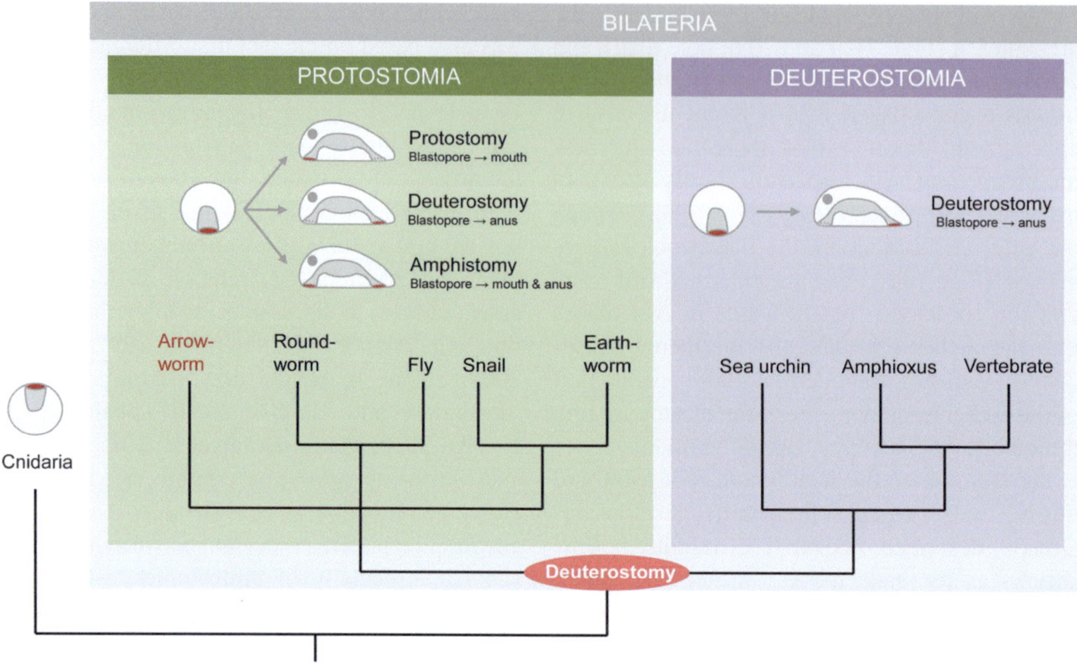

Fig. 10.8 Diversity of blastoporal fates in the Bilateria (Modified from Martín-Duran et al. 2012). While in Deuterostomia the blastopore forms the anus, the Protostomia exhibit a diversity of blastoporal fates. This situation may be best explained when considering deuterostomy to be part of the bilaterian ground pattern which subsequently was modified convergently in the various protostome lineages. Chaetognaths are here considered the sister group to all other protostomia (Compare Perez et al. 2014) and have retained deuterostomy as a plesiomorphic character from the bilaterian ground pattern

Peterson and Eernisse (2001) advocated that traditional characters placing the lophophorates into the deuterostomes are plesiomorphies of bilaterians. Chaetognaths also exhibit such plesiomorphies: a complete gut with a mouth not arising from the blastopore and coelomic cavities forming by inward folding of the endoderm (Papillon et al. 2004). What is more, based on embryological analyses of priapulid development, Martín-Duran et al. (2012) suggested that deuterostomy was the ancestral developmental programme in bilaterians (Fig. 10.8). Along these lines, Chaetognatha may have simply retained deuterostomy as a plesiomorphic character from the bilaterian ground pattern so that this developmental mode remains uninformative about their phylogenetic position (Fig. 10.8). Interestingly, such a pivotal position of Chaetognatha between deuterostomes and protostomes is compatible with the report of Shimotori and Goto (2001) that the fates of the first four blastomeres in chaetognath embryos may have some similarities to spiralians and with an analysis of the expression pattern of the *brachyury* gene in a chaetognath species, *Paraspadella gotoi* (Takada et al. 2002). These authors conclude that the expression pattern of this gene in the embryonic chaetognath blastopore and mouth resembles that of hemichordates and echinoderms, whereas the pattern in the region of the new mouth opening of the hatchling appears to be novel.

LATE DEVELOPMENT

Ontogeny of Mesoderm-Derived Tissues and Organisation of the Internal Body Cavities: Are Chaetognaths Bipartite Animals?

Chaetognatha was traditionally placed within Deuterostomia mainly because of their mode of coelom formation and the tripartite body organisation with three distinct coelomic cavities, a situation which recalls the archimeric condition of basal deuterostomes and lophophorates. However, chaetognaths are supposed to pass through a peculiar developmental process, known as "heterocoely" (Kapp 2000). This aberrant way by which the mesoderm and body cavities arise as well as the questionable persistence and histological anatomy of these cavities throughout ontogeny has led to controversies as to whether or not chaetognaths are deuterostomes. Furthermore, it has been debated for decades how their peculiar coelomogenesis can be compared to other coelomates (Hyman 1959; Willmer 1990; Backeljau et al. 1993; Schram and Ellis 1994; Ghiradelli 1995; Kapp 2000; Brusca and Brusca 2003; see Jenner 2004 for latest review). For example, cladistic analyses based on morphological and embryological data placed the chaetognaths among the aschelminths (Meglitsch and Schram 1991; Backeljau et al. 1993; Schram and Ellis 1994; Nielsen et al. 1996). Furthermore, Meglitsch and Schram (1991); also questioned the enterocoelic nature of the body cavities and considered the coelomic compartments not to be homologous to those of the typical archimeric animals.

Nowadays, it has been accepted that, as a working definition relevant to assess the mesoderm organisation and nature of body cavities in coelomates, a coelom always consists of a body cavity which is lined by a specialised epithelium, the coelothel, and not by an extracellular matrix (ECM; e.g., Schmidt-Rhaesa 2007; Koch et al. 2014). The apicolateral junctional complex adhering neighboured coelothelial cells thereby faces the coelomic cavity, if present and clearly discernible. In this definition, ultrastructural data (Duvert and Salat 1979; Welsch and Storch 1982; Shinn 1994b, 1997; Shinn and Roberts 1994) unambiguously support the hypothesis that the chaetognath ground pattern comprises true coelomic cavities and coelothelia at least in the trunk and tail of the adult. Extensive studies on this topic were conducted by Shinn (1994b, 1997) and Shinn and Roberts (1994) in *Ferosagitta hispida*. These authors did not only analyse the mesodermal epithelial arrangement in adults but also in hatchlings, thus providing valuable insights into the formation and (functional) transformation of the chaetognath coelom. In hatchlings, the trunk and tail mesoderm consist of stereotypically arranged myoepithelial cells corresponding in position to specific adult tissue, e.g., the lateral fields, the longitudinal body wall muscles, medial cells forming the dorsal and ventral mesenteries, and peri-intestinal muscular cells. Only the coelothelial cells (= peritoneocytes)

overlying the adult longitudinal muscles in adult animals do not have any equivalents in hatchlings. These specialised adult peritoneocytes are supposed to descend from lateral or medial cells observed in hatchlings. This indicates that in chaetognaths the coelothel initially is made up of a layer of myoepithelial cells, whereas formation of the peritoneum which lines the musculature is delayed to few days after hatching. Furthermore, only numerous small triangular spaces situated between the hatchling's mesodermal cells are visible. These are supposed to coalesce to form the coelomic cavities in the adult (Shinn 1994b; Shinn and Roberts 1994). This contradicts previous descriptions of the hatchling's morphology in Sagittidae and Spadellidae, arguing that the coelom is obliterated during embryogenesis (Doncaster 1903; John 1933). Interestingly, it seems that chaetognaths show a singular posterior-anterior gradient in tissue differentiation. Indeed, Shinn and Roberts (1994) stated that the mesoderm cavitation is relatively more advanced in the posterior part of the tail and just anterior to the posterior septum.

Recent studies of the authors of the present chapter on *Spadella cephaloptera* provide some further insight into this issue (Rieger et al. 2011) and show that in hatchlings the development of the brain is delayed when compared to the ventral nerve centre. In *Spadella cephaloptera*, the mesodermal tissues of hatchlings look compact and do not display any coelomic cavities in the analysed body regions (Fig. 10.9A). Those mesodermal cells the vast majority of which later become muscle cells have close lateral and apical contact to each other (Fig. 10.9B). They rest on the ECM of the body wall and lack intercellular junctions at their apical membrane's centre. Their cytoplasm contains numerous electron-dense yolk bodies, mitochondria, and various amounts of rough endoplasmic reticulum (RER). Cross sections through the mid-trunk region reveal that the differentiation of the longitudinal muscle cells is poorly advanced, with each quadrant consisting of only three or four semicircularly arranged fibres (Fig. 10.9D). A thin vertical sheet of ECM continuous with the ECM of the body wall separates the adjacent muscle cells belonging to the cephalic and trunk mesodermal compartments (Fig. 10.10A, B). This boundary matches the location of the anterior septum which separates the head and trunk regions. The anterior septum

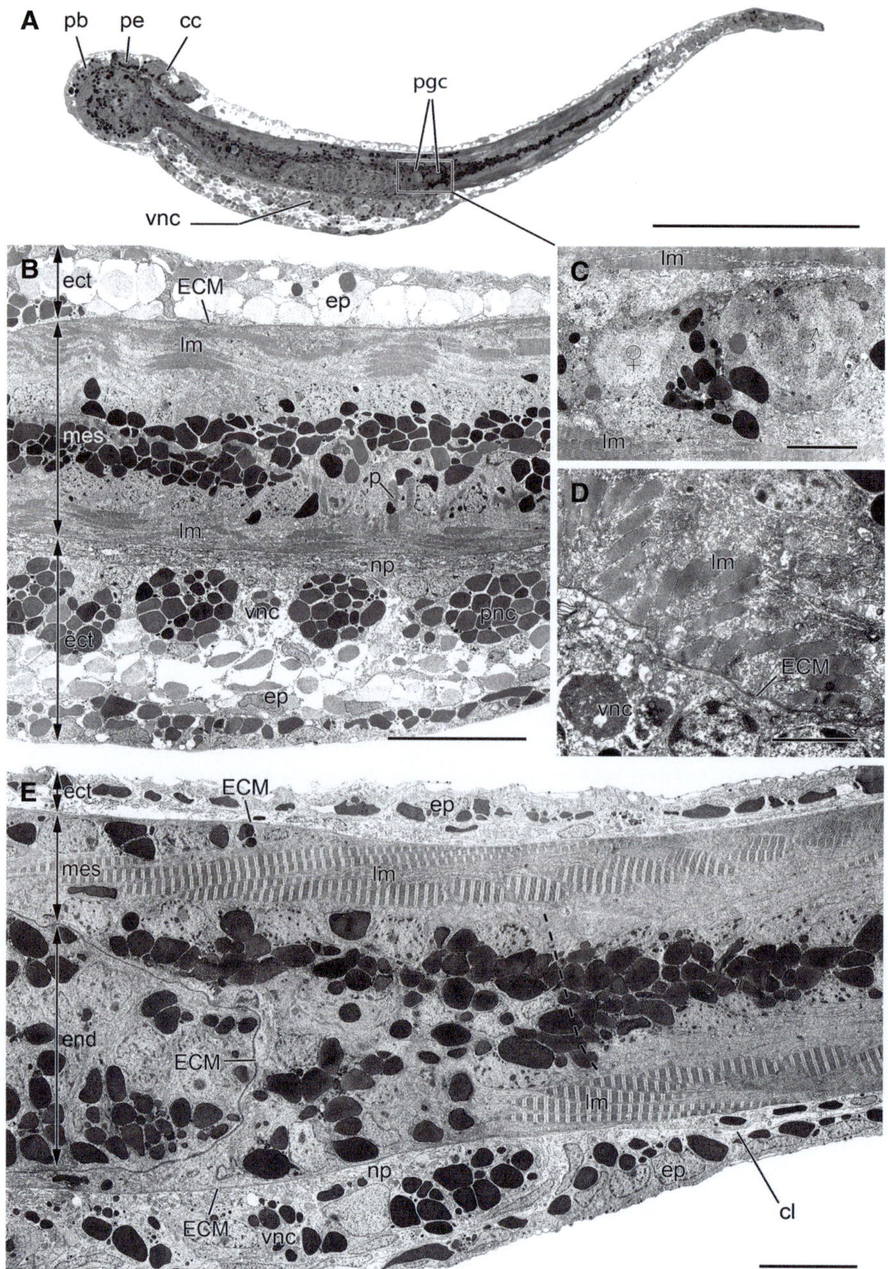

Fig. 10.9 Light micrographs (**A**) and transmission electron micrographs (**B–E**) from sections through a hatchling of *Spadella cephaloptera*. (**A**) Longitudinal section through the whole body. The region where the primordial germ cells stop their migration coincides with the position of the future posterior septum between the trunk and tail in the adult. (**B**) Longitudinal aspect of the mid-trunk region of the same specimen as shown in (**A**). Note the serial organisation of the neural progenitor cells which appear in a segmental fashion, although there is no segmentation in the trunk. (**C**) Detail of the female and male primordial germ cells. (**D**) Cross section through the mid-trunk region showing the poorly developed longitudinal muscles. (**E**) Detail of the transition of the trunk and tail (paramedian section). *Dotted line* indicates the position of the future posterior septum which results in the final segregation of the male and female primordial germ cells shown in (**C**). Abbreviations: *cc* corona ciliata, *cl* caudal loop, *ect* ectoderm, *ECM* extracellular matrix, *ep* epidermis, *hm* head mesoderm, *lm* longitudinal muscle, *mes* mesoderm, *np* neuropil, *npc* neural progenitor cells, *p* phragm (transverse muscle), *pb* primordial brain, *pe* primordial eye, *pgc* primordial germ cell, *tm* trunk mesoderm, *vnc* ventral nerve centre, *yo* yolk, ♂ male primordial germ cell, ♀ female primordial germ cell. Originals: Y. Perez and C.H.G. Müller. Scale bars: A = 0.25 mm; B = 25 µm; C, D = 4 µm; E = 8 µm

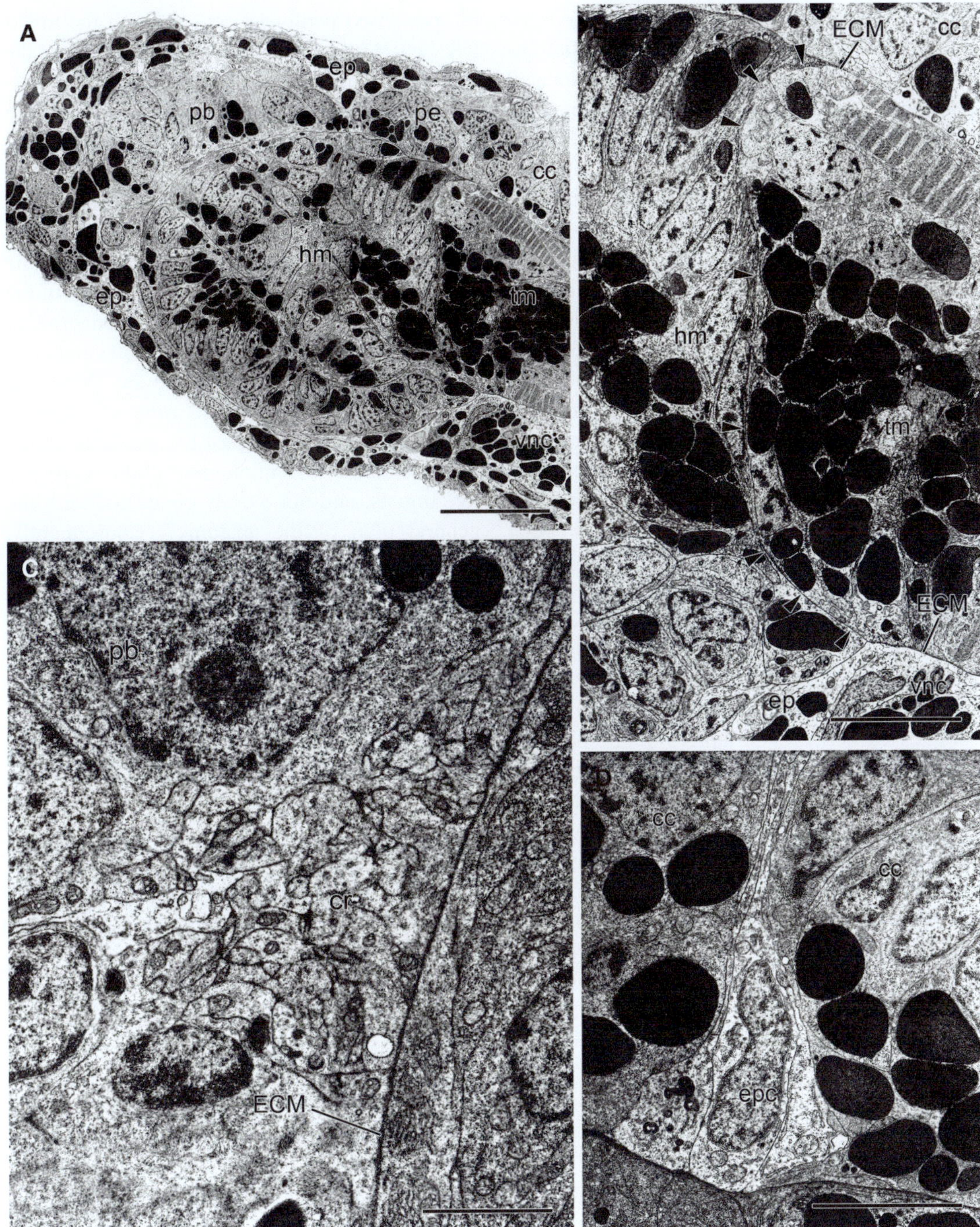

Fig. 10.10 Transmission electron micrographs from (para)median sections through a hatchling of *Spadella cephaloptera*. (**A**) Head and neck region. Note the low level of differentiation of the cephalic muscles in comparison to those of the trunk. (**B**) Detail of the transition of head and trunk. *Arrowheads* indicate contacts between developing muscle cells from the head and trunk mesodermal pouches. (**C**) Primordial brain showing the cephalic nerve ring which consists of a bundle of neurites (cross section) in basiepidermal position. The cephalic nerve ring sits on the extracellular matrix of the body wall, spins all around the head, and connects the primordial brain to the ventral nerve centre. (**D**) Detail of a neuronal cell from the epidermal plexus at the basis of the corona ciliata. Abbreviations: *cc* corona ciliata, *cr* cephalic ring, *ECM* extracellular matrix, *pb* primordial brain, *pe* primordial eye, *ep* epidermis, *hm* head mesoderm, *tm* trunk mesoderm, *vnc* ventral nerve centre, *epc* epidermal plexus cell. Originals: Y. Perez and C.H.G. Müller. Scale bars: A = 20 μm; B = 10 μm; C = 2 μm; D = 4 μm

is formed by the direct juxtaposition of the muscle cells and does not exhibit the same histological organisation as the posterior one located between the trunk and tail which is formed by two layers of specialised mesodermal cells (Shinn and Roberts 1994). The region in which the posterior septum appears is marked anteriorly by the two female PGCs and posteriorly by the two male PGCs (Fig. 10.9A, C). It has been proposed on the basis of histological data that the completion of the posterior septum occurs several days after hatching (Doncaster 1903; John 1933). However, according to ultrastructural observations (Shinn and Roberts 1994), the rudiment of the posterior septum is already formed at hatching and consists of two layers of specialised lateral and peri-intestinal cells but without any trace of ECM or junctional complexes between these cells. Considering the lack of ECM separating the specialised mesodermal cells which enfold the PGCs, the haemal system located in the adult posterior septum is not complete. Therefore, the trunk and tail coelomic cavities in hatchlings, although remaining very narrow and compressed between the apices of mesodermal cells, are continuous and must therefore be considered as a single cavity extending from the anterior septum to the posterior end of the body.

The full adult morphology develops 2 days after hatching (Fig. 10.11A). At this time, the young *Spadella* exhibits the typical segmentation of the body into the head, trunk, and tail. In the cephalic region, the chitinous structures (hooks, teeth, ventral epidermis, lateral and ventral plates) appear. The head muscles, the brain, the eyes, and the corona ciliata are also complete by this time. The mouth and anus are opened. The stomodeum is connected to the intestine at the level of the anterior septum. In the trunk, the mesoderm- and endoderm-derived tissues are well differentiated. The posterior septum which lies just behind the anus is now complete and splits the trunk and tail coelomic cavities (Fig. 10.11A, B). Importantly, completion of the posterior septum comes along with a continuous transition from trunk to tail, as the posterior septum does not interrupt the longitudinal muscles, which extend up from the beginning of the trunk to the end of the tail segment. There are two layers of specialised peritoneal cells forming the trunk/tail septum (Fig. 10.11C). Each cell layer lines the coelomic cavities of the trunk and tail with its apical region. Towards their proximal regions, both tightly adjoined peritoneal epithelia secrete the ECM, the actual posterior septum, which displays two basal laminae (from either peritoneal epithelium opposing) enclosing a thin, fibrous interlayer.

Taken together, these suggest that the establishment of the tripartite organisation of the adult body plan is delayed and is concomitant with the final segregation of the female and male PGCs. As previously suggested by Doncaster (1903) "From its mode and time of origin it seems reasonable to regard the posterior transverse septum as essentially part of the reproductive organs, and not closely connected with the general plan of the anatomy". Consequently, chaetognaths must be fundamentally regarded as being bipartite animals. The high organisation level of the coelomic lining and the longitudinal muscles and the early formation of a rudimentary posterior septum observed in *Ferosagitta hispida* (Shinn and Roberts 1994) could be a taxon and/or species-specific feature, revealing some differences in the level of tissue differentiation, especially when the studied species exhibit different ecological features such as benthic (Spadellidae) versus pelagic (Sagittidae) lifestyles. The experimental breeding conditions could also explain such variations. Accordingly, the body cavities observed in chaetognaths are likely formed just as the coelomic cavities of recognised schizocoelic protostomes are formed, e.g., from a compact band of mesodermal cells that transforms to an epithelial organisation and then surround the continuously widening coelomic cavity (Turbeville 1986; Schmidt-Rhaesa 2007). Interestingly, in small annelids, the development of a coelom is arrested at the stage when an epithelial but still compact mesodermal cell mass is present (for review, see Schmidt-Rhaesa 2007), a situation highly similar to the hatchling condition in chaetognaths. Accordingly, the chaetognaths' coelomogenesis, by nature of its precursor mesodermal cells and the late differentiation and cavitation of the mesoderm, appears like a derived, postembryogenic variant of

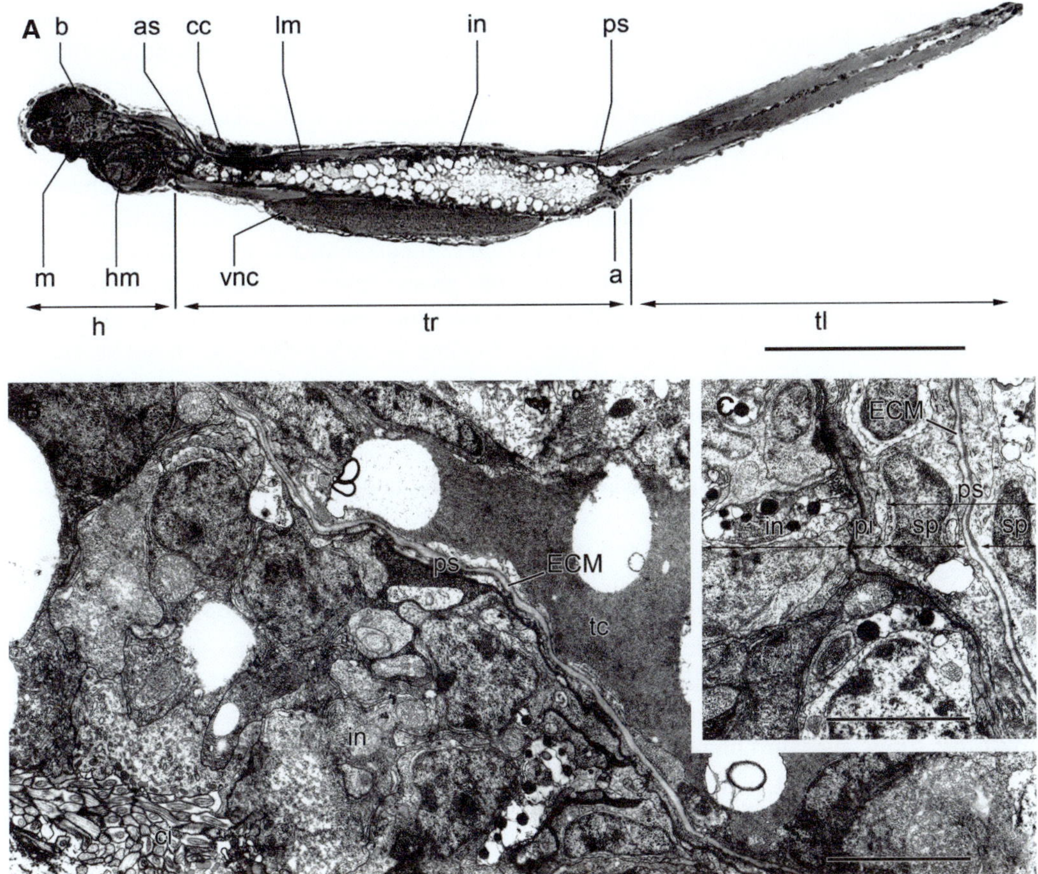

Fig. 10.11 Light (**A**) and transmission electron micrographs (**B**, **C**) from longitudinal sections through a 2-day-old specimen of *Spadella cephaloptera*. (**A**) Longitudinal section through the whole body. Note the completion of the posterior septum (*ps*) without interruption of the longitudinal muscles (*lm*). (**B**) Longitudinal section showing the posterior septum. (**C**) Detail of the posterior septum constituted by two layers of specialised peritoneal cells. The posterior septum has a distinct ECM secreted from basal domains of opposing peritoneal cells. Abbreviations: *a* anus, *as* anterior septum, *b* brain, *cc* corona ciliata, *ci* cilia, *ECM* extracellular matrix, *h* head, *hm* head muscles, *in* intestine, *m* mouth, *pi* peri-intestinal cells, *ps* posterior septum, *sp* specialised peritoneocyte, *st* stomodeum, *tc* tail coelom, *tl* tail, *tr* trunk, *vnc* ventral nerve centre. Originals: Y. Perez and C.H.G. Müller. Scale bars: A=0.25 mm; B=4 μm; C=0.5 μm

schizocoely which, however, is not necessarily homologous to other coelomates. Indeed, it is important to note that coelomogenesis by schizocoely starts when fluid accumulates between desmosomes that connect opposing epithelio-muscle cells of the somatic and visceral muscle (Koch et al. 2014). This is apparently not the case in chaetognaths, since no specialised junctional complexes occur between the longitudinal muscle cells and the peri-intestinal cells of the hatchling (Shinn and Roberts 1994). If chaetognaths are not true enterocoelic animals, even more doubt arises as to whether the head, trunk, and tail coelomic cavities are homologous to the protocoel, mesocoel, and metacoel of classic archimeric animals. According to the latest data on their embryology and the new consensus on their pivotal position between deuterostomes and protostomes, the apomorphic chaetognath's coelomogenesis could represent the first attempt to modify the typical enterocoely observed in deuterostomes and some protostomes towards the derived schizocoely of lophotrochozoan lineages.

Neurogenesis

The general organisation of the adult nervous system of chaetognaths has been described by several authors, notably by Bone and Pulsford (1984), Goto and Yoshida (1984, 1987), and Shinn (1997; recently summarised in Perez et al. 2014). The recent studies by Harzsch and Müller (2007) and Harzsch et al. (2009) on the ventral nerve centre, by Rieger et al. (2010) on the brain, and by Müller et al. (2014) on ciliated sense organs have contributed new facets to this picture. Chaetognath neurogenesis was first examined histologically by Doncaster (1903), who described the initial stages of neural development in some detail. According to this author, when the head coelom has formed, the ventral ectoderm of the trunk and the ectoderm above the mouth develop thickenings that become the rudiments of the ventral nerve centre and the brain, respectively. Cell proliferation occurs along two ventrolateral bands of somata in the trunk, which, just after hatching, are clearly marked off from the surrounding epidermis. These form the primordia of the ventral nerve centre (VNC). At this point, the general organisation of the VNC is well established. Beyond these basic aspects, embryonic neurogenesis is poorly understood in chaetognaths.

Recently, Perez et al. (2013) have used S-phase-specific proliferation markers to analyse the mitotic activity of neuronal progenitor cells in chaetognath hatchlings. Furthermore, Goto et al. (1992) and Rieger et al. (2011) have characterised the postembryonic development of some peptidergic and aminergic neurons with immunofluorescence methods. The ventral nerve centre is a dominant organ in hatchlings (Figs. 10.3 and 10.9), while the brain is still rather rudimentary. A comparison of the system of neurons that express RFamide-related neuropeptides in the ventral nerve centre showed that this pattern in the hatchlings in many respects already resembles that in adults. The number of somata with RFamide-like immunoreactivity increases very little from hatching onwards, but the system of their neurites becomes more complex as development proceeds (Harzsch et al. 2009; Rieger et al. 2011). However, experiments using the S-phase-specific mitosis

marker bromodeoxyuridine (BrdU) provided evidence for a high level of mitotic activity in the ventral nerve centre for approx. 3 days after hatching (Perez et al. 2013). Neurogenesis in the hatchlings is carried by presumptive neuronal progenitor cells that cycle rapidly and most likely divide asymmetrically (Figs. 10.12 and 10.13). These progenitors are arranged in a distinct grid-like geometrical pattern including about 35 transverse rows and most likely are also actively dividing in earlier embryos. In adults, the VNC controls swimming by initiating contractions of the body wall musculature and coordinating mechanosensory input from the numerous ciliary fence receptors in the epidermis. For motor control, the VNC closely interacts with the peripheral, exclusively epidermal plexus that innervates the muscles (reviewed in Bone and Pulsford 1984; Goto and Yoshida 1984, 1987; Shinn 1997; Perez et al. 2014). Similar to the situation in the adult, the VNC most likely modulates body movements in hatchlings, enabling them to escape predators and control their attachment to seaweed, the preferred substrate for attachment in this species. It appears that the hatchlings are already equipped with some functional "fence receptor organs", sensory organs that perceive hydrodynamic stimuli, and an array of papillae on the ventral surface of the body that probably mediate substrate attachment (Müller and Perez, unpublished observations). The experiments described above provide evidence that the VNC is far from being completely differentiated at hatching, but that instead new neurons are added and existing neuronal systems continue to differentiate. Likewise, unpublished results suggest that new fence receptor organs are added on the body surface and provide new sensory input to the VNC that needs to be integrated. In conclusion, the newly hatched, non-feeding animals mostly rely on a set of embryonic fence receptors that perceive hydrodynamic stimuli for navigating in their habitat, predator avoidance, and attachment to their preferred substrate. Because these fence receptors feed into the VNC and because the VNC is essential for swimming behaviour, this major neuronal centre for sensory-motor integration must be functional to a certain degree at hatching.

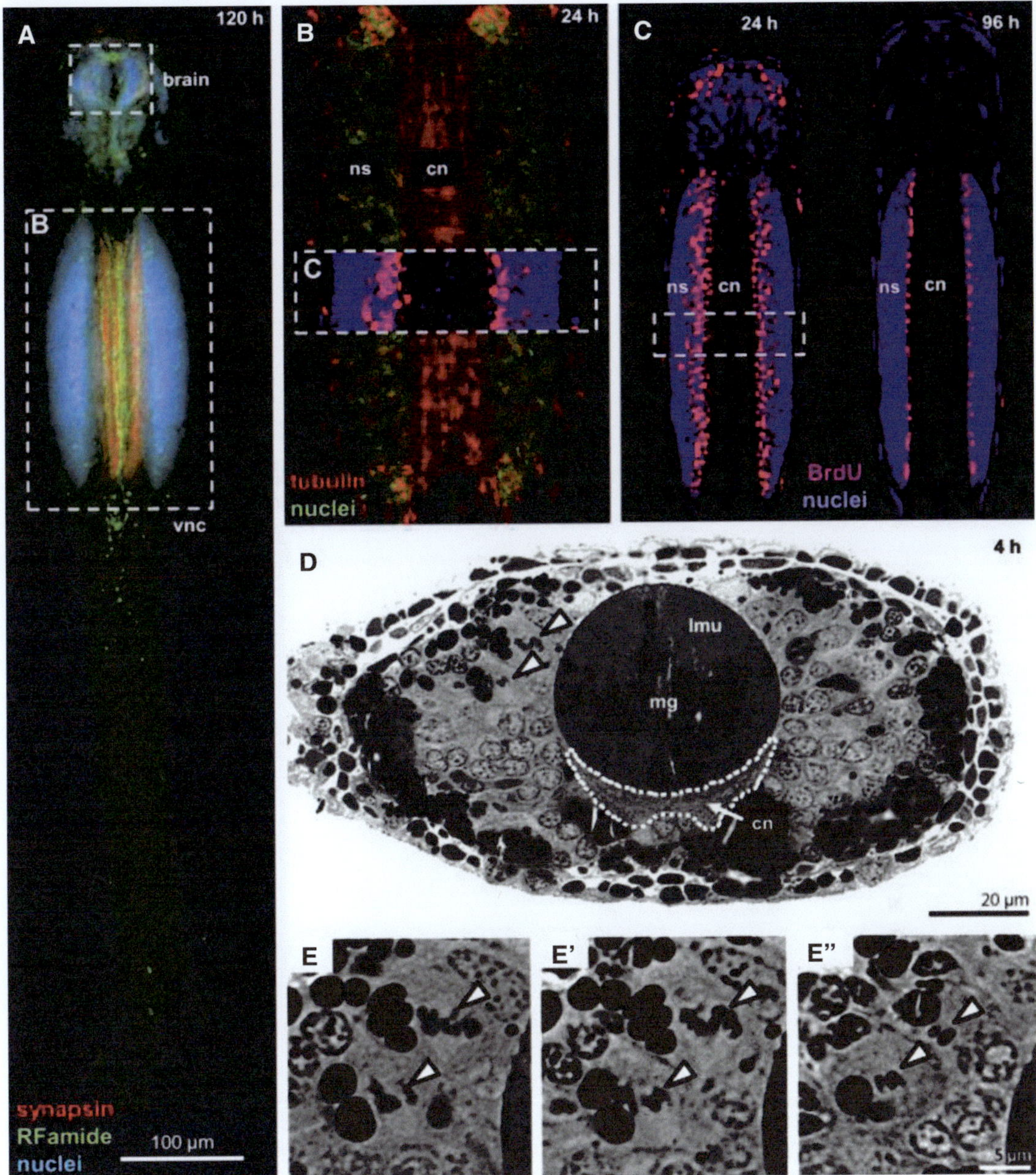

Fig. 10.12 The ventral nerve centre of *Spadella cephaloptera* (From Rieger et al. 2011 and Perez et al. 2013). (**A**) Hatchling at 120 h triple labeled for nuclei (*blue*), synaptic proteins (*red*), and the neuropeptide RFamide (*green*). *Boxes* identify the brain and the ventral nerve centre (*vnc*). (**B**) Higher magnification of the ventral nerve centre (as indicated in **A**) of a 24-h-old hatchling double labeled for tubulin (*red*) to show the central neuropil that is flanked by neuronal somata labeled by a nuclear marker (*green*). The inset shows a higher magnification of the boxed area in (**C**). (**C**) Hatchlings at 24 h (*left*) and 96 h (*right*) after 4 h BrdU labeling. The cells in S-phase (*red*) are located at the interface of the central neuropil and the peripheral layers of neuronal somata (*blue*). Note the decreasing mitotic activity in the older hatchling. (**D**) Semithin cross section (toluidine blue staining) through the ventral nerve centre of a hatchling. The hatchling was fixed shortly after hatching (age about 4 h). *Arrowheads* identify two nuclei of dividing cells (see higher magnifications in (**E**)). The *dotted line* surrounds the central neuropil of the ventral nerve centre. (**E–E''**) Higher magnifications of three consecutive sections (1 µm) showing mitotic cells in M-phase (*arrowheads*). Abbreviations: *cn* central neuropil, *lmu* longitudinal muscles, *mg* midgut, *ns* neuronal somata, *vnc* ventral nerve centre

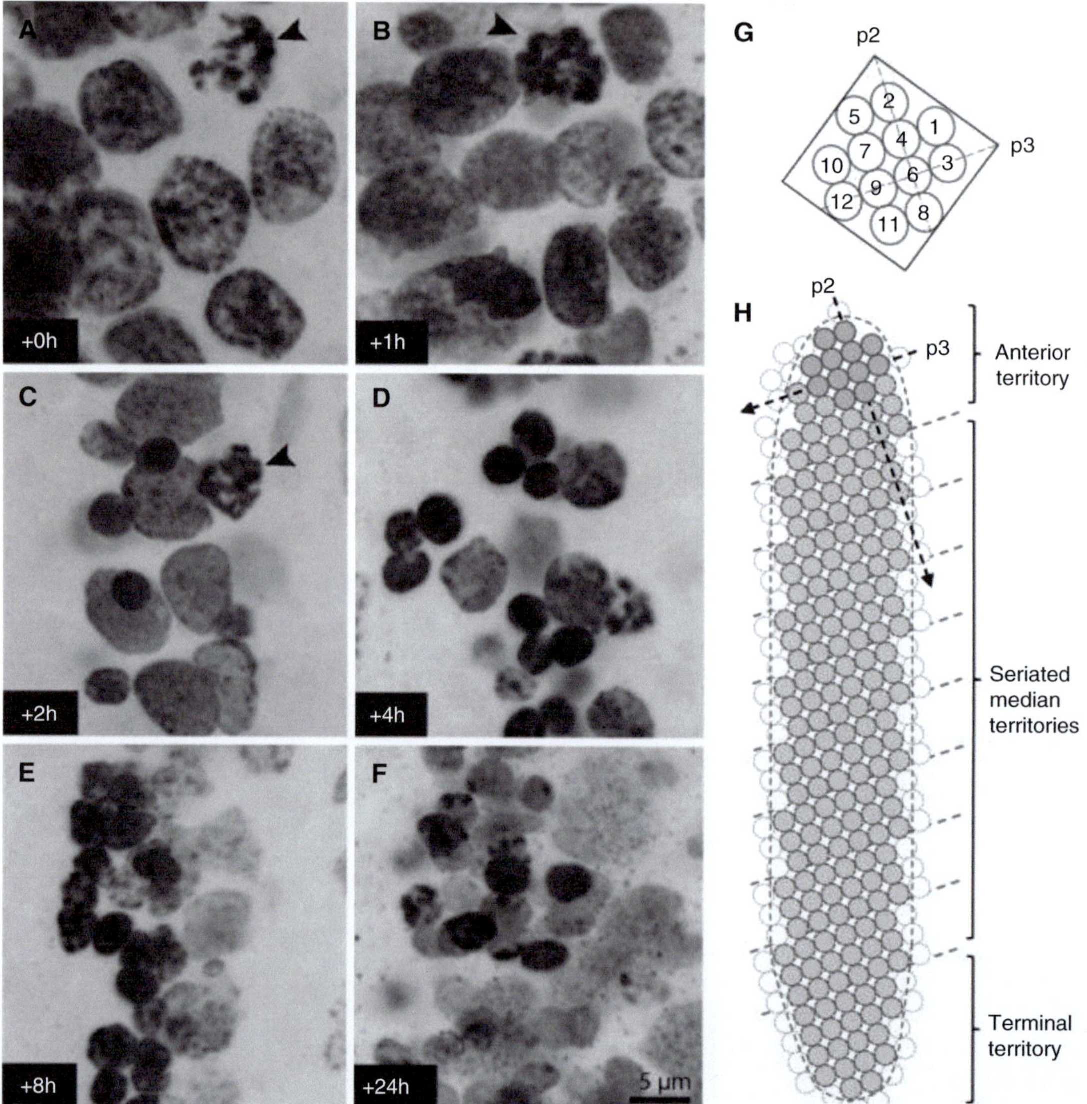

Fig. 10.13 Pulse-chase experiments to track the mode of cell division in the ventral nerve centre of hatchlings of *Spadella cephaloptera* (From Perez et al. 2013). A 10-min bromodeoxyuridine pulse was applied to a group of hatchlings at 48 h. Specimens from these groups were fixed and processed after a period of chase as follows: (**A**) No chase. (**B**) 1 h. (**C**) 2 h. (**D**) 4 h. (**E**) 8 h. (**F**) 24 h. The images are single optical sections (confocal laser scanning microscopy) that were *black-white* inverted. The excerpts chosen are located *left* of the neuropil; anterior is towards the *top*. The *arrowheads* indicate nuclei in pro-metaphase. (**G**) Packed circles pattern on an unrolled cylinder surface. On the unrolled surface the two to three parastichies (*p2, p3*) through point 0 are shown. (**H**) Schematic diagram representing the organisation of neuronal stem cells in a frontal plane of the ventral nerve centre (*left cluster*) after iteration of the two to three patterns. Note the serially organised domains

The brain in adult arrow worms is subdivided into an anterior and a posterior domain as described by Rieger et al. (2010). The posterior domain receives input from the sensory organs and hence, according to Rieger et al. (2010), is probably involved in the modulation of motor behaviour in response to changing sensory input. The anterior domain is associated with the stomatogastric nervous system and therefore likely assists in controlling the activity of the grasping

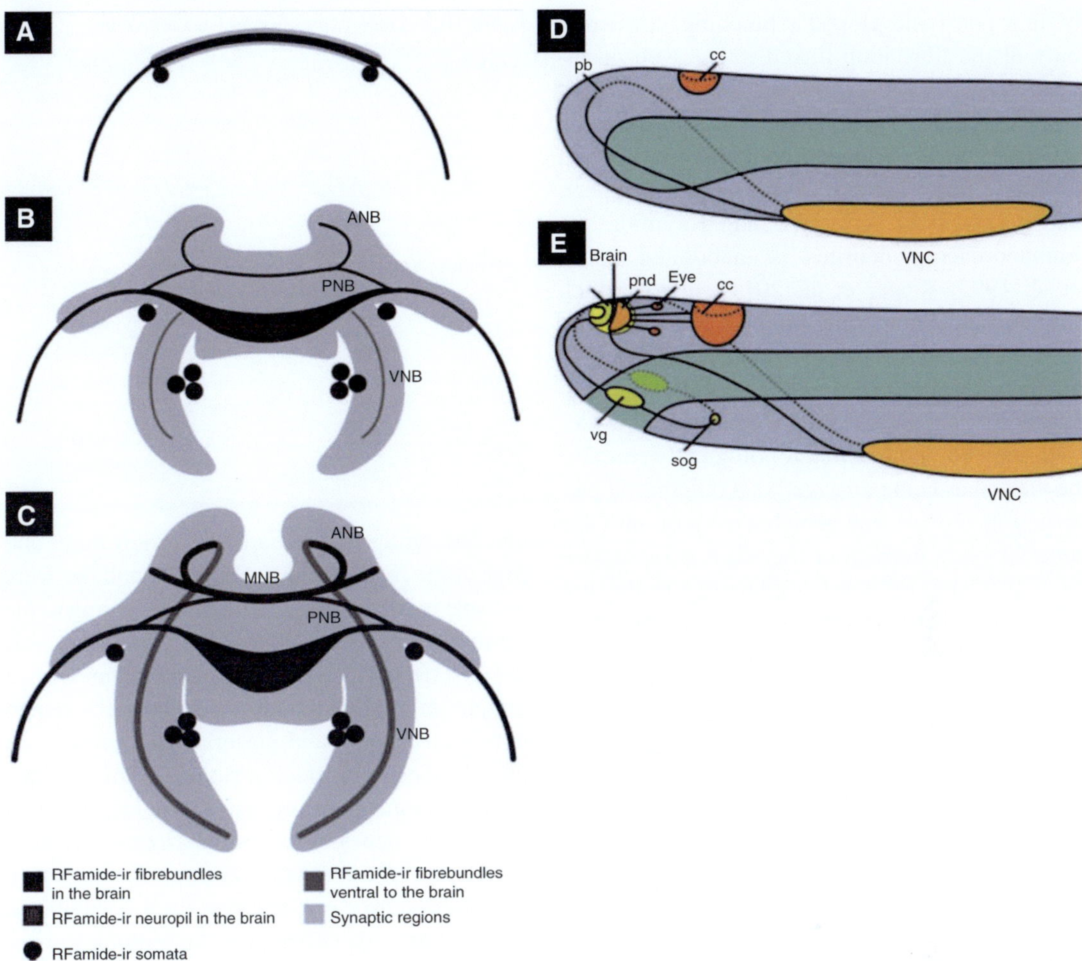

Fig. 10.14 (**A–C**) Schematic representations comparing the development of the brain and of RFamide-like immunoreactivity in hatchlings of *Spadella cephaloptera* (From Rieger et al. 2011). The schemes represent stages of (**A**) 24 h, (**B**) 72 h, and (**C**) 120 h after hatching. Abbreviations: *ANB* anterior neurite bundle, *MNB* median neurite bundle, *PNB* posterior neurite bundle, *vg* vestibular ganglion, *VNB* ventral neurite bundle. (**D, E**) Schematic representation comparing the brain in a newly hatched (**D**) and an adult (**E**) chaetognath (From Rieger et al. 2011). Abbreviations: *and* anterior neuropil domain of the brain, *cc* corona ciliata, *pb* primordial brain, *pnd* posterior neuropil domain of the brain, *sog* subesophageal ganglion, *vg* vestibular ganglion, *VNC* ventral nerve centre

spines and the musculature responsible for opening and closing the mouth. As for development, Rieger et al. (2011) could show that the brain of chaetognath hatchlings is delayed when compared to the VNC. At hatching, the brain is represented simply by a fibre loop that extends anteriorly from the VNC and surrounds the area, where a few days later the new mouth opening and its connection to the gut will form (Figs. 10.3, 10.10C, and 10.14). Neurons belonging to the epidermal plexus are also visible in the vicinity

of the corona ciliata (Fig. 10.10D). This "primary brain", which topologically can be viewed as a circumoral nerve ring, will develop into the posterior brain domain of the adult. Acting as a kind of "sensory brain", it receives input from the sensory eyes and corona ciliata and may be involved in the modulation of motor behaviours in response to changing sensory input (Rieger et al. 2010, 2011). Considering that the hatchlings neither feed nor are equipped with the adult set of sensory organs, it is not surprising that the brain is poorly developed at hatching, consisting only of the fibre loop linked to the well-developed VNC. However, during the first 48 h of postembryonic development, zones of synaptic neuropil emerge in the brain, new serotonin- and RFamide-like immunoreactive neurons appear, and a system of serotonin- and RFamide-like immunoreactive neurites is elaborated (Goto et al. 1992; Rieger et al. 2011), all of which implies that a functional brain is present by the time the hatchlings switch from feeding on yolk supplies to active predation. Postembryonic neurogenesis also includes the emergence of a second brain component that topologically can also be viewed as being arranged in a circumoral pattern (Fig. 10.14) and which develops into the anterior brain domain of the adult nervous system. This anterior domain gives rise to the stomatogastric nervous system which, in the adult, is involved in controlling the activity of the grasping spines and the innervation of the musculature responsible for opening and closing the mouth (Rieger et al. 2010). The circumesophageal arrangement of the adult cephalic nervous system including, in addition to the brain, the vestibular and the subesophageal ganglion (see Fig. 10.11) has already been recognised by Nielsen (2001). However, the situation in the hatchlings clearly shows that we do not only face one but two brain components that have a basically circumoral arrangement (Fig. 10.14; Rieger et al. 2011).

GENE EXPRESSION

The expression of investigated genes in Chaetognatha is summarised in Table 10.1. As there appear to be specific diagnostic amino acid motifs in the Hox genes of the three main lineages of bilaterians, this family of homeotic genes was an important and informative field of investigations for Chaetognatha, too. This aspect was recently reviewed by Perez et al. (2014) and this section is reprinted from their contribution. Papillon et al. (2003) isolated the homeodomains of six Hox genes from *Spadella cephaloptera*,

Table 10.1 Gene expression in Chaetognatha

Gene	Reference	Site of expression
Brachyury	Takada et al. (2002)	Embryonic blastopore and mouth
SceMed4	Papillon et al. (2005)	Two lateral stripes in a restricted region of the developing ventral nerve centre
Actin isoforms	Yasuda et al. (1997)	
PgAct1		Adults: oocytes, neurons, spermatocytes
PgAct2		Adults: head muscle, spermaduct
PgAct3		Adults: trunk muscle, oocytes, spermatocytes

one belonging to the paralogy group three and four to the median class. Yet, these authors were not able to identify any sequence belonging to the posterior Hox genes. One important result was the discovery of a new homeodomain with a unique set of signature amino acid motifs shared both with median and posterior Hox proteins of protostomes and deuterostomes. This unique mosaic organisation suggests that at least some of the median genes in extant metazoans may have derived from tandem duplication of an ancestral median/posterior one and that the Chaetognatha might be an early offshoot of the triploblastic lineage that predates the deuterostome-protostome split. However, the authors also noted carefully that this mosaic gene could be highly derived and only present in Chaetognatha. In 2007, eight Hox genes and one ParaHox gene were isolated from *Flaccisagitta enflata* (Matus et al. 2007). The presence of the mosaic median/posterior gene was confirmed in this pelagic species and, additionally, two posterior Hox genes were isolated. The finding of posterior Hox genes in Chaetognatha supports the hypothesis that the mosaic median/posterior gene is actually an apomorphy of all extent chaetognaths inherited from their last common ancestor. Yet, the careful analysis of homeodomains showed that the posterior Hox genes of chaetognaths possess both ecdysozoan and lophotrochozoan signature amino acid motifs, while the central class Hox genes lack the diagnostic amino acid motifs used to assign lophotrochozoan or protostome affinities (e.g., Lox5 spiralian parapeptide and Ubd-A peptide found in both ecdysozoans and lophotrochozoans). Thus, this last study did not allow the authors to decide between the tested two hypotheses of a sister group relationship of Chaetognatha to all other Protostomia versus to all Lophotrochozoa/Spiralia (Matus et al. 2007).

The expression pattern of only one median Hox gene has been analysed in Chaetognatha. Papillon et al. (2005) investigated the expression pattern of *SceMed4* (a putative ortholog to the *Scr*/*Hox5* or *Antp* orthology groups) in late embryos and in hatchlings and juveniles of *Spadella cephaloptera*. These authors showed a position-specific expression pattern of *SceMed4* in the ventral nerve centre typical of the coordinated expression of Hox genes observed in other Bilateria, suggesting a potential role in regionalisation of the nervous system. Because *SceMed4* expression starts earlier (before hatching) than the detection of serotonin and RFamide neurons, it is likely that this gene contributes to the diversity of neuronal subpopulations and to the establishment of distinct axon projection patterns.

Brachyury is one of the key transcription factors in the determination of mesoderm in vertebrates, but the comparative analysis of its expression in a variety of metazoans including the cnidarians suggests that an ancestral function is likely to specify the blastoporal region (Technau 2001). Takada et al. (2002) analysed the expression of *brachyury* (*Pg-Bra*) in *Paraspadella gotoi* and showed that *Pg-Bra* is expressed in two specific domains, around the blastopore and on the opposite side of the early embryo and then around the mouth opening region at the time of hatching. The expression of *Pg-Bra* in the early embryo resembles that of basal deuterostomes such as hemichordates, whereas that in the mouth opening region in the hatchling appears to be a chaetognath novelty. Once again, these results not only suggest the conservation but also the drift of chaetognath developmental features.

OPEN QUESTIONS

- Cleavage patterns beyond four-cell stage with respect to the question of spiral versus radial cleavage
- Gastrulation with respect to the emergence of the mesoderm
- Neurogenesis in chaetognath embryos
- The relation of the embryonic blastopore and mouth, which then close again, in comparison to the newly emerging mouth and anus of the hatchlings
- The expression of key regulatory genes (Hox, ParaHox, other homeobox genes, etc.) controlling embryogenesis

Acknowledgements We would like to thank Verena Rieger for her contribution to the experimental work on nervous system development reported here and Tina Kirchhoff for editing the reference list. We gratefully acknowledge the exchange with Thurston Lacalli and Andreas Hejnol on metazoan gastrulation patterns, Carolin and Joachim Haug on fossil chaetognaths, and Günter Purschke on photoreceptor structure. Research on arrow worms by SH was supported by grants HA 2540/7-1, 2, 3 in the DFG focus programme "Metazoan Deep Phylogeny".

References

Alvarino A (1992) Chaetognatha. In: Adiyodi KG, Adiyodi RG (eds) Reproductive biology of invertebrates. Oxford & IBH Publishing Co. PVT. LTD., New Delhi, pp 425–470, Chapter 22

Arendt D, Nübler-Jung K (1997) Dorsal or ventral: similarities in fate maps and gastrulation patterns in annelids, arthropods and chordates. Mech Dev 61:7–21

Ax P (2001) Das System der Metazoa III. Spektrum Akademischer Verlag, Hamburg

Backeljau T, Winnepenninckx B, De Bruyn L (1993) Cladistic analysis of metazoan relationships: a reappraisal. Cladistics 9:167–181

Bieri R (1991) Systematics of the Chaetognatha. In: Bone Q, Kapp H, Pierrot-Bults AC (eds) The biology of Chaetognaths. Oxford University Press, New York, pp 122–136

Bone Q, Goto T (1991) The nervous system. In: Bone Q, Kapp H, Pierrot-Bults AC (eds) The biology of Chaetognaths. Oxford University Press, New York, pp 18–31

Bone Q, Pulsford A (1984) The sense organs and ventral ganglion of *Sagitta* (Chaetognatha). Acta Zool (Stockh) 65:209–220

Bone Q, Kapp H, Pierrot-Bults AC (1991) The biology of Chaetognaths. Oxford University Press, New York, pp 1–173

Brusca RC, Brusca GJ (2003) The invertebrates, 2nd edn. Sinauer Associates, Sunderland

Burfield ST (1927) Sagitta. Liverpool Mar Biol Com Mem 28:1–104

Carré D, Djediat C, Sardet C (2002) Formation of a large vasa-positive germ granule and its inheritance by germ cells in the enigmatic chaetognaths. Development 129:661–670

Conway MS (2009) The Burgess Shale animal *Oesia* is not a chaetognath: a reply to Szaniawski. Acta Palaeontol Pol 54:175–179

Darwin C (1844) Observations on the structure and propagation of the genus Sagitta. Ann Mag Nat Hist Lond 13:1–6

Doncaster L (1903) On the development of *Sagitta*, with notes on the anatomy of adult. Q J Microsc Sci 46:351–398

Duvert M, Salat C (1979) Fine structure of muscle and other components of the trunk of *Sagitta setosa* (Chaetognatha). Tissue Cell 11:217–230

Edgecombe GD, Giribet G, Dunn CW, Hejnol A, Kristensen RM, Neves RC, Rouse GW, Worsaae K, Sørensen MV (2011) Higher-level metazoan relationships: recent progress and remaining questions. Org Divers Evol 11:151–172

Elpatiewsky VW (1909) Die Urgeschlechtszellenbildung bei *Sagitta*. Anat Anz 35:226–239

Feigenbaum DL (1976) Development of the adhesive organ in *Spadella schizoptera* (Chaetognatha) with comments on growth and pigmentation. Bull Mar Sci 26:600–603

Ghiradelli E (1995) Chaetognaths: two unsolved problems: the coelom and their affinities. In: Lanzavecchia G, Valvassori R, Candia Carnevali MD (eds) Body cavities: function and phylogeny. Selected Symposia and Monographs, vol 8. UZI, Modena, Italy, pp 167–185

Ghirardelli E (1953) L'accopiamento in *Spadella cephaloptera* Busch. Pubbl Staz Zool Napoli XXIV, Fasc 3°: 3–12

Ghirardelli E (1968) Some aspects of the biology of the chaetognaths. Adv Arch Biol 6:271–357

Goto T (1999) Fertilization process in the arrow worm *Spadella cephaloptera* (Chaetognatha). Zool Sci 16:109–114

Goto T, Yoshida M (1984) Photoreception in Chaetognatha. In: Ali MA (ed) Photoreception and vision in invertebrates. Plenum Publishing Corporation, New York, pp 727–742

Goto T, Yoshida M (1985) The mating sequence of the benthic arrowworm *Spadella schizoptera*. Biol Bull 169:328–333

Goto T, Yoshida M (1987) Nervous system in Chaetognatha. In: Ali MA (ed) Nervous systems in invertebrates. Plenum Publishing Corporation, New York, pp 461–481

Goto T, Yoshida M (1997) Growth and reproduction of the benthic arrowworm *Paraspadella gotoi* (Chaetognatha) in laboratory culture. Invert Repro Dev 32: 210–207

Goto T, Katayama-Kumoi Y, Tohyama M, Yoshida M (1992) Distribution and development of the serotonin- and RFamide-like immunoreactive neurons in the arrowworm, *Paraspadella gotoi* (Chaetognatha). Cell Tissue Res 267:215–222

Harzsch S, Müller CHG (2007) A new look at the ventral nerve centre of *Sagitta*: implications for the phylogenetic position of Chaetognatha (arrow worms) and the evolution of the bilaterian nervous system. Front Zool 4:14

Harzsch S, Wanninger A (2009) Evolution of invertebrate nervous systems: the Chaetognatha as a case study. Acta Zool (Stockh) 91:35–41

Harzsch S, Müller CHG, Rieger V, Perez Y, Sintoni S, Sardet C, Hansson B (2009) Fine structure of the ventral nerve centre and interspecific identification of individual neurons in the enigmatic Chaetognatha. Zoomorphology 128:53–73

Hejnol A, Martindale MQ (2009) The mouth, the anus and the blastopore – open questions about questionable openings. In: Telford MJ, Littlewood DTJ (eds) Animal evolution: genes, genomes, fossils and trees. Oxford University Press, Oxford, pp 33–40

Hertwig O (1880) Die Chaetognathen. Monatl Jenaer Z Med Nat 14:196–311

Hyman LH (1959) Phylum Chaetognatha. Smaller coelomate groups. The invertebrates, vol 5. McGraw Hill Book Company, New York, pp 1–71

Jenner RA (2004) Towards a phylogeny of the Metazoa: evaluating alternative phylogenetic positions of Platyhelminthes, Nemertea, and Gnathostomulida, with a critical reappraisal of cladistic characters. Contr Zool 73:3–163

John CC (1933) Habits, structure and development of *Spadella cephaloptera*. Q J Microsc Sci 75: 625–696

Kapp H (1991) Morphology and anatomy. In: Bone Q, Kapp H, Pierrot-Bults AC (eds) The biology of Chaetognaths. Oxford University Press, New York, pp 5–17

Kapp H (2000) The unique embryology of Chaetognatha. Zool Anz 239:263–266

Kapp H (2007) Chaetognatha, Pfeilwürmer. In: Westheide W, Rieger R (eds) Spezielle Zoologie. Teil 1. Einzeller und Wirbellose Tiere. Gustav Fischer Verlag, Stuttgart, pp 898–904

Koch M, Quast B, Bartolomaeus T (2014) Coeloms and nephridia in annelids and arthropods. In: Wägele JW, Bartolomaeus T (eds) Deep metazoan phylogeny: the backbone of the tree of life. New insights from analyses of molecules, morphology, and theory of data analysis. De Gruyter, Berlin, pp 173–284

Kotori M (1975) Morphology of *Sagitta elegans* (Chaetognatha) in Early Larval Stages. J Oceanographical Society of Japan 31:139–144

Kuhl W (1938) Chaetognatha. In: Bronn HG (ed) Klassen und Ordnungen des Tierreiches, vol IV. Akademische Verlagsgesellschaft, Leipzig, Abt. IV, Buch 2, Teil 1

Kuhl W, Kuhl G (1965) Die Dynamik der Frühentwicklung von *Sagitta setosa*. Helgol Mar Res 12:260–301

Lacalli TC (2010) The emergence of the chordate body plan: some puzzles and problems. Acta Zool (Stockh) 9:4–10

Martindale MQ, Heijnol A (2009) A developmental perspective: changes in the position of the blastopore during bilaterian evolution. Dev Cell 17:162–174

Martín-Duran JM, Janssen R, Wennberg S, Budd GE, Hejnol A (2012) Deuterostomic development in the protostome *Priapulus caudatus*. Curr Biol 22:2161–2166

Matus DQ, Halanych KM, Martindale MQ (2007) The *Hox* gene complement of a pelagic chaetognath, *Flaccisagitta enflata*. Int Comput Biol 47:854–864

Meglitsch PA, Schram FR (1991) Invertebrate Zoology. Oxford University Press, Oxford

Müller CHG, Rieger V, Perez Y, Harzsch S (2014) Immunohistochemical and ultrastructural studies on ciliary sense organs of arrow worms (Chaetognatha). Zoomorphology 133:167–189

Nielsen C, Scharff N, Eibye-Jacobsen D (1996) Cladistic analyses of the animal kingdom. Biol J Linn Soc 57:385–410

Nielsen C (2001) Animal evolution: interrelationships of the living phyla, 2nd edn. Oxford University Press, Oxford

Nielsen C (2005a) Larval and adult brains. Evol Dev 7:483–489

Nielsen C (2005b) Trochophora larvae: cell-lineages, ciliary bands and body regions. 2. Other groups and general discussion. J Exp Zool 304B:401–447

Nielsen C (2010) Some aspects of spiralian development. Acta Zool (Stockh) 91:20–28

Nielsen C (2012) Animal evolution: interrelationships of the living phyla, 3rd edn. Oxford University Press, Oxford

Papillon D, Perez Y, Fasano L, Le Parco Y, Caubit X (2003) *Hox* gene survey in the chaetognath *Spadella cephaloptera*: evolutionary implications. Dev Genet Evol 213:142–148

Papillon D, Perez Y, Caubit X, Le Parco Y (2004) Identification of chaetognaths as protostomes is supported by the analysis of their mitochondrial genome. Mol Biol Evol 21:2122–2129

Papillon D, Perez Y, Fasano L, Parco YL, Caubit X (2005) Restricted expression of a median *Hox* gene in the central nervous system of chaetognaths. Dev Genet Evol 215:369–373

Pearre S (1991) Growth and reproduction. In: Bone Q, Kapp H, Pierrot-Bults AC (eds) The biology of Chaetognaths. Oxford University Press, New York, pp 61–75

Perez Y, Rieger V, Martin E, Müller C, Harzsch S (2013) Neurogenesis in an early protostome relative: progenitor cells in the ventral nerve centre of chaetognath hatchlings are arranged in a highly organized geometrical pattern. J Exp Zool 320:179–193

Perez Y, Müller CHG, Harzsch S (2014) The Chaetognatha: an anarchistic taxon between Protostomia and Deuterostomia. In: Wägele JW, Bartolomaeus T (eds) Deep metazoan phylogeny: the backbone of the tree of life. Walter De Gruyter GmbH, Berlin, pp 49–74

Peterson KJ, Eernisse DJ (2001) Animal phylogeny and the ancestry of bilaterians: inferences from morphology and 18s rDNA gene sequences. Evol Dev 3:170–205

Reeve MR (1970) Complete cycle of development of a pelagic chaetognath in culture. Nature 227:381

Reeve MR, Lester B (1974) The process of egg-laying in the chaetognath *Sagitta hispida*. Biol Bull 147:247–256

Reeve MR, Walter MA (1972) Conditions of culture, food-size selection and the effects of temperature and salinity on growth rate and generation time in *Sagitta hispida* Conant. J Exp Mar Biol Ecol 9:191–200

Rieger V, Perez Y, Müller CHG, Lipke E, Sombke A, Hansson BS, Harzsch S (2010) Immunohistochemical analysis and 3D reconstruction of the cephalic nervous system in Chaetognatha: insights into an early bilaterian brain? Invertebr Biol 129:77–104

Rieger V, Perez Y, Müller CHG, Lacalli T, Hansson BS, Harzsch S (2011) Development of the nervous system in hatchlings of *Spadella cephaloptera* (Chaetognatha), and implications for nervous system evolution in Bilateria. Dev Growth Differ 53:740–759

Schram FR, Ellis WN (1994) Metazoan relationships: a rebuttal. Cladistics 10:331–337

Schmidt-Rhaesa A (2007) The evolution of organ systems. Oxford University Press, Oxford

Shimotori T, Goto T (1999) Establishment of axial properties in the arrow worm embryo, *Paraspadella gotoi* (Chaetognatha): developmental fate of the first two blastomeres. Zool Sci 16:459–469

Shimotori T, Goto T (2001) Developmental fates of the first four blastomeres of the chaetognath *Paraspadella gotoi*: relationship to protostomes. Dev Growth Differ 43:371–382

Shinn GL (1994a) Ultrastructural evidence that somatic "accessory cells" participate in chaetognath fertilization. In: Wilson WH, SA Stricker, Shinn GL (eds) Reproduction and development of marine invertebrates. Johns Hopkins Univ Press, London, pp 96–105

Shinn GL (1994b) Epithelial origin of mesodermal structures in arrowworms (*Phylum Chaetognatha*) Amer Zool 34:523–532

Shinn GL (1997) Chaetognatha. In: Harrison FW, Ruppert EE (eds) Microscopic anatomy of invertebrates, vol 15, Hemichordata, Chaetognatha, and the invertebrate chordates. Wiley-Liss, New York, pp 103–220

Shinn GL, Roberts ME (1994) Ultrastructure of hatchling chaetognaths (*Ferosagitta hispida*): epithelial arrangement of mesoderm and its phylogenetic implications. J Morphol 219:143–163

Szaniawski H (1982) Chaetognath grasping spines recognized among Cambrian protoconodonts. Journal Paleontol 56:806–810

Szaniawski H (2002) New evidence for the protoconodont origin of chaetognaths. Acta Palaeontol Pol 47:405–419

Szaniawski H (2005) Cambrian chaetognaths recognized in Burgess Shale fossils. Acta Palaeontol Pol 50:1–8

Szaniawski H (2009) Fossil chaetognaths from the Burgess Shale: a reply to Conway Morris. Acta Palaeontol Pol 54:361–364

Takada N, Goto T, Satoh N (2002) Expression pattern of the *brachyury* gene in the arrow worm *Paraspadella gotoi* (Chaetognatha). Genesis 32:240–245

Technau U (2001) *Brachyury*, the blastopore and the evolution of the mesoderm. Bioessays 23:788–794

Terazaki M, Miller CB (1982) Reproduction of meso and bathypelagic chaetognaths in the genus *Eukrohnia*. Mar Biol 71:193–196

Turbeville JM (1986) An ultrastructural analysis of coelomogenesis in the hoplonemertine *prosorhochmus americanus* and the *Polychaete Magelona* sp. J Morphol 187:51–60

Valentine JW (1997) Cleavage patterns and the topology of the metazoan tree of life. Proc Natl Acad Sci USA 94:8001–8005

Vannier J, Steiner M, Renvoise E, Hu S-X, Casanova J-P (2007) Early Cambrian origin of modern food webs: evidence from predator arrow worms. Proc R Soc Lond B 274:627–633

Welsch U, Storch V (1982) Fine structure of the coelomic epithelium of *Sagitta elegans*. Zoomorphology 100:217–222

Willmer P (1990) Invertebrate relationships: patterns in animal evolution. Cambridge University Press, New York

Yasuda E, Goto T, Makabe KW, Satoh N (1997) Expression of actin genes in the arrow worm *Paraspadella gotoi* (Chaetognatha). Zool Sci 14:953–960

Index

A. Wanninger (ed.), *Evolutionary Developmental Biology of Invertebrates 1:*
Introduction, Non-Bilateria, Acoelomorpha, Xenoturbellida, Chaetognatha
DOI 10.1007/978-3-7091-1862-7, © Springer-Verlag Wien 2015

 MIX
Papier aus verantwortungsvollen Quellen
Paper from responsible sources
FSC® C105338
FSC
www.fsc.org

If you have any concerns about our products,
you can contact us on
ProductSafety@springernature.com

In case Publisher is established outside the EU,
the EU authorized representative is:
Springer Nature Customer Service Center GmbH
Europaplatz 3, 69115 Heidelberg, Germany

Printed by Libri Plureos GmbH
in Hamburg, Germany